T0418814

OPTIMAL OPERATION OF INTEGRATED MULTI-ENERGY SYSTEMS UNDER UNCERTAINTY

OPTIMAL OPERATION OF INTEGRATED MULTI-ENERGY SYSTEMS UNDER UNCERTAINTY

QIUWEI WU
Associate Professor, Technical University of Denmark, Denmark

JIN TAN
PhD candidate, Technical University of Denmark, Denmark

MENGLIN ZHANG
Postdoctoral researcher, Technical University of Denmark, Denmark

XIAOLONG JIN
Postdoctoral researcher, Technical University of Denmark, Denmark

ANA TURK
PhD candidate, Technical University of Denmark, Denmark

ELSEVIER

Elsevier
Radarweg 29, PO Box 211, 1000 AE Amsterdam, Netherlands
The Boulevard, Langford Lane, Kidlington, Oxford OX5 1GB, United Kingdom
50 Hampshire Street, 5th Floor, Cambridge, MA 02139, United States

Copyright © 2022 Elsevier Inc. All rights reserved.

No part of this publication may be reproduced or transmitted in any form or by any means, electronic or mechanical, including photocopying, recording, or any information storage and retrieval system, without permission in writing from the publisher. Details on how to seek permission, further information about the Publisher's permissions policies and our arrangements with organizations such as the Copyright Clearance Center and the Copyright Licensing Agency, can be found at our website: www.elsevier.com/permissions.

This book and the individual contributions contained in it are protected under copyright by the Publisher (other than as may be noted herein).

Notices
Knowledge and best practice in this field are constantly changing. As new research and experience broaden our understanding, changes in research methods, professional practices, or medical treatment may become necessary.

Practitioners and researchers must always rely on their own experience and knowledge in evaluating and using any information, methods, compounds, or experiments described herein. In using such information or methods they should be mindful of their own safety and the safety of others, including parties for whom they have a professional responsibility.

To the fullest extent of the law, neither the Publisher nor the authors, contributors, or editors, assume any liability for any injury and/or damage to persons or property as a matter of products liability, negligence or otherwise, or from any use or operation of any methods, products, instructions, or ideas contained in the material herein.

British Library Cataloguing-in-Publication Data
A catalogue record for this book is available from the British Library

Library of Congress Cataloging-in-Publication Data
A catalog record for this book is available from the Library of Congress

ISBN: 978-0-12-824114-1

For Information on all Elsevier publications visit our website at
https://www.elsevier.com/books-and-journals

Publisher: Joe Hayton
Acquisitions Editor: Graham Nisbet
Editorial Project Manager: Sara Valentino
Production Project Manager: Kamesh Ramajogi
Cover Designer: Christian J. Bilbow

Working together
to grow libraries in
developing countries

www.elsevier.com • www.bookaid.org

Typeset by Aptara, New Delhi, India

Contents

Biography

Qiuwei Wu received the PhD degree in Power System Engineering from Nanyang Technological University, Singapore, in 2009. He was a senior R&D engineer with Vestas Technology R&D Singapore Pte Ltd from Mar. 2008 to Oct. 2009. He has been working at Department of Electrical Engineering, Technical University of Denmark (DTU) since Nov. 2009 (PostDoc Nov. 2009-Oct. 2010, Assistant Professor Nov. 2010-Aug. 2013, Associate Professor since Sept. 2013). He was a visiting scholar at the Department of Industrial Engineering & Operations Research (IEOR), University of California, Berkeley, from Feb. 2012 to May 2012. He was a visiting scholar at the School of Engineering and Applied Sciences, Harvard University from Nov. 2017 Oct. 2018. His research interests are operation and control of power systems with high penetration of renewables, including wind power modelling and control, active distribution networks, and operation of integrated energy systems.

Jin Tan received the M.S. degree from the Department of Electrical Engineering, Wuhan University, Wuhan, China, in 2018. She is working toward the Ph.D. degree in electrical engineering from Technical University of Denmark, Kongens Lyngby, Denmark. Her research interests include the optimal operation of integrated electricity and heating system and renewable energy integration. Currently, she is involved in the project of Using Flexible District Heating with Heat Pumps for Integrated Electricity and Heat Dispatch with Renewables. She focuses on modeling the integrated electricity and heating system, investigating the flexibility that district heating system could provide to the electric power system, and the optimal operation for the integrated electricity and heat system considering wind power uncertainty.

Menglin Zhang received the B.S. degree in electrical engineering from Southwest Jiaotong University (SWJTU), Chengdu, China, in 2011, and the Ph.D. degree in electrical engineering from Wuhan University (WHU), Wuhan, China, in 2017. She was with the Department of Electrical Engineering, Huazhong University of Science and Technology (HUST), Wuhan, China from 2017 to 2019. Currently, she is a Post-Doctoral Researcher with the Center for Electric Power and Energy, Technical University of Denmark (DTU). Her current research interests include the modeling of temporal-spatial correlation of renewables in stochastic programming and advanced uncertainty set to reduce conservativeness in robust optimization, the modeling of optimal operation of integrated electricity and heat system considering flexibility, and the accelerated solving algorithm for the bulk system.

Xiaolong Jin obtained the Ph.D. degree from the School of Electrical and Information Engineering, Tianjin University, Tianjin, China, in 2019. He is now a Postdoc researcher with Technical University of Denmark (DTU). His research interests include energy management of multi-energy systems and multi-energy buildings. Specifically, his research focuses on improving energy efficiency and reducing operating cost of multi-energy systems and multi-energy buildings with designed energy management frameworks, which uses the flexibilities from three aspects: 1) Use the demand-side flexibility by dispatching the flexible multi-energy loads in smart buildings; 2) Use the network-side flexibility by coordinating the multi-vector energy networks; 3) Use the supply-side flexibility by scheduling the various generations in the energy stations and the distributed energy resources connected with multi-energy systems and multi-energy buildings.

Ana Turk received the B.S. degree from the Faculty of Electrical Engineering and Computer Science at University of Maribor in Slovenia and MSc degree in Energy Engineering from Faculty of Engineering and Science at Aalborg University in Denmark in 2018. She is currently pursuing a Ph.D. at the Center of Electric Power and Energy (CEE) at the Department of Electrical Engineering at the Technical University of Denmark (DTU). Her research interest include integration and modeling of multi-energy systems (district heating, natural gas and electric power system), stochastic programming and optimal operation and scheduling of multi-energy systems. In particular, special focus is on optimal operation and real time control of integrated energy systems by using model predictive control.

CHAPTER 1

Introduction of integrated energy systems

1.1 Introduction

Energy plays an important role in the development of society. Before the industrial revolution, biomass (i.e., wood) was the world's main primary energy source. Since 1900, most primary energy came from wood and coal [1], but with the advent of the automobile and airplanes in the early 20th century, oil became the dominant fuel. In 2018, most of the world's energy was generated from fossil fuels (81%). The rest came from bioenergy, including traditional solid biomass (9.4%), nuclear (5%), hydro (2.5%), and other renewables such as wind, solar, and geothermal (2.1%) [2]. In recent years, with growing environmental concerns, especially over global climate change and local pollution, attention to international agreements for reducing greenhouse gas emissions and cleaning air, with a consequential increase of renewable energy technologies, has increased. The European Union is committed to reducing greenhouse gas emissions to be 80% to 95% below 1990 levels by 2050, and about two thirds of the energy should be from the renewable sources [3]. In the United States, a renewable energy transition is under way, led by communities and states [4]. Denmark has been a pioneer in implementing renewable energy, and the Danish energy system has undergone a transformational change, whereas China ranked first in the world in terms of cumulative and new installations of onshore wind power by 2018 [5].

Since major renewable sources like wind and solar can easily be turned into electricity, and electric power can easily be transmitted, transformed, and used, such resources are expected to become the dominant energy carriers in the future. In recent years, wind power and photovoltaic (PV) power have been growing rapidly in many counties. Fig. 1.1 shows the installed wind power capacity from 2010 to 2019, which reached about 622 GW in 2019 [6]. China's cumulative installed wind power capacity from 2004 to 2019 is illustrated in Fig. 1.2, which reached about 236 GW in the latter year, accounting for almost 36% of total installed wind capacity worldwide [8]. It is estimated that more than 25% of new offshore wind power capacity will be

Optimal operation of integrated multi-energy systems under uncertainty.
DOI: https://doi.org/10.1016/B978-0-12-824114-1.00006-8 Copyright © 2022 Elsevier Inc.
All rights reserved.

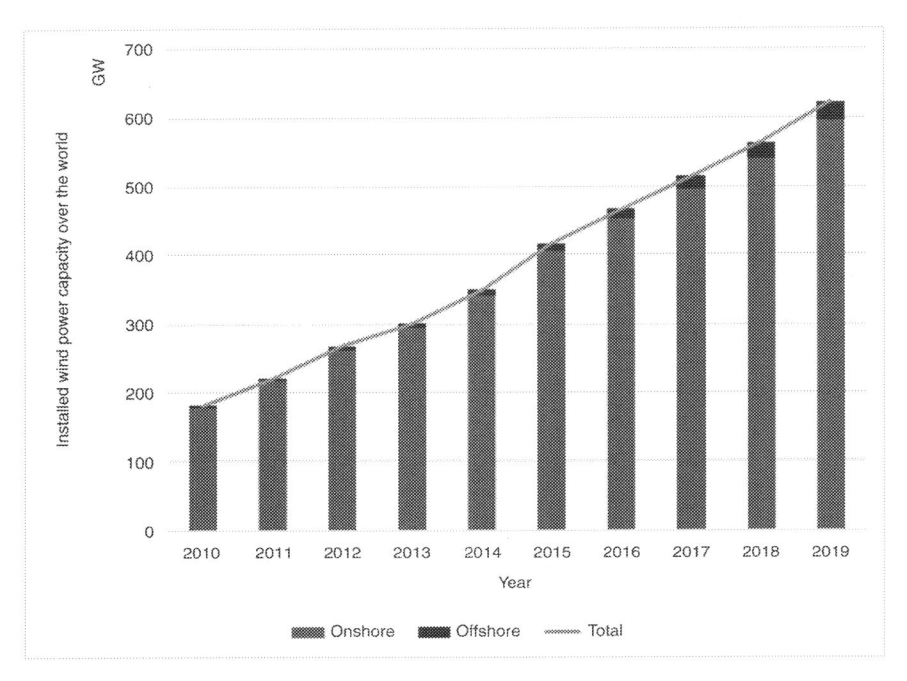

Figure 1.1 Installed wind power capacity over the world from 2010 to 2019. From International Renewable Energy Agency [6].

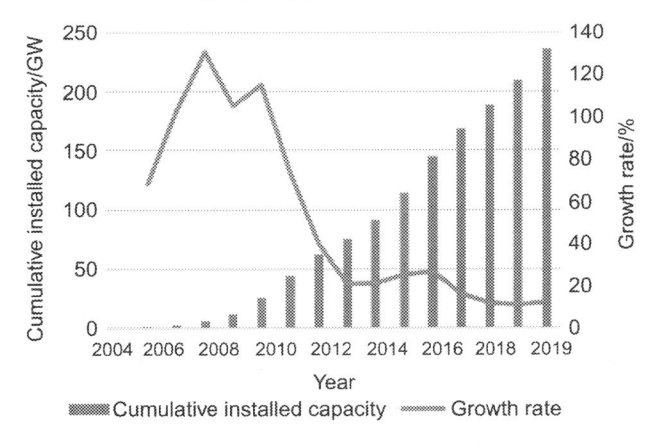

Figure 1.2 Cumulative installed wind power capacity in China. From He et al. [7].

added in China by 2030 [9]. In Denmark, cumulative wind power capacity was 6.13 GW in 2019, with onshore and offshore wind turbine capacities reaching 4.43 GW and 1.70 GW, respectively [10]. In addition, wind power production accounted for 47.2% of Denmark's domestic electricity supply.

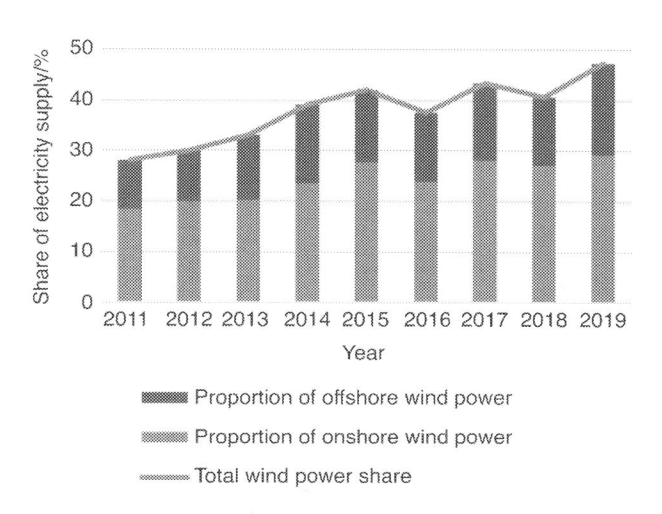

Figure 1.3 Proportion of onshore and offshore wind power and the total wind power share for electricity supply in Denmark.

Fig. 1.3 shows the proportion of onshore and offshore wind power, as well as the total wind power share of electricity supply in Denmark from 2011 to 2019. Meanwhile, solar power and solar thermal energy have experienced strong growth in the past two decades. Fig. 1.4 shows the installed solar power and solar energy capacity from 2010 to 201, which reached about 585 GW in 2019 [11].

It should be noted that the power output of renewable technologies like wind and solar PV fluctuates due to rapidly changing meteorological conditions. Because they have a zero marginal cost, renewable power from wind and PV is replacing conventional thermal power plants, which conventionally have been responsible for providing many electrical power system services, such as reserves, voltage control, frequency control, stability services, and black start restoration [11]. The increasing penetration of these renewable power sources is therefore posing substantial challenges to the planning, secure and reliable operation, and control of power systems [12], [13]. Hence, the requirements for flexibility to accommodate large amounts of naturally fluctuating renewable energy are increasing.

In several countries and regions, parts of the gas, heating, cooling, and transportation systems have responded to these flexibility requirements by means of a deep coupling of multiple energy sectors. Indeed, due to the intrinsic storage capabilities of, for example, the thermal inertia of district heating pipes and buildings [14], the storage of electric vehicle batteries [15],

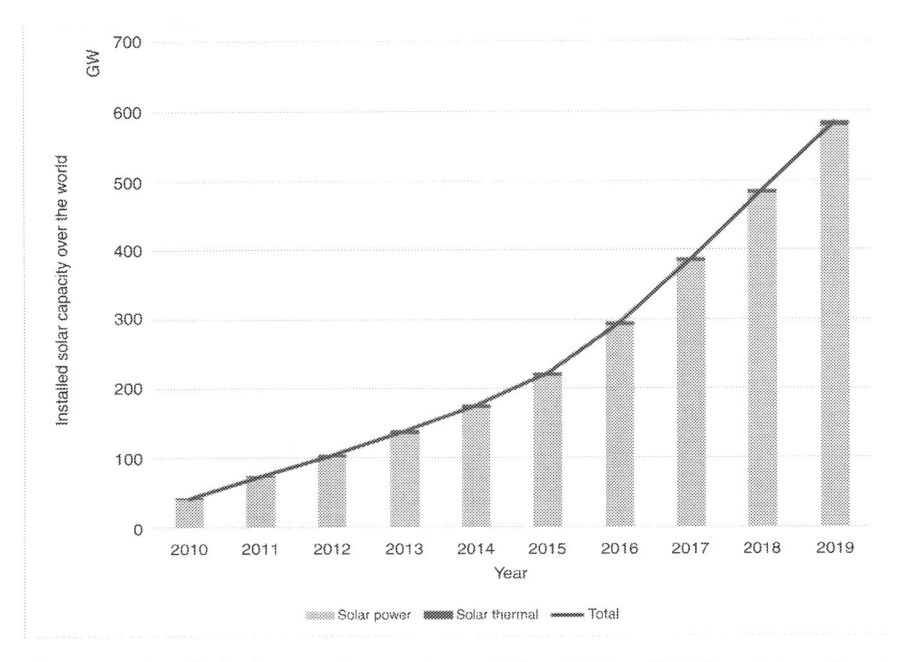

Figure 1.4 Installed solar capacity over the world from 2010 to 2019. From International Renewable Energy Agency [11].

and various energy conversion techniques (e.g., power-to-heat, power-to-gas [P2G], combined heat and power [CHP] units), the heating and gas sectors can provide extra flexibility to the electric power system (EPS). Different energy sectors are coupled at the production or demand side.

The integration of different energy sectors can solve some challenges to the stable and reliable operation of electric power systems with high penetrations of renewable energy [16]. This will not only facilitate the integration of renewable energy but also can improve the cost efficiency of the whole energy system if done properly. In Denmark, Energinet is the transmission system operator of the country's electricity and gas transmission grids [11], which, by controlling both wholesale grid systems, basically demonstrates the integration of two different energy systems. In addition, the Danish Partnership Smart Energy Networks was established in 2014 to bring together Danish energy companies, industry, and knowledge institutions within electricity, heating, cooling, and gas. This promises to be an effective approach for achieving the ambitious Danish climate and energy goal of a fully 100% renewable-based energy system by 2050.

In addition to the integration of infrastructural technology, coordinated operation and control of an integrated energy system are necessary [17], [18]. At present, in most countries, the regulation and management of different energy sectors are still separate both for historical reasons and due to the different sets of rules based on diverse principles. This lack of uniform standardization impedes the development of efficient integration and optimal solutions for the whole system. Well-functioning and efficient integrated energy systems should be based on an integrated energy system approach that incorporates novel digital solutions, including sensors and actuators embedded in the system, various Internet technologies, platforms with service-based designs, and novel business models [19], but this can only work effectively if the cross-sector regulations are compatible.

During the past few decades, the concept of the smart grid has emerged. It involves new concepts and technologies in the EPS. The European Union Commission Task Force for Smart Grids defines the smart grid as "an electricity network that can cost-efficiently integrate the behavior and actions of all users connected to it—generators, consumers, and those that do both—in order to ensure an economically efficient, sustainable power system with low losses and high levels of quality and security of supply and safety." The idea of the smart grid can be extended to smart energy, whereby information and communication technology also play an important role in enhancing the performance of the coordinated operation and control of all of the coupled energy sectors [20].

1.2 Integrated energy system

An *integrated energy system* is defined as a cost-effective, sustainable, and secure energy system in which renewable energy production, infrastructure, and consumption are integrated and coordinated through energy services, active users, and enabling technologies. Fig. 1.5 gives an overview of a Danish integrated energy system providing flexibility for the cost-effective integration of renewable energies. The different characteristics of the coupled electricity, heating, and gas energy sectors in integrated energy systems are listed in Table 1.1.

1.2.1 Electricity sector

There will be more naturally fluctuating power generation in the EPS, which can flow bi-directionally, from large-scale generators via the grid to the

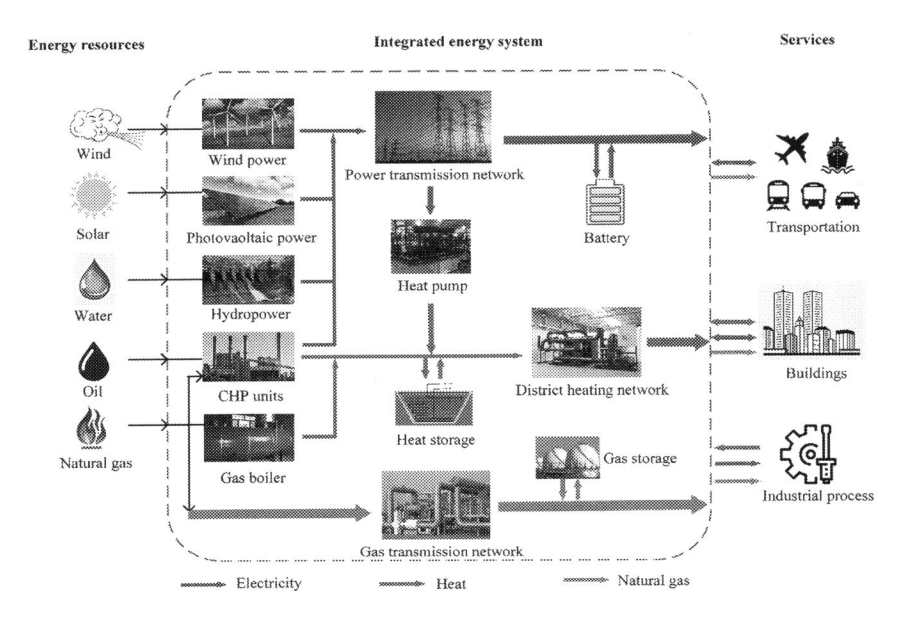

Figure 1.5 Overview of an integrated energy system.

Table 1.1 Different characteristics of various energy sectors.

Energy Sector	Properties	Intrinsic Flexibility	Flexibility Need
Electricity	Long–distance transport Low losses Easy to generate from renewable energy sources Easy conversion to other energy carriers	Very low (seconds)	High
Heating	Local/district Medium losses Difficult to convert to other energy carriers (if low temperature)	Medium (days)	Medium
Gas	Long–distance transport Low transmission losses Intrinsic losses during conversion at the point of use Easy to convert to heat and electricity	High (months)	Low

From Lund et al. [27].

consumer, and be reinjected by prosumers into the grid. In addition, over time, small-scale distributed generation and fluctuating renewable generation will gradually replace conventional central power plants. Since the traditional (large-scale) synchronous electricity generation units that provide inertia response are being replaced by nonsynchronous renewable energy technologies (effectively via power electronics inverters), the total inertia in the system is being reduced, leading in turn to adverse impacts on the frequency security of the EPS. This considerably increases the requirement for flexibility, especially frequency control reserves, to maintain system frequency security. To obtain additional flexibility to support the operation and control of the EPS, new energy conversion techniques, demand response, and new power generation scheduling strategies are being introduced into the electricity sector. The conversion techniques include the still conventional gas-fired electric power plants (gas to power) and cogeneration plants (gas to heat and electrical power), as well as heat pumps (electric power to heat) and future technologies that convert electrical energy into molecules such as hydrogen and methane (P2G) [21]. By promoting appropriate interaction between electric power generation and active consumers (including commercial, industry, and residential), demand response can offer great benefits to operation of the system [22]. The electrical loads are controlled by intelligent management systems participating in the electricity markets. As an example, in the European Union–supported project EcoGrid EU, flexibility on the consumer side is supposed to originate mainly from local heating systems in buildings [23].

1.2.2 Heating and cooling sector

With the further development of low-energy buildings, residential and office energy consumption, including heating and cooling demand, will fall correspondingly. By developing more district heating and cooling systems where appropriate and justified, it is possible to move toward a more sustainable energy system based on renewable energy [24]. In this regard, the concept of a 4th Generation District Heating System (4GDH) was proposed in Denmark [25], whereas the 5th Generation District Heating System (5GDH) was developed further, also known as Cold District Heating Networks [26]. These systems are based on the idea of low-temperature and ultra-low-temperature district heating systems (DHSs) respectively, which can reuse the waste heat from industry and buildings, as well as reduce heat loss. The heating networks in 4GDH are characterized by normal distribution temperatures of 50°C (supply pipe) and 20°C (return pipe) as annual averages, whereas the temperatures in pipes with 5GDH are around 5 to 30°C, which keeps heat loss to a minimum and reduces the need for extensive insulation. In 5GDH, electrical heat boosters are usually installed at the building side for heating hot tap water. In addition, heat storage is playing an increasingly important role in the heating sector, which can enhance the flexibility of CHP units and integrate fluctuating wind power better through the conversion of electrical energy into heat.

1.2.3 Natural gas sector

Due to the low cost of the energy carrier, low environmental emissions, and high efficiency of natural gas–based technologies, natural gas has become the second largest source of the world energy consumption [5]. On the one hand, gas can easily be converted into electricity and heat by gas-fired power generation, such as combined cycle gas turbines, high–efficiency condensing boilers, and CHP. CHP (or co-generation) intensifies the coupling between natural gas and electricity power systems. On the other hand, with the hopefully successful future development of P2G technology, electric power can also be converted into gas (hydrogen and methane), and then the converted gas can be injected into the natural gas system together with biogas [28]. It should be noted, however, that although the overall efficiency of this process is quite low, it may nevertheless be a necessary building block in achieving the required system integration between sectors to ensure long-term storage. The P2G route can help decrease the curtailing of renewable energies and provide more flexibility for the EPS. Furthermore, the natural gas system has large-scale storage capabilities due to the pressure flexibility and the large volumes in pipelines and caverns [11].

1.2.4 Transportation sector

The European Environmental Agency, which keeps track of worldwide final energy consumption, has found that the transport sector is responsible for about a third of overall final energy consumption [29]. Thus, because of the accompanying CO_2 emissions and local pollution, it is crucial that the transportation sector replaces fossil fuels with renewable-based energy carriers [30]. The electrification of transportation through battery electric vehicles (BEVs) and fuel cell hybrid electric vehicles are promising technologies, since they can reduce fossil fuel consumption, as well as enhance the integration of naturally fluctuating renewable energies. For instance, BEVs can be charged and discharged at different times and locations. Thus, it is treated as a flexible load (G2V) and storage in the power system, which can change the load both in time and space [31]. Meanwhile, BEVs can discharge electric power to the power system like generation units through vehicle to grid (V2G) technology. With the proper design and control strategies, BEVs can provide multiple ancillary services to the power system, such as frequency response [32].

1.2.5 Operation of integrated energy systems

The optimal operation and smart control of integrated energy systems can improve the sustainability, reliability, and cost efficiency of the whole system. Taking into account diverse energy conversion technologies and the coordination of different energy sectors, the energy services required by customers or system operators can be provided in many different ways. With centralized control, the entire smart energy system is generally managed by a single operator, and overall, the appropriate operation constitutes a large-scale centralized problem, which is more complicated than with an individual system. To improve computation efficiency and protect information privacy, distributed or decentralized solutions are desired to achieve independent yet coordinated operation [33]. In addition, there will be quite a lot of local control via integrators and aggregators, which also requires distributed operation and control. Apart from the aforementioned technical aspects, a proper market design with the right incentives and clear (i.e., stimulating and no mutually opposing) regulations will be required to ensure the effective operation of integrated smart energy systems as well.

1.3 Current status of integrated energy systems in China and Denmark

During the period of the 12th Five-Year Plan (2011–2015), coal consumption in China fell by 5.2% and the consumption of non-fossil fuels increased by 2.6% [34]. Before 2006, electricity generation came mainly from conventional thermal power units and hydropower units. To reduce CO_2 emissions, power generation from renewable sources of energy such as wind, solar, and hydro has developed rapidly in China during the past decade. It is expected that CO_2 emissions will peak at around 2030 and that the non-fossil-fuel share of primary energy will increase by 20% by the same year [35]. Moreover, the Chinese government has recently announced a target of achieving carbon neutrality by 2060. In addition, due to its higher conversion efficiency and lower environmental emissions, natural gas has attracted increasing attention, expecting to reach 15% of total fuel consumption in the whole energy sector by 2030. In northern China, the DHS is being adopted to supply heat to consumers. The CHP units and heat boilers cover 62.9% and 35.7% of heating production, respectively, with the rest mainly being supplied by industry waste heat and geothermal. However, the electric power and heat generation of CHP units depends on heat loads, which limits the operational region of the CHP units. CHP units must run when the heat is needed, leading to a high curtailment of wind power in the winter.

Since 2015, the National Energy Administration of China has issued several policies to support the development of integrated energy systems, including microgrids with high renewable penetration and an overall integrated energy system, referred to as the Energy Internet (http://www.nea.gov.cn/). The State Grid Tianjin Electricity Power Company is the first company to conduct demonstration projects of integrated energy systems, which would achieve coordinated management and control of the electricity, heating, and cooling fluxes and flows. The integrated energy system, if done properly, improves the cost efficiency of the whole system and reduces CO_2 emissions. The State Grid Jiangsu Electricity Power Company has completed a demonstration project of a district smart energy system with 70% penetration of renewables, which incorporates the electricity, heating, cooling, and transportation energy sectors. In addition, the China Southern Power Grid has investigated how to design and operate a smart energy system that includes the electricity, heating, gas, and transportation energy sectors. However, at present, the EPS, DHS, and gas systems are operated by different entities in China and are thus planned individually.

In 2018, electricity from renewables accounted for 60% of Denmark's domestic electricity supply, and wind power accounted for 40% [36]. In particular, the transition from fossil fuels to renewable energy for district heating is significant in Denmark. The percentage of renewables covered 60% of district heating production in 2018 [36]. Apart from securing adequate capacity through the connection with neighboring countries, the heating sector in Denmark plays a major role to provide flexibility for the EPS in integrating fluctuating wind power. The heating and electricity sectors are coupled through CHP plants, which generate around 70% of thermal energy in the Danish DHS. Since the electricity tax is being reduced gradually over time, electric boilers and heat pumps have attracted increasing attention. Combined with the electric boilers, heat pumps, and heat storage, CHP units can provide more flexibility to the EPS.

To facilitate the integration of wind power, Denmark has conducted numerous research projects on future integrated energy systems. For example, the EnergyLab Nordhavn project is a demonstration project for a dense and integrated future energy system. It demonstrates how electricity and heating, energy-efficient buildings, and electric transportation with the innovative use of data and analytics can be integrated into an intelligent, flexible, and optimized energy system [37]. A low-temperature district heating system incorporating smart energy network technologies, heat storage, energy-flexible buildings, decentralized supply options, and fuel-shift solutions has been developed. In the Copenhagen Nordhavn area, active participation by occupants of the low-energy buildings acting as agile consumers and users, and therefore becoming active energy-flexible elements, has been investigated. Another project, Centre for IT-Intelligent Energy Systems (CITIES), has developed methodologies and digital solutions for the analysis, operation, and development of integrated urban energy systems, with the ultimate aim of achieving independence from fossil fuels by utilizing the flexibility of the energy system through intelligence, integration, and planning [38]. The EnergyPlan tool has been developed by Aalborg University to design a 100% renewable energy system that includes electricity, heating, cooling, transportation, and industrial sectors. The EnergyPlan tool is investigating the modeling of all relevant energy generation units, energy storage, and energy conversion technologies [39].

1.4 Recommendations for further development of integrated energy systems

An efficient transition to a smart energy system requires intensive research and development efforts regarding the integration of various energy conversion techniques, system operation frameworks, digitalization, and communication systems, among others. The following suggestions for further research into and development of integrated energy systems are recommended:

- *Investigate new optimal operation frameworks and control strategies for multiple energy systems.* Given the fact that various energy sectors are managed by different entities and that the coordination of different energy sectors is insufficient at present, research should be conducted to coordinate various energy sectors with different operational timescales and characteristics while respecting the privacy of different entities. The development of integrated energy systems should focus on providing secure and reliable energy services to end users.

- *Design multi-energy carrier markets and develop new business model frameworks.* To distribute smart energy system costs and benefits across energy sectors and services efficiently, new regulations and business models should be developed. A corresponding demonstration acting as operational platforms for new business models is needed. In addition, the incentives needed for energy consumers and building management to adopt flexible consumption should be explored.
- *Develop solutions for the more efficient integration of energy storage and advanced energy conversion technologies to accommodate the growth in fluctuating renewable energy.* Optimal operation and smart control of the various energy infrastructures should be investigated in depth, enabling additional flexibility across these infrastructures to efficiently balance and utilize renewable energy, mainly integrated into the power system.
- *Design and develop low-energy buildings for a green transition.* Buildings play an important role as the main consumers in cities. Together with indoor climates and thermal inertia, the potential flexibility of buildings can be utilized. Advanced building energy management and control systems should be developed to interact with the (external) smart energy system and increase energy flexibility.
- *Develop integrated design and planning methods across energy sectors for integrated energy systems.* At present, there are no national policies and regulations regarding integrated energy systems in either China or Denmark. The coordinated design and planning should evolve to remove the barriers between the different energy sectors and facilitate the deployment of smart energy solutions.

1.5 Conclusion

Cross-sector integrated energy systems will be developed over the world to cope with the fluctuations and uncertainty from renewables in an efficient way. The integrated energy system will be the most efficient solution to increase the energy efficiency of systems and reduce environmental emissions. This chapter described the concept of the integrated energy system, integrating the electricity, heating, cooling, gas, and transportation sectors with high renewable energy penetration. The different timescales and characteristics of the different energy sectors create challenges for the coordinated operation of the different energy sectors. Many projects have been conducted to demonstrate the economic and environmental benefits of integrated energy systems. Further research and development are required to deal with the challenges to the green transition toward a smart energy system, including advanced technologies, novel market designs and business models, and consistent national regulations to remove the barriers between the different energy sectors.

References

[1] V Smil, Energy Transitions: History, Requirements, Prospects, Praeger Publishers, West Port, CT, 2010.

[2] International Energy Agency, World energy outlook 2019, 2019. https://www.iea.org/reports/world-energy-outlook-2019. (Accessed 16 April 2021).

[3] European Union. Energy roadmap 2050, 2012. https://ec.europa.eu/energy/sites/ener/files/documents/2012_energy_roadmap_2050_en_0.pdf. (Accessed 16 April 2021).

[4] UCLA Luskin Centre for Innovation, Progress toward 100% clean energy: in cities and states across the U.S., 2019. https://innovation.luskin.ucla.edu/wp-content/uploads/2019/11/100-Clean-Energy-Progress-Report-UCLA-2.pdf. (Accessed 16 April 2021).

[5] Y He, M Yan, M Shahidehpour, Z Li, C Guo, L Wu, et al., Decentralized optimization of multi-area electricity-natural gas flows based on cone reformulation, IEEE Trans Power Syst 33 (4) (2018) 4531–4542.

[6] International Renewable Energy Agency, Wind energy. https://www.irena.org/wind. (Accessed 16 April 2021).

[7] Z He, D Drozdov, J Wang, W Shen, C Li, W Li, Competitiveness of the wind power industry in China: an analysis based on the extended Diamond Model, J Renew Sustain Energy 12 (2020) 052701.

[8] Global Wind Energy Council, Global wind report 2019. https://gwec.net/. (Accessed 16 April 2021).

[9] Reve, China to account for over 25% global offshore wind power capacity by 2030, 2020. https://www.evwind.es/2020/08/25/china-to-account-for-over-25-global-offshore-wind-power-capacity-by-2030/76767. (Accessed 16 April 2021).

[10] Wind Europe, Wind energy in Europe in 2019, 2019. https://windeurope.org/wp-content/uploads/files/about-wind/statistics/WindEurope-Annual-Statistics-2019.pdf. (Accessed 16 April 2021).

[11] International Renewable Energy Agency, Solar energy. https://www.irena.org/solar. (Accessed 16 April 2021).

[12] HH Larsen, LS Petersen, DTU International Energy Report: Energy System Integrations for the Transition to Non-Fossil Energy Systems, Technical University of Denmark, Lyngby, 2015..

[13] C Tang, J Xu, Y Sun, J Liu, X Li, D Ke, et al., A versatile mixture distribution and its application in economic dispatch with multiple wind farms, IEEE Trans Sustain Energy 8 (4) (2017) 1747–1762.

[14] N Zhang, C Kang, Q Xia, J Liang, Modeling conditional forecast error for wind power in generation scheduling, IEEE Trans Power Syst 29 (3) (2014) 1316–1324.

[15] G Liu, T Jiang, TB Ollis, X Zhang, K Tomsovic, Distributed energy management for community microgrids considering network operational constraints and building thermal dynamics, Appl Energy 239 (2019) 83–95.

[16] BV Mathiesen, H Lund, P Nørgaard, Integrated transport and renewable energy systems, Util Policy 16 (2) (2008) 107–116.

[17] P Pinson, L Mitridati, C Ordoudis, J Ostergaard, Towards fully renewable energy systems: experience and trends in Denmark, CSEE J Power Energy 3 (1) (2017) 26–35.

[18] Z Li, W Wu, J Wang, B Zhang, T Zheng, Transmission-constrained unit commitment considering combined electricity and district heating networks, IEEE Trans Sustain Energy 7 (2) (2015) 480–492.

[19] Y He, M Yan, M Shahidehpour, Z Li, C Guo, L Wu, Y Ding, Decentralized Optimization of Multi-Area Electricity-Natural Gas Flows Based on Cone Reformulation, IEEE Trans Power Systems 33 (4) (2018) 4531–4542 In this issue, doi:10.1109/TPWRS.2017.2788052.

[20] H Lund, PA Østergaard, D Connolly, BV Mathiesen, Smart energy and smart energy systems, Energy 137 (2017) 556–565.

[21] X Yu, X Xu, S Chen, J Wu, H Jia, A brief review to integrated energy system and energy Internet, Diangong Jishu Xuebao/Trans China Electrotech Soc 31 (1) (2016) 1–13.

[22] F Teng, V Trovato, G Strbac, Stochastic scheduling with inertia-dependent frequency regulation, IEEE Trans Power Syst 31 (2) (2015) 1557–1566.

[23] E Guelpa, A Bischi, V Verda, M Chertkov, H Lund, Towards future infrastructures for sustainable multi-energy systems: a review, Energy 184 (2019) 2–21.

[24] P Siano, Demand response and smart grids: a survey, Renew. Sustain. Energy Rev. 30 (2014) 461–478.

[25] L Zhang, N Good, P Mancarella, Building-to-grid flexibility: modelling and assessment metrics for residential demand response from heat pump aggregations, Appl Energy 233–234 (2019) 709–723.

[26] Y Zong, S You, J Wang, Z.Y Dong, H Cai, C Traeholt, Investigation of real-time flexibility of combined heat and power plants in district heating applications, Appl Energy 237 (2019) 196–209.

[27] H Lund, S Werner, R Wiltshire, S Svendsen, J Eric, F Hvelplund, et al., 4th Generation District Heating (4GDH): integrating smart thermal grids into future sustainable energy systems, Energy 68 (2014) 1–11.

[28] M Wirtz, L Kivilip, P Remmen, D Muller, 5th Generation District Heating: a novel design approach based on mathematical optimization, Appl Energy 260 (2020) 114158.

[29] LH Hvidtfeldt, SP Leif, DTU International Energy Report 2015 : Energy Systems Integration for the Transition to Non-Fossil Energy Systems, Technical University of Denmark, Lyngby, 2015. https://backend.orbit.dtu.dk/ws/portalfiles/portal/119583507/DTU_International_Energy_Report_2015_rev.pdf. (Accessed 16 April 2021).

[30] Q Zeng, B Zhang, J Fang, Z Chen, A bi-level programming for multistage co-expansion planning of the integrated gas and electricity system, Appl Energy 200 (2017) 192–203.

[31] Shell International BV. Shell World Energy Model: a view to 2100, 2017. https://www.shell.com/energy-and-innovation/the-energy-future/scenarios/shell-scenarios-energy-models/world-energy-model/_jcr_content/par/textimage.stream/1510344160326/2ee82a9c68cd84e572c9db09cc43d7ec3e3fafe7/shell-world-energy-model.pdf. (Accessed 16 April 2021).

[32] T Qian, C Shao, X Li, X Wang, M Shahidehpour, Enhanced coordinated operations of electric power and transportation networks via EV charging services, IEEE Trans Smart Grid 11 (4) (2020) 3019–3030.

[33] K Knezović, M Marinelli, A Zecchino, PB Andersen, C Traeholt, Supporting involvement of electric vehicles in distribution grids: lowering the barriers for a proactive integration, Energy 134 (2017) 458–468.

[34] RJ Askjar, PB Andersen, A Thingvad, M Marinelli, Demonstration of a technology neutral control architecture for providing frequency control using unidirectional charging of electric vehicles, in: Proc. 2020 55th International Universities Power Engineering Conference (UPEC), 2020.

[35] J Tan, Q Wu, W Wei, F Liu, C Li, B Zhou, Decentralized robust energy and reserve co-optimization for multiple integrated electricity and heating systems, Energy 205 (2020) 118040.

[36] The development of energy in China, 2018 (in Chinese), 2018. http://www.chyxx.com/industry/201802/611045.html. (Accessed 16 April 2021).

[37] International Energy Agency, World energy outlook 2017, 2017. https://webstore.iea.org/download/direct/1055?fileName=World_Energy_Outlook_2017.pdf. (Accessed 16 April 2021).

[38] Danish Energy Agency, Energy statistics 2018, 2018. https://ens.dk/sites/ens.dk/files/Statistik/energy_statistics_2018.pdf. (Accessed 16 April 2021).

[39] EnergyLab Nordhavn, A smart city energy lab, 2019, 2019. http://www.energylabnordhavn.com/. (Accessed 16 April 2021).

CHAPTER 2

Mathematical model of multi-energy systems

2.1 Introduction

In past decades, with growing environmental concerns such as global climate change and local pollution, the international agreement to reduce greenhouse gas (GHG) emissions has attracted increasing attention, which encourages striving for cleaner air by developing renewable energy technologies [1]. Thus, many countries are making efforts toward shifting from fossil fuels to renewable energy sources (RESs).

Electric power and heat, as two main forms of energy carrier, play important roles in the development of human society [2]. Since electric power can easily be generated by primary energy sources (e.g., fossil fuels, gas, hydro, wind, and solar), and can be easily transmitted, transformed, and used, it has become the dominant energy carrier in society. Heat is another major form of energy demand in a human's life. In the global final energy consumption in 2018, heat accounted for around 50% as the largest end use, which was mainly used for industrial processes and for space and water heating in buildings [3].

Since the integrated electricity and heating system (IEHS) can improve energy efficiency and reduce CO_2 emissions, it has become a promising paradigm of the future energy system. Electric power systems (EPSs) and district heating systems (DHSs) are interconnected through coupling devices in both the energy source side and energy consumption side. A combined heat and power (CHP) plant generates electricity and heat simultaneously, which is a high-energy-efficiency technology that reuses the waste heat along with the electricity production [4]. Heat pumps and electric boilers are considered as electric loads that produce heat by consuming electric power. Heat pumps usually transfer heat from low-temperature heat sources (e.g., surrounding air sources and seawater) to high temperatures using electricity [5]. Electric boilers use a resistance heater to convert electricity to heat. In addition, the circulation pumps in DHSs also need to consume electricity to drive the water flow in DHSs.

Optimal operation of integrated multi-energy systems under uncertainty. Copyright © 2022 Elsevier Inc.
DOI: https://doi.org/10.1016/B978-0-12-824114-1.00009-3 All rights reserved.

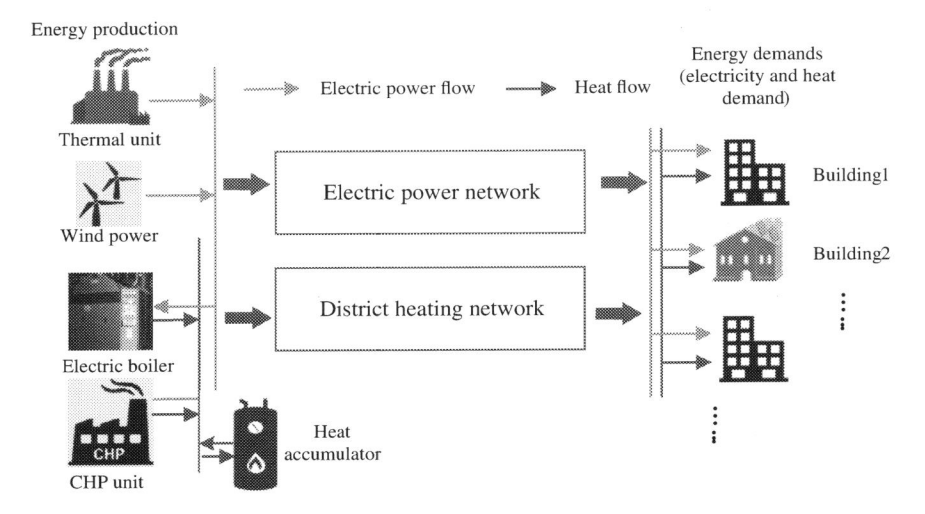

Figure 2.1 Schematic structure of an IEHS.

Fig. 2.1 shows a schematic structure of an IEHS. As can be seen, electric power is produced by generators, and transmitted from power sources to electric demands through electric power networks. Heat energy produced by heat sources is carried by hot water and then delivered to consumers through district heating networks. Electric energy storages and heat accumulators are also deployed to store excess power and heat energies in the system, which can provide substantial flexibility to integrate more fluctuating RESs [6]. The electric power network usually consists of two parts: the transmission network and distribution network. A large amount of power produced by large generators is transmitted to substations with a high-voltage level, and then distribution networks with a low-voltage level deliver the power from transmission networks to end users via step-down transformers. Similarly, the heat energy produced by large-scale heat sources is delivered to heat exchange stations via primary heat pipelines and then supplied to consumers via secondary heat pipelines. The district heating network consists of supply pipelines with a temperature of 70 to 120°C delivering heat to load nodes and return pipes with a temperature of 35 to 50°C carrying mass flow back to heat source nodes [7].

Electric power and heat energy have different characteristics, and the mathematical models corresponding to the two energy systems are different [8]. The most obvious feature of electricity is that the produced electricity should be consumed immediately (i.e., the power production and consumption in EPSs need to be balanced all the time). Electric power can easily be transmitted through a long distance and the power loss is relatively low, and thus generators generally are located far from end users. Electricity is expensive to store, whereas heat energy is easy to store. In addition, the flow rate of heat energy is much lower and there are many heat losses during the heat delivery; therefore, heat demands are locally supplied by DHSs. Taking into account the thermal inertia of district heating networks and buildings, a certain amount of thermal energy storages from district heating networks and buildings can be utilized [9]. To summarize, this chapter introduces the mathematical model of multi-energy systems, which consists of electric power system, district heating system and natural gas system. With the rapid development of energy conversion technologies, e.g. power-to-heat and power-to-gas, the coupling among these energy systems are intensified these years. First, the steady-state models of different kinds of coupling devices are introduced in Section 2.2, including CHP plants, heat pumps, and electric boilers. The mathematical model of the district heating network is presented in Section 2.3, which comprises hydraulic model and thermal models. Then, Section 2.4 shows the mathematical model of the electric power network. Finally, the overview and mathematical model of the natural gas system is given in Section 2.5.

2.2 Modeling of coupling devices

The implementation of coupling devices connects the EPS and DHS, which enables the interaction between these two systems. In this chapter, the detailed models of coupling devices, including CHP units, heat pumps, electric boilers, and water pumps, are introduced in the following chapters.

2.2.1 CHP unit model

The CHP unit is the component that produces electric power while capturing available heat energy produced during the process. According to different prime movers, CHP units can be classified into different types, including the reciprocating engine, gas turbine, steam turbine, fuel cell, and microturbines [4]. Different types of CHP units have different operation constraints. In this chapter, the reciprocating engine, gas turbine, and steam turbine are introduced and applied on the IEHS.

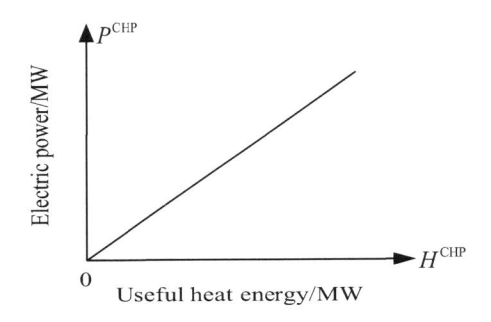

Figure 2.2 Electric power and heat production of CHP units with reciprocating engines or gas turbines.

Reciprocating engine and gas turbine. The relationship between heat and electricity produced by CHP units with reciprocating engines and gas turbines can be simplified as a linear heat to power ratio [10]. The mathematical expression for such relationship between heat and electricity is shown in Eq. (2.1):

$$H^{\mathrm{CHP}} = \eta^{\mathrm{CHP}} P^{\mathrm{CHP}}, \tag{2.1}$$

where H^{CHP} is the available heat production of the CHP unit, P^{CHP} is the electric power production of the CHP unit, and η^{CHP} is the power-to-heat ratio.

Fig. 2.2 shows the relationship between the heat and electricity production of CHP units with reciprocating engines or gas turbines.

Steam turbine. CHP units with steam turbines can be divided into two main types: back-pressure and extraction [4]. Back-pressure turbines gather the exhaust steam at atmospheric pressures and above. Low-pressure steam is usually used for district heating, whereas high-pressure steam is often used in industrial processes. An extraction turbine has one or more openings to extract a portion of the steam at some intermediate pressure, which is shown in Fig. 2.3.

By adjusting the extraction ratio, the extraction turbine can operate from full condensing to full extraction modes, and thus the ratio of heat output for district heating and electric power output can be adjusted. The adjustment of the power–to–heat ratio is realized by controlling the opening degree of valves in the district heating network [11]. The operation region of a CHP unit with extraction steam turbine is illustrated in Fig. 2.4. Line AB shows the fully condensing operation mode of the steam turbine without heat output. Line CD in Fig. 2.4 represents the back-pressure mode. Lines

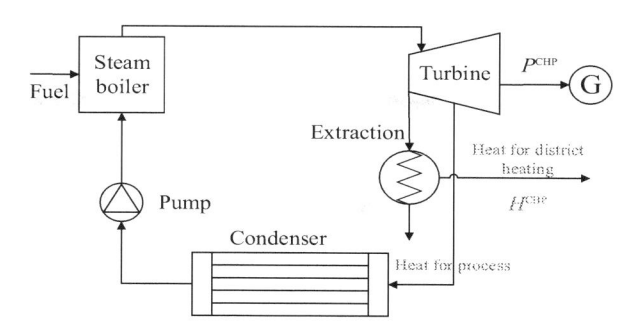

Figure 2.3 Schematic structure of steam turbine–based CHP units with extraction.

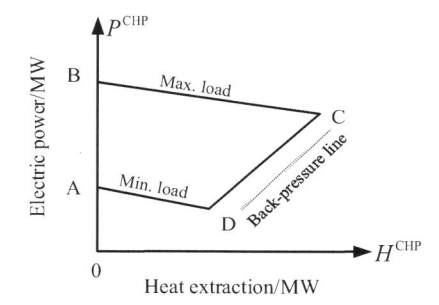

Figure 2.4 Convex operation region of extraction CHP unit.

BC and AD show the operation of the steam turbine with the maximum and minimum load.

With regard to the extraction CHP unit with a convex feasible operation region as shown in Fig. 2.4, the corresponding mathematical model of the electricity and heat production is formulated as a convex combination of the extreme points of the feasible operation region [12], which is modeled as follows:

$$P^{\mathrm{CHP}} = \sum_{k=1}^{M_i} \alpha_k \mathrm{P}_k^{\mathrm{CHP}}, \tag{2.2}$$

$$H^{\mathrm{CHP}} = \sum_{k=1}^{M_i} \alpha_k \mathrm{H}_k^{\mathrm{CHP}}, \tag{2.3}$$

$$\sum_{k=1}^{M_i} \alpha_k = 1, \ 0 \le \alpha_k \le 1, \ \forall k \in \mathrm{K}, \tag{2.4}$$

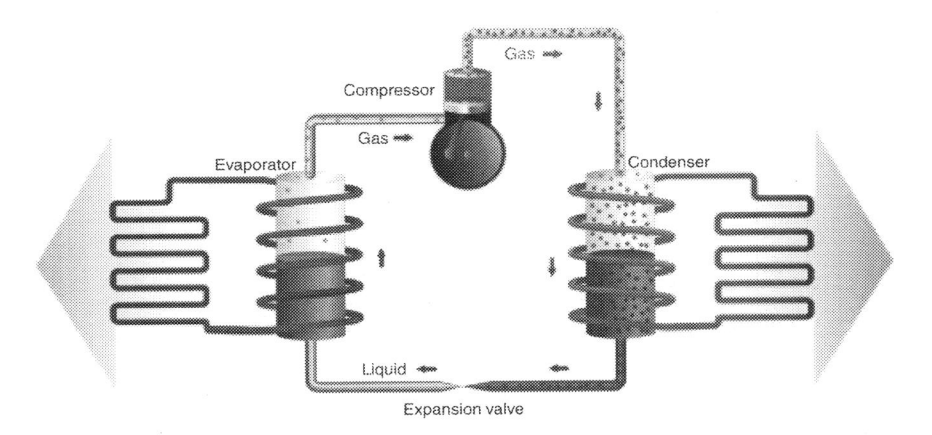

Figure 2.5 Schematic structure of a heat pump.

where $P_{i,k}^{CHP}$ and $H_{i,k}^{CHP}$ are the electric power and heat production on the kth vertex of the feasible operation region of the CHP unit. $\alpha_{i,k}^{CHP}$ is the combination coefficient.

2.2.2 Heat pump model

Heat pumps are able to provide both required heating and cooling to the internal space by consuming external electricity for compression [13]. When working in a heating mode, the heat pump extracts heat energy from a low-temperature heat source and releases it to heat sinks with higher temperatures. Fig. 2.5 shows the schematic structure of a heat pump, including an evaporator, compressor, condenser, and expansion valve. On the heat source side, the refrigerant that is a volatile evaporating fluid enters the evaporator and then absorbs heat from the low-temperature heat sources during evaporation. On the heat sink side, the refrigerant vapor is compressed and condensed into liquid, and the heat is thus released to the heat sink during condensation [14]. According to various heat sources, the main types of heat pumps are air-to-air, waste source, and geothermal. The applications of heat pumps in the smart grid can be found in the work of Fischer and Hatef [5].

The available heat from a heat pump depends on its coefficient of performance (COP), which is defined as the ratio of heat output and consumed electricity:

$$H^{\text{HP}} = \eta^{\text{HP}} P^{\text{HP}}, \tag{2.5}$$

where H^{HP} is the available heat production of the heat pump, P^{HP} is the electric power consumption of the heat pump, and η^{HP} is the COP.

The COP of heat pumps is generally around 3 to 5, which is higher compared with the COP around 1 of conventional electrical resistance heaters. The COP of heat pumps is influenced by the temperature difference between the heat source and heat sink [15]. The larger the temperature difference, the lower the efficiency (i.e., COP) of the heat pump.

2.2.3 Electric boiler model

Similar to heat pumps, the electric boiler is also a kind of power-to-heat component, which usually heats water through resistance heating elements. The relationship between the heat output and electricity consumption of the electric boiler is linear, which is depicted as follows:

$$H^{EB} = \eta^{EB} P^{EB} \qquad (2.6)$$

where H^{EB} and P^{EB} are the available heat production and electric power consumption of electric boiler, respectively. η^{EB} is the COP of the electric boiler, which is usually around 1.

2.2.4 Water pump model

In the district heating network, water pumps are installed to compensate for the accumulated pressure drops along pipelines due to friction [16]. The consumed electric pumping power is proportional to the mass flow at the water pump m_{pump} and total pressure losses in critical pipelines, which is calculated as follows:

$$P_{pump} = \frac{m_{pump} \sum\limits_{p \in \Phi_c^P} hl_p}{\eta_{pump}\rho} \qquad (2.7)$$

where hl_p is the pressure loss along pipeline p, which is introduced in detail in Section 2.2.1. η_{pump} is the conversion efficiency of the water pump, and ρ is the water density.

2.3 Mathematical model of the district heating network

Similar to transmission and distribution networks in the EPS, there are primary and secondary district heating networks in the DHS [17], which are connected by heat exchange stations. The primary district heating networks

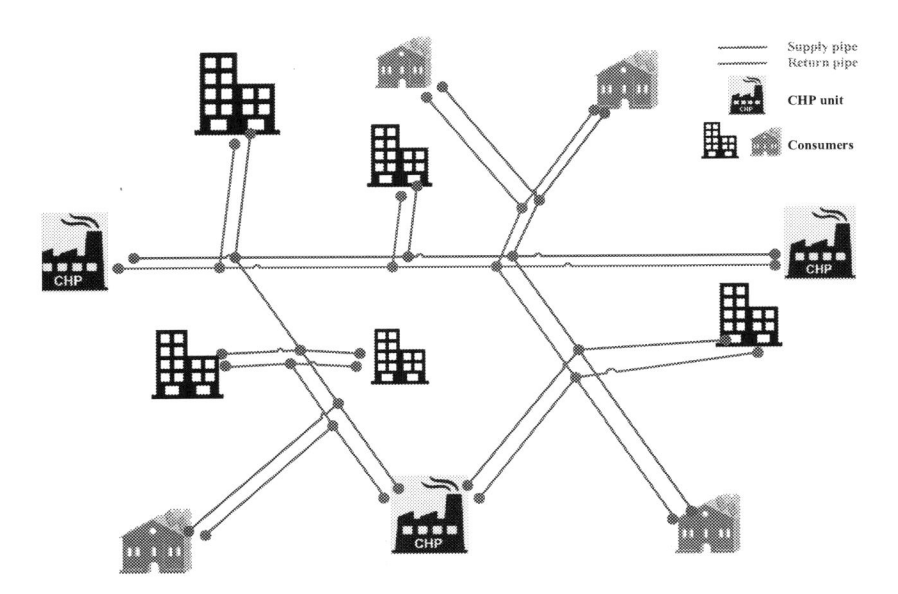

Figure 2.6 Schematic figure of a district heating network.

are responsible for delivering heat to heat exchange stations in a large area, whereas the secondary district heating networks distribute heat to end users at a local or building level. The pipe network in DHSs transports heat energy in the form of steam or hot water from heat sources to the consumers. It includes supply pipes with high temperature delivering heat energy to load points and return pipes with low temperature carrying mass flow back to heat sources. Fig. 2.6 illustrates a simplified structure of the DHS.

During operation of the district heating network, the mass flow conservation and heat energy conservation are satisfied. In addition, the principles between fluid friction and energy dissipation are also taken into account. Generally, the steady-state model of district heating networks includes two parts: the hydraulic model and the thermal model [18]. The goal of the hydraulic model is to obtain mass flows within each pipe and the mass flow flowing in or out of each node. There are three basic rules in the analysis of the hydraulic model—continuity of flow, loop pressure equation, and head loss equation, which are introduced in detail in Section 2.3.1, whereas the aim of the thermal model is to calculate the temperature at each node in supply pipelines and return pipelines. The temperature mixture at heating nodes and temperature drop along pipelines are modeled, which

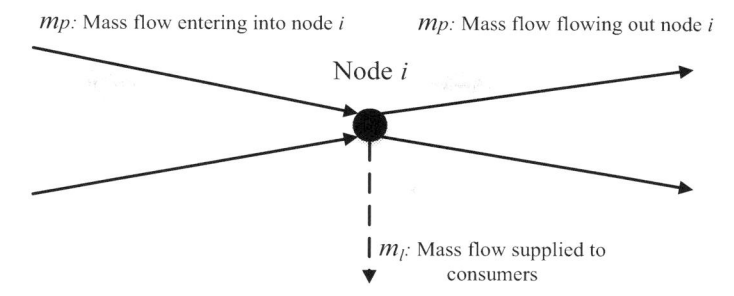

Figure 2.7 Mass flows entering into and flowing out node *i*.

are shown in Section 2.3.2. Typically, the DHS has different operation modes, including constant mass flow and constant temperature (CF–CT), constant mass flow and variable temperature (CF–VT), variable mass flow and constant temperature (VF–CT), and variable mass flow and variable temperature (VF–VT) [19].

2.3.1 Hydraulic model

Continuity of mass flow. The continuity of mass flow rate means that for each node in the district heating network, the mass flows entering the node are equal to the mass flows leaving the node. Fig. 2.7 depicts the relationship when there are multiple pipes connected to the node simultaneously.

The continuity of mass flow in both the supply network and return network can be expressed as follows, and the steady-state energy flow model is considered here:

$$\sum_{j\in\Omega_{nd}^{HS}} m_j^{HS} + \sum_{p\in\Omega_{nd}^{S,pipe}} a_{nd,p} m_p^{S,pipe} - \sum_{l\in\Omega_{nd}^{HL}} m_l^{HL} = 0, \ \forall nd \in \Phi^{HN}, \qquad (2.8)$$

$$\sum_{l\in\Omega_{nd}^{HL}} m_l^{HL} - \sum_{p\in\Omega_{nd}^{R,pipe}} a_{nd,p} m_p^{R,pipe} - \sum_{j\in\Omega_{nd}^{HS}} m_j^{HS} = 0, \ \forall nd \in \Phi^{HN}, \qquad (2.9)$$

where $a_{nd,p}$ is defined in the following matrix **A** in Eq. (2.10), denoting the network matrix that connects heating nodes with all supply pipelines [18]. The possible values of $a_{nd,p}$ are +1, −1, and 0 according to the connection of node nd and pipeline p. The +1 and −1 represent the mass flow entering or leaving the node nd through pipeline p, respectively. The 0 represents that there is no connection between pipeline p and node nd.

$$\mathbf{A} = \begin{bmatrix} a_{1,1} & a_{1,2} & \cdots & a_{1,p} \\ a_{2,1} & a_{2,2} & \cdots & a_{2,p} \\ \vdots & \vdots & \ddots & \vdots \\ a_{nd,1} & a_{nd,2} & \cdots & a_{nd,p} \end{bmatrix} \tag{2.10}$$

Head loss and loop pressure. To ensure that the hot water is pumped and delivered to end users, the pressure at nodes in the heating network should be maintained at a suitable level. In the hydraulic model, the pressure in pipelines decreases along with the direction of mass flow due to friction, which is referred to as head losses. The head loss along pipeline p connecting two nodes is described as follows:

$$hl_p = K_p m_p |m_p|, \tag{2.11}$$

where hl_p is the head loss along pipeline p. K_p is the resistance coefficient of pipeline p, and the detailed derivation of K_p can be found in the work of Larock et al. [20]. m_p is the mass flow rate in pipeline p.

The loop pressure equation has to be applied when there is a loop or more than one loop in the heating network. The loop pressure equation for one closed loop means that the sum of all head loss along the pipes belonging to the chosen loop is equal to 0. Given Φ_l^P denoting a set of pipelines belonging to the closed loop $l, l \in \Phi^L$, the mathematical expression of loop l is given as follows:

$$\sum_{p \in \Phi_l^P} hl_p = \sum_{p \in \Phi_l^P} K_p m_p |m_p| = 0, \ \forall l \in \Phi^L. \tag{2.12}$$

2.3.2 Thermal model

The temperature at each node in the district heating network is obtained using the thermal model. To guarantee the comfort of consumers, the temperatures of mass flows in supply networks need to be maintained at a specific level. After consumers extract heat energy, the temperature of mass flow decreases and the mass flow returns to heat stations and is heated again. This circulation corresponds to the hydraulic model mentioned in Section 2.3.1. According to the heat energy conservation, the quantity and direction of mass flow in the pipelines will affect the temperature distribution in the overall network [21]. Fig. 2.8 shows an overview of the temperature distribution in a radial network. The temperature changes

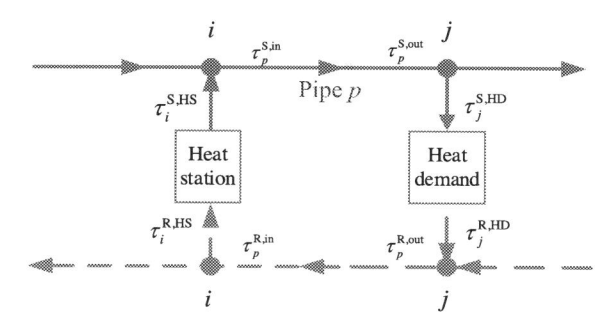

Figure 2.8 Temperature related to pipelines and nodes.

in the district heating network mainly depends on three parts. First, the temperature of mass flow increases (decreases) at the source (demand) node by charging (discharging) heat energy. Second, the temperature of mass flow decreases due to energy losses, which radiate from pipelines to the ambient environment. Third, the temperature at each node is determined by the mixture of mass flow with multiple incoming pipes connected to the node. The detailed models of these temperature changes are described in the following.

Heat source and heat demand. As introduced in Section 2.2, CHP units, heat pumps, and electric boilers are effective heat sources. The heat energy generated from these heat sources is used to heat water or steam to fulfill heat demands. The temperature of mass flow in the heat source is raised from $\tau_i^{R,HS}$ in return pipelines to $\tau_i^{S,HS}$ in supply pipelines, and the heat production from heat sources is described by Eq. (2.13):

$$H_i^{HS} = c \cdot m_i^{HS} \cdot \left(\tau_i^{S,HS} - \tau_i^{R,HS} \right), \forall i \in \Phi^{HS}, \tag{2.13}$$

where c is the specific heat of water and m_i^{HS} is the mass flow at the heat source.

However, the heat energy is extracted by consumers at heat demand nodes. After heat extraction, the temperature of mass flow in the heat demand is reduced from $\tau_i^{S,HD}$ in supply pipelines to $\tau_i^{R,HD}$ in return pipelines. Denoting the mass flow at heat demand as m_i^{HD}, the consumed heat demand is expressed as follows:

$$H_i^{HD} = c \cdot m_i^{HD} \cdot \left(\tau_i^{S,HD} - \tau_i^{R,HD} \right), \quad \forall i \in \Phi^{HD} \tag{2.14}$$

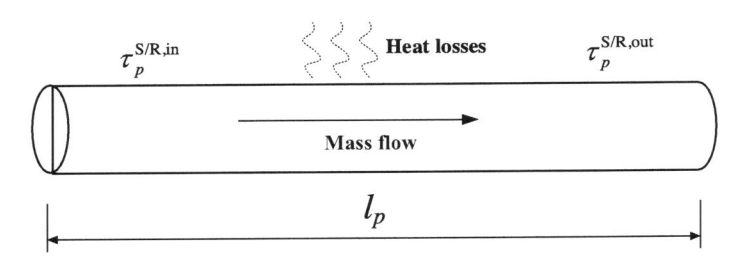

Figure 2.9 Temperature and heat losses along pipelines.

Usually, the supply temperature at the source node is specified and the return temperature at the demand node depends on the heat energy consumed by consumers.

Temperature drop. Fig. 2.9 presents the temperatures distributed along a pipeline. Since the temperature of mass flow is generally higher than the ambient temperature, there will be heat losses (i.e., radiation of heat energy to the ambient environment) when the mass flow passes through the pipelines. In this regard, temperature drop occurs in the supply and return pipelines caused by heat losses.

Denoting the temperature of mass flow in pipeline p at starting and ending points as $\tau_p^{S/R,in}$ and $\tau_p^{S/R,out}$, respectively, the temperature drops in supply and return pipelines are given in Eqs. (2.15) and (2.16):

$$\tau_p^{S,out} - \tau^a = \left(\tau_p^{S,in} - \tau^a\right) e^{\frac{-\lambda l_p}{c \cdot m_p^{S,pipe}}}, \forall p \in \Phi^{S,pipe} \tag{2.15}$$

$$\tau_p^{R,out} - \tau^a = \left(\tau_p^{R,in} - \tau^a\right) e^{\frac{-\lambda l_p}{c \cdot m_p^{R,pipe}}}, \forall p \in \Phi^{R,pipe} \tag{2.16}$$

where τ^a is the ambient temperature, λ is the heat transfer coefficient of pipelines, and l_p is the length of pipeline p.

As can be seen, the temperature drop along pipelines is affected by the mass flow rate, the length of pipelines, and the temperature difference between the temperature of mass flow and the ambient environment. Even though increasing the mass flow rate can reduce temperature drops, it will consume more electricity to pump hot water. Thus, it is reasonable to increase the mass flow rate ultimately to reduce the temperature drop. However, it is possible to reduce a certain amount of heat losses adopting a lower supply temperature in district heating networks. The supply and return temperatures in low–temperature district heating are usually set at 50 to 70°C and 30 to 40°C, respectively. The conception and development of low-temperature district heating can be found in the work of Lund et al. [22].

Temperature mixture. As shown in Fig. 2.7, it is common that multiple pipelines connect to the same node, where the mass flows enter into or flow out of one node simultaneously. There are three temperature variables related to each node: the temperature of mass flow $\tau_p^{S/R,in}$ at the starting point in pipeline p, the temperature of mass flow $\tau_p^{S/R,out}$ at the ending point in pipeline p, and the temperature at node nd $\tau_{nd}^{S/R,ND}$.

When there are multiple pipelines flowing into the same node, the water temperature of all pipelines are mixed. The temperature at each heat node is calculated as the mixture temperature of mass flows flowing into the node. The mathematical models of the temperature mixture in the supply and return networks are given by Eqs. (2.17) and (2.18) according to the energy conservation law.

$$\sum_{p\in\Omega_{nd}^{S,E,pipe}}\left(m_p^{S,pipe}\cdot\tau_p^{S,out}\right)=\tau_{nd}^{S,ND}\sum_{p\in\Omega_{nd}^{S,B,pipe}}m_p^{S,pipe},\quad\forall nd\in\Phi^{HN}\qquad(2.17)$$

$$\sum_{p\in\Omega_{nd}^{R,E,pipe}}\left(m_p^{R,pipe}\cdot\tau_p^{R,out}\right)=\tau_{nd,}^{R,ND}\sum_{p\in\Omega_{nd}^{R,B,pipe}}m_p^{R,pipe},\quad\forall nd\in\Phi^{HN}\qquad(2.18)$$

Moreover, all of the water in the pipelines flowing out of the same node has the same temperature, which is equal to the mixed temperature at this node. The mathematical model is given by Eqs. (2.19) and (2.20).

$$\tau_p^{S,in}=\tau_{nd}^{S,ND},\forall nd\in\Phi^{HN},\forall p\in\Omega_{nd}^{S,B,pipe}\qquad(2.19)$$

$$\tau_p^{R,in}=\tau_{nd}^{R,ND},\forall nd\in\Phi^{HN},\forall p\in\Omega_{nd}^{R,B,pipe}\qquad(2.20)$$

The water temperature in both the supply and return networks is constrained by the following limits.

$$\tau^{S,min}\leq\tau_{nd}^{S,ND}\leq\tau^{S,max},\forall nd\in\Phi^{HN}\qquad(2.21)$$

$$\tau^{R,min}\leq\tau_{nd}^{R,ND}\leq\tau^{R,max},\forall nd\in\Phi^{HN}\qquad(2.22)$$

Heat accumulator. The heat accumulator can store and release heat energy in an insulated heat storage tank, which connects the supply network and return network. It can improve the flexibility of DHSs by storing excess heat energy and then releasing it when needed [23]. The high-temperature water from supply pipelines is injected into heat accumulator, and the low-temperature water in the heat accumulator flows out through return pipelines during the charging process. During the discharging process, the high-temperature water is released from heat accumulators and injected into supply networks. The mathematical model of heat accumulators includes the charging/discharging heat ΔH_i^{HA} and the heat energy level H_i^{HA}, which are shown next.

$$H_{i,t+1}^{\mathrm{HA}} = H_{i,t}^{\mathrm{HA}} + \Delta H_{i,t}^{\mathrm{HA}}, \ \forall i \in \Phi^{\mathrm{HA}}, \forall t \in T \tag{2.23}$$

$$-\Delta H_i^{\mathrm{HA,max}} \leq \Delta H_{i,t}^{\mathrm{HA}} \leq \Delta H_i^{\mathrm{HA,max}}, \ \forall i \in \Phi^{\mathrm{HA}}, \forall t \in T \tag{2.24}$$

$$H_i^{\mathrm{HA,min}} \leq H_{i,t}^{\mathrm{HA}} \leq H_i^{\mathrm{HA,max}}, \ \forall i \in \Phi^{\mathrm{HA}}, \forall t \in T \tag{2.25}$$

2.4 Mathematical model of the electric power network

Power flow is a basic concept in power system analysis. Power flow calculation determines the steady-state power flow across the overall electric power network. Generators and loads are generally connected to the busbar in the electric power network, and different busbars are connected with transmission lines or distribution lines. In reality, each line has resistance and reactance resulting in active power loss and reactive power loss. There are four parameters or variables at each bus: active power (P), reactive power (Q), voltage angle (θ), and voltage magnitude (V). The original mathematical model of power flow for AC systems is nonlinear with complex numbers. To simplify the power flow calculation, the original nonlinear equations are approximated by linearized DC power flow [24]. We have the following several assumptions: (1) the line resistance is much smaller than its reactance, so the line resistance is ignored, as well as the reactive power; (2) the voltage at each bus is kept at nominal value; and (3) the angle differences are small such that $\sin(\theta_1 - \theta_2) = \theta_1 - \theta_2$.

For each bus, the power flow balance should be maintained—that is, the electricity production injected into the bus should be equal to the total electric power flow out of the bus. The angle of the reference bus is set as 0:

$$\sum_{g \in \Omega_n} P_g^G - \sum_{d \in \Omega_n} P_d^D - \sum_{m \in \Omega_n} B_{nn}(\theta_n - \theta_m) = 0, \ \forall n \tag{2.26}$$

$$\theta_{ref} = 0 \tag{2.27}$$

where P_g^G is the power production, P_d^D is the electricity demand, and B_{nn} is the line susceptance.

For each transmission line, the power flow is limited by line capacity P_{nm}^{cap}.

$$-P_{nm}^{cap} \le B_{nm}(\theta_n - \theta_m) \le P_{nm}^{cap}, \quad \forall n, m \in \Omega_n \qquad (2.28)$$

For each generator and demand, the power production and consumption is limited by their minimum and maximum outputs P_{nm}^{cap}.

$$P_g^{G,\min} \le P_g^G \le P_g^{G,\max}, \quad \forall g \qquad (2.29)$$

$$0 \le P_d^D \le P_d^{D,\max}, \quad \forall d \qquad (2.30)$$

2.5 Modeling of the natural gas system

An overview of the natural gas system is given in Section 2.5.1. The modeling of the natural gas system is presented in Section 2.5.2. Then, Sections 2.5.3 and 2.5.4 demonstrate the conversion to the unity system and linearization procedure. Finally, integration of the natural gas system with the EPS and DHS is demonstrated in Section 2.5.5.

2.5.1 Natural gas system overview

Natural gas is extracted from the gas fields and processed before being used for pipeline transportation or commercial use. The natural gas system resembles the EPS [25]. The similarity between the natural gas system and the EPS is shown in Fig. 2.10. Both the EPS and natural gas system have four main components: suppliers, transmission and distribution system, operators, and consumers. The main difference between the two systems is that the electric power is not the primary form of energy as natural gas is. To generate electricity, a primary fuel is required, and a primary fuel can be natural gas. Thus, natural gas can be used as a fuel to produce electricity. Another difference worth mentioning is the ability of storing electricity and natural gas. Natural gas can be efficient and easy to store compared to electricity [25].

The natural gas system is divided into the transmission and distribution level [26]. The natural gas transmission system is operated at high pressures, whereas the natural gas distribution system is operated under lower pressures [27]. The transmission and distribution system are connected through meter and regulating (M/R) stations [26]. The M/R stations regulate the pressure down to the distribution system pipelines. As the natural gas is an odorless,

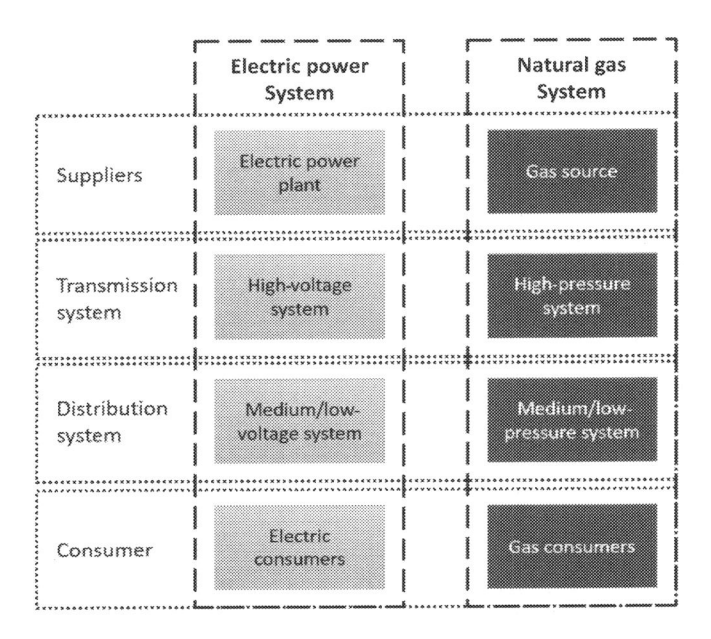

Figure 2.10 Comparison and resemblance of the EPS and natural gas system.

colorless, and tasteless gas, for safety measures, odorant is added at M/R stations that lead toward the distributions system. Thus, gas leaks can be detected.

Transmission system operators are responsible for assessment of the natural gas offtake at each M/R station based on the reporting of distribution system operators. Through such assessment, the required capacity is provided to the distribution system [26]. As the gas is supplied from different sources, either gas fields, imports from neighboring countries, storages, or upgraded biomethane plants, different gas qualities can be expected. According to transmission system operators for natural gas, the specifications are set for the quality of natural gas in the pipelines and the composition must match the requirements and specifications. Therefore, the natural gas supplied from the transmission level must ensure a certain quality of natural gas. The natural gas quality is usually measured by the Wobbe index, calorific value, and relative density based on the composition of natural gas. Further reading on gas quality specifications can be found in the work of Energinet [28], and a summary is given in the following.

Natural gas composition varies. However, it is mostly formed of methane. The composition often includes ethane, propane, nitrogen, CO_2, and hydrogen, among others [29]. The composition of the natural gas system is measured at different locations to further calculate the calorific value and Wobbe index. The following definitions are introduced for a better understanding of the modeling of the natural gas system. The definitions are summarized based on other works [26, 28–31].

The calorific value is the quantity of heat generated during combustion of natural gas. Considering that the payments are done for the amount of energy delivered and not for the amount of gas delivered, the energy a consumer receives is calculated by multiplying the delivered amount of gas by the calorific value. Therefore, natural gas is expressed in kWh.

The gross calorific value is the quantity of heat generated during combustion of $1m^3$ of natural gas at constant pressure when the air and gas have a combustion temperature of 25°C. Such combustion results in water formed in the liquid state. The gross calorific value is expressed in kWh/m^3 or MJ/m^3.

The specific gravity, also called *relative density*, is calculated as the ratio of natural gas density and air density at the same pressure and temperature. Specific gravity indicates the amount of hydrocarbons in the natural gas composition (i.e., how heavy the gas is compared to the air at the same conditions). Usually, the absolute pressure of 1.01325bar and 0°C is used.

The Wobbe index is the gross calorific value divided by a square root of the relative density and often is expressed in kWh/m^3 or MJ/m^3. It indicates a heat input the burner is exposed to during combustion of the fuel and a safety measure. A very high Wobbe index can result in carbon monoxide production and thermal overload.

2.5.2 Mathematical model of the natural gas system

In this section, the mathematical model for the natural gas system will be introduced. In addition to division into transmission and distribution systems, the natural gas system includes the following components. The nodes are divided into supply nodes, demand nodes, and intermittent nodes [25]. At the supply nodes, the gas is either injected from the gas source, imported from the neighboring countries, or injected from an upgrading biogas power plant (i.e., biomethane input). The demand nodes are nodes where the gas flows out of the system toward gas-fired power plants and other natural gas consumers.

The connection between two nodes is a pipeline. The pipelines can be divided into active and passive pipelines [25]. Active pipelines include a compressor compared to passive pipelines excluding the compressor. Compressors increase the pressure difference between two nodes of the pipeline and therefore increase the gas flow in the pipelines. In addition to the gas source, load, and compressor, the natural gas system can also include units such as gas storages and power–to–gas (P2G) units.

Hence, first, the pipeline flow rate will be demonstrated through two equations. It must be noted that there are several equations expressing the flow dependency on the gas properties, pipeline length and diameter, temperature, and so forth. The pipeline flow expressions are demonstrated in Section 2.5.2.1. Second, the nodal gas balance equation are introduced and the limitations are illustrated for the variables in Sections 2.5.2.2 and 2.5.2.3, respectively. Then, compressor modeling is demonstrated in Section 2.5.2.4. Furthermore, the modelling of the gas storages is given in Section 2.5.2.5. Finally, linepack modeling is presented in Section 2.5.2.6. The equations are given in SI units. The sets used for a mathematical model are defined next.

Sets and indices

$(nm) \in \Omega^{LP}$	represents the set of pipelines linepack storages
$(nm) \in \Omega^{NGS}$	notes pipeline from node n to node m
$n, m \in \Omega^{NGS}$	notes node n and node m
$d \in \Omega^{GD}$	stands for gas demand
$g \in \Omega^{GS}$	notes gas source
$ic \in \Omega^{IC}$	notes interconnections
$n \in \Omega^{NGS}$	notes node n
$p \in \Omega^{P2G}$	notes power–to–gas unit
$s \in \Omega^{ST}$	notes gas storages
$t \in \Omega^{T}$	notes time

2.5.2.1 Pipeline gas flow equation

Before continuing to the flow equation, the following simplifications are introduced. First, the equations are presented in steady state [27, 32]. Furthermore, viscosity (i.e., the fluid resistance to flow in the pipeline) highly depends on the pressure and temperature. When a gas temperature increases, the viscosity increases, and consequently the gas flow quantity decreases. Therefore, it is important to operate at a lower gas temperature to have higher injection of the gas in the pipeline. However, it is computationally

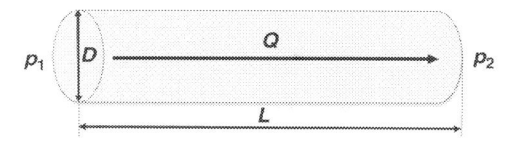

Figure 2.11 Pipeline parameters.

hard to take the gas and ambient temperatures, as well as thermal conductivities, into account. Hence, constant gas temperature along the pipeline, called *isothermal flow*, is assumed. The approximation is valid for long pipelines, as the gas temperature will reach a constant value [32].

A gas pipeline with important parameters is illustrated in Fig. 2.11. The pressures are noted as p, diameter is noted as D, length is noted as L, and gas flow is noted as Q. Notation p_1 is often called *inlet or upstream pressure*, whereas p_2 is *outlet or downstream pressure*. By considering isothermal flow, when the gas flows through the pipeline, the gas pressure drops along the pipeline due to friction and the gas expands [27, 32]. Therefore, to have the gas flow in the pipeline, it is important to have a pressure difference between two ends of the pipeline. The pressure difference acts as a force for the gas flow in the pipeline. In the situation of upstream and downstream pressure being equal, there will be no gas flow in the pipeline.

The flow equation for the gas pipeline, based on the explanations given, is stated as follows [33, 32]:

$$\left((p_n)^2 - (p_m)^2\right)C_{nm}^2 = (Q_{nm})^2, \quad \forall(nm) \in \Omega^{NGS} \tag{2.31}$$

where p_n and p_m are inlet and outlet pressures in the pipeline at a certain time t, respectively. The pipeline parameters are expressed through parameter C_{nm}. Q_{nm} represents a gas flow in the pipeline n-m. Hence, Eq. (2.31) illustrates the relationship between gas flow, gas pressures, and pipeline parameters.

Parameter C_{nm} can be expressed by several equations [29, 32]. According to the general flow equation, C_{nm} is expressed as follows:

$$C_{nm} = C\frac{T_b}{p_b}D_{nm}^{2.5}\left(\frac{1}{L_{nm}\gamma_g T_a Z_a f}\right)^{0.5}\eta_{p,nm}, \quad \forall(nm) \in \Omega^{NGS} \tag{2.32}$$

where

$\quad C = 47.8917 \times 10^{-6}$,

$\quad T_b =$ temperature based on normal cubic meter conditions (K),

$\quad p_b =$ pressure based on normal cubic meter conditions (MPa),

D_{nm} = pipeline diameter (mm),

L_{nm} = pipeline length (km),

γ_g = relative density of natural gas,

T_a = average absolute gas temperature (K),

Z_a = average compressibility factor,

f = friction constant, and

$\eta_{p,nm}$ = efficiency of the pipeline.

Based on the preceding units, the constant C is dependent on the system unit. Hence, the unit for gas flow Q_{nm} in Eq. (2.31) is m^3/h and the units for pressure is MPa. Consequently, units of C_{nm} is (m^3/h)/MPa. A few conclusions can be drawn based on Eqs. (2.31) and (2.32). Based on Section 2.5.1, the specific gravity illustrates how heavy the gas is. Therefore, if the gas is heavier, the flow is decreasing. Similarly, as the pipe length is larger, the flow decreases. On the contrary, if the diameter of the pipelines is larger, the flow increases. In Eq. (2.31), it is assumed that there is no elevation between two ends of the pipeline. In the situation in which the elevation is known, the reader is advised to follow the work of Menon [32].

Normal cubic meter conditions are at 0°C and absolute pressure of 1.01325 bar [26]. The relative density ranges from 0.55 to 0.7 [26]. γ_g can be found as reported by the Danish transmission system operator Energinet [30]. Temperatures can range from −5 to 50°C according to information obtained from Energinet. Z_a can be obtained from explanations given in the work of Mokhatab et al. [29]. The friction factor can be determined based on the Moody diagram depending on the type of flow (i.e., laminar or turbulent). The Moody diagram can be found in the work of Mokhatab et al. [29] and Menon [32]. In addition to general flow in the work of Menon [32], the efficiency of the pipeline is considered to realistically represent the long pipelines. Efficiency is usually close to 1, and typical values are from 0.6 to 0.92 [29]. In the situation when a parameter is known, it is advised to add the efficiency of the pipeline. As an example, the energy loss in the Danish natural gas system is around 0.05% of the total gas consumption [34].

Parameter C_{nm} resembles the Weymouth equation for the calculation of gas flow. The Weymouth equation is yet another among several equations to describe gas flow. It has been used for high-pressure pipelines with large diameters [32]. The parameter C_{nm} used for the Weymouth equation is as follows:

$$C_{nm} = C\frac{T_b}{p_b}D_{nm}^{2.667}\left(\frac{1}{L_e\gamma_g T_a Z_a}\right)^{0.5}\eta_{p,nm} \ , \ \forall n, m \in \Omega^{NGS} \tag{2.33}$$

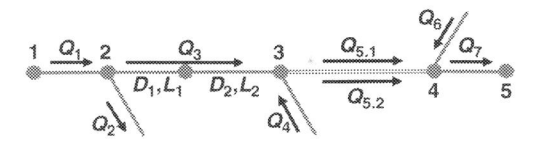

Figure 2.12 Simplified gas network: pipelines connected in series and parallel.

where C equals 155.979×10^{-6} and L_e represents the equivalent pipeline length in kilometers. Eq. (2.33) does not include the friction constant. However, it always includes the pipeline efficiency. According to Menon [32], it is considered as the most conservative equation for the calculation of the gas flow, as it obtains the highest pressure drop for a given gas flow.

2.5.2.2 Nodal flow equation and pipelines

The nodal gas balance equation demonstrates the balance of the gas flow entering or exiting the node. Fig. 2.12 illustrates a simplified gas network. Before expressing the gas flows for such a network, one can notice that there are pipelines connected in series and pipelines connected in parallel. To further simplify the network, equivalent pipeline parameters are found for series and parallel pipelines.

Under assumption, there is an equal flow through the pipelines connected in series as shown in Fig. 2.12 from node 2 to node 3, and an equivalent length concept is demonstrated to obtain one equivalent pipeline [32]. In case the diameters of two sections are equal, the length will be the sum of the length of the two sections. On the contrary, under assumption that the diameters differ, the equivalent length is calculated based on the constant diameter. The pressure drop of the equivalent length pipeline should match the pressure drop of the original pipeline. The procedure is explained in detail by Menon [32]. Thus, only the final equation for the equivalent length is given as follows.

$$L_e = L_1 + L_2 \left(\frac{D_1}{D_2}\right)^5 + L_3 \left(\frac{D_1}{D_3}\right)^5 \tag{2.34}$$

Therefore, an equivalent pipeline is obtained with flow Q_3, length L_e, and diameter D_1 as demonstrated in Fig. 2.12 and Fig. 2.13.

In Fig. 2.12, two parallel pipelines are situated between node 3 and node 4. According to flow conservation, the flow entering the node equals the flow exiting the node. Thus, the total flow from node 3 to node 4 is as

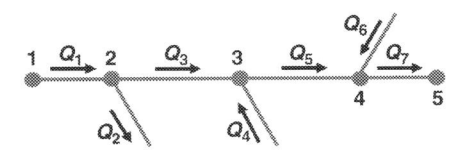

Figure 2.13 Simplified gas network: pipelines and nodes.

follows.

$$Q_5 = Q_{5.1} + Q_{5.2} \tag{2.35}$$

Moreover, the pressure drop in both pipelines connected in parallel is the same, and therefore the following equation gives the relationship between parameters of the two pipelines in parallel.

$$\frac{Q_{5.1}}{Q_{5.2}} = \left(\frac{L_2}{L_1}\right)^{0.5} \left(\frac{D_1}{D_2}\right)^{2.5} \tag{2.36}$$

It can be assumed that the equivalent length is one of the original lengths, and the equations can be expressed as follows [32].

$$\frac{L_1 Q_{5.1}^2}{D_1^5} = \frac{L_2 Q_{5.2}^2}{D_2^5} = \frac{L_1 Q_5^2}{D_e^5} \tag{2.37}$$

From Eq. (2.37), the equivalent diameter can be expressed as follows:

$$D_e = D_1 \sqrt[5]{\left(\frac{1+A}{A}\right)^2} \tag{2.38}$$

where

$$A = \sqrt{\left(\frac{D_1}{D_2}\right)^5 \left(\frac{L_2}{L_1}\right)} \tag{2.39}$$

Finally, the simplified gas network with equivalent pipelines is demonstrated in Fig. 2.13.

The gas flows are calculated as follows.

$$Q_3 = Q_1 - Q_2$$
$$Q_5 = Q_3 + Q_4$$
$$Q_7 = Q_5 + Q_6$$

For the purpose of illustrating the conservation of the gas flow at the nodes, several units are included, as shown in Fig. 2.14.

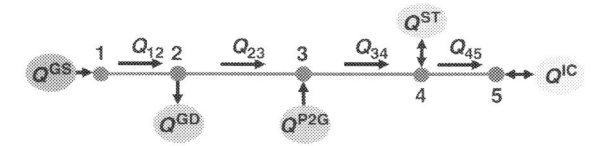

Figure 2.14 Simplified gas network: gas suppliers and consumers.

The gas flow is calculated for each node as demonstrated in the following. Q^{EXP} and Q^{IMP} represent the exit of natural gas to a neighboring country and entry of natural gas from a neighboring country, respectively. Q^{GD} denotes gas demand and Q^{GS} denotes gas source production, whereas $Q^{ST,IN}$ and $Q^{ST,OUT}$ denote injection and extraction to and from the gas storage, respectively. Q^{P2G} denotes the gas production by P2G.

Node 1: $\quad Q^{GS} = Q_{12}$

Node 2: $\quad Q_{12} = Q^{GD} + Q_{23}$

Node 3: $\quad Q_{23} = Q_{34} - Q^{P2G}$

Node 4: $\quad Q_{34} = Q^{ST,IN} - Q^{ST,OUT} + Q_{45}$

Node 5: $\quad Q_{45} = -Q^{IMP} + Q^{EXP}$

Finally, a compact equation for nodal gas balance can be expressed as follows.

$$\sum_{g\in\Omega_n^{GS}} Q_g^{GS} + \sum_{ic\in\Omega_n^{IC}} \left(Q_{ic}^{IMP} - Q_{ic}^{EXP} \right)$$
$$+ \sum_{s\in\Omega_n^{ST}} \left(Q_s^{ST,OUT} - Q_s^{ST,IN} \right) + \sum_{p\in\Omega_n^{P2G}} Q_p^{P2G}$$
$$- \sum_{d\in\Omega_n^{GD}} Q_d^{GD} = \sum_{m\in\Omega_n^{NGS}} Q_{nm}, \quad \forall n, m \in \Omega^{NGS} \qquad (2.40)$$

2.5.2.3 Gas network consumers and suppliers, and its limitations

In this section, the limitations of the gas source and interconnection to neighboring countries are given. Operational limits for gas pressure and gas flow are presented as well.

The gas source is often considered as an injection of the gas from the gas field. The injection has its maximum operational limits as follows.

$$Q_g^{GS,min} \leq Q_g^{GS} \leq Q_g^{GS,max}, \quad \forall g \in \Omega^{GS} \qquad (2.41)$$

Accordingly, the gas import and export to and from the neighboring countries are limited with its maximum entry and exit capacity limits, respectively, as follows.

$$0 \leq Q_{ic}^{IMP} \leq Q_{ic}^{IMP,max} x_{ic}^{IMP}, \quad \forall ic \in \Omega^{IC} \tag{2.42}$$

$$0 \leq Q_{ic}^{EXP} \leq Q_{ic}^{EXP,max} x_{ic}^{EXP}, \quad \forall ic \in \Omega^{IC} \tag{2.43}$$

However, either export or import can happen at a certain time period. Hence, additional binary variables for import and export are included as x_{ic}^{IMP} and x_{ic}^{EXP}, respectively, as presented in Eqs. (2.45) and (2.46). The constraint allowing either import or export is shown in Eq. (2.44).

$$x_{ic}^{IMP} + x_{ic}^{EXP} \leq 1, \quad \forall ic \in \Omega^{IC} \tag{2.44}$$

$$x_{ic}^{IMP} \in \{0, 1\}, \quad \forall ic \in \Omega^{IC} \tag{2.45}$$

$$x_{ic}^{EXP} \in \{0, 1\}, \quad \forall ic \in \Omega^{IC} \tag{2.46}$$

The natural gas system also includes constraint for the reference pressure node as follows.

$$\left(p_m\right)^2 = p_{ref} \tag{2.47}$$

The gas flow rate and pressure are limited by its maximum limits as shown in Eqs. (2.48) and (2.49), respectively.

$$-Q_{nm}^{max} \leq Q_{nm,t} \leq Q_{nm}^{max}, \quad \forall (nm) \in \Omega^{NGS} \tag{2.48}$$

$$\left(p_m^{min}\right)^2 \leq \left(p_{m,t}\right)^2 \leq \left(p_m^{max}\right)^2, \quad \forall m \in \Omega^{NGS} \tag{2.49}$$

2.5.2.4 Gas compressor

As mentioned earlier, the pressure drop occurs when the gas flows through the pipeline. To control the flow in the pipeline and compensate for the energy losses, the gas compressors are integrated in the pipelines [27, 29]. The advantages of the gas compressors are that a higher amount of gas flow can flow in the pipeline and energy losses are lower due to the increase in pressure [29]. The compressors are consuming either electricity or gas. The important parameter to control the gas flow is the compression ratio. Compression ratio, denoted as CR, is the ratio between discharge and suction pressures. The constraint to control the gas flow is as follows.

$$\left(p_m\right)^2 \leq CR^2\left(p_n\right)^2, \quad \forall (nm) \in \Omega^{NGS} \tag{2.50}$$

The compression ratio is always greater than 1 and lower than 2. Very often in the literature, a compression ratio of 1.4 is used due to the relative density ranging from 0.58 to 0.65 [29]. The following equation presents the gas compressor consumption, D_{gc}^{DGC}.

$$D_{gc}^{DGC} = \lambda_{gc}^{GC} Q_{gc}^{GC}, \quad \forall gc \in \Omega^{GC} \tag{2.51}$$

The gas compressor consumption is calculated for each pipe containing a gas compressor where the flow through the gas compressor is denoted as Q_g^{GC}. For every passive pipeline, the compression ratio is given as unity. λ_{gc}^{GC} is a parameter used in the calculation of the consumption of the gas compressor as follows:

$$\lambda_{gc}^{GC} = C^{GC} Z_a \frac{T_s}{E^{GC} \eta_{gc}^{GC}} \frac{c_k}{c_k - 1} \left(CR^{\frac{c_k-1}{c_k}} - 1 \right), \quad \forall gc \in \Omega^{GC} \tag{2.52}$$

where
C^{GC} = constant for the gas compressor dependent on units,
Z_a = average compressibility factor,
T_s = suction temperature (°R),
E^{GC} = parasitic efficiency,
c_k = specific heat constant for the gas compressor, and
η_g^{GC} = compression efficiency.

The gas compressor consumption highly depends on the compressor ration and gas flow in the pipeline. Initially, the gas flow can be represented by a volumetric gas flow rate in million standard cubic feet of gas per day, and the consumption of the compressor is given in brake horsepower, as shown in Eq. (2.52). The consumption of the gas compressor converted to SI units is as follows:

$$D_{gc}^{DGC} = C^{GC} Z_a \frac{T_s}{E^{GC} \eta_{gc}^{GC}} \frac{c_k}{c_k - 1} \left(CR^{\frac{c_k-1}{c_k}} - 1 \right) Q_{gc}^{GC}, \quad \forall gc \in \Omega^{GC} \tag{2.53}$$

where Q_g^{GC} and D_{gc}^{DGC} are the volumetric flow rate of gas in MW and gas compressor consumption in Mm^3/day, respectively. C^{GC} equals to 4.0639×10^{-3}, and Z_a is approximately 0.95. E^{GC} and η_g^{GC} depend on the type of the compressor. The usual range of E^{GC} for a reciprocal compressor is 0.72 to 0.85, whereas a centrifugal compressor equals to 0.99. η_g^{GC} equals to 1 for a reciprocal unit, whereas it ranges from 0.8 to 0.87 for a centrifugal unit. c_k ranges from 1.3 to 1.31 for relative density of 0.55 [29]. The mechanical and pressure losses account for parasitic efficiency. Often, the parasitic efficiency is represented as mechanical efficiency, whereas compression efficiency is represented by adiabatic efficiency [32]. T_s is expressed in K.

2.5.2.5 Gas storages

Most common gas storages are underground. Gas storages can be classified into three groups: depleted gas storages, aquifer reservoirs, and salt caverns. Depleted gas storages are the most common and prominent. They are less expensive technology for storing gas compared to the other types. Aquifer reservoirs are porous rock formations and require high investments and evaluation of the aquifers' sustainability to store natural gas. Last, salt caverns have high deliverability and a long life span. Due to the strong walls, the gas cannot leak from the storage. The gas storages have input gas and output gas. Input gas is usually pure methane, whereas the output is similar with the addition that it must be cleaned before being used (i.e., water should be removed) [35]. The volume of gas that can be extracted from the storage (i.e., working gas volume) is very high in depleted and aquifer types of storage, whereas it is relatively low in salt caverns. On the contrary, the withdrawal and injection rates are very high in salt caverns compared to depleted and aquifer storages. All mentioned storages must have a certain amount of cushion gas. Cushion gas is a base or residual gas that must be maintained to ensure appropriate pressure in the storage. The percentage of cushion gas is the highest in aquifer storages and lowest in salt cavern storages [36].

Hence, gas storages can participate in the regulation and operation of the natural gas system. The regulation ability is often long term. The storages can provide primary regulation and be a backup for the security of the gas supply [36].

The gas storage state of energy (denoted as SOE) is calculated based on the injection or withdrawal to and from the gas storage as follows:

$$SOE_{s,t}^{ST} = SOE_{s,t-1}^{ST} + Q_{s,t}^{ST,IN} - Q_{s,t}^{ST,OUT}, \quad \forall s \in \Omega^{ST}, \forall t \in T \quad (2.54)$$

Therefore, in the next timestep, the gas stock in the gas increases for the injection input gas storage in the current timestep or decreases for the withdrawal from the gas storage in the current timestep. The gas storage capacity is limited as shown next.

$$SOE_{s,t}^{ST,min} \leq SOE_{s,t}^{ST} \leq SOE_{s,t}^{ST,max}, \quad \forall s \in \Omega^{ST}, \forall t \in T \quad (2.55)$$

The gas storage has maximum injection and withdrawal rates. Similar to import and export, the binary variables are introduced to provide either injection or extraction to and from the gas storage as shown in Eqs. (2.56) through (2.60). $x_{s,t}^{ST,IN}$ and $x_{s,t}^{ST,OUT}$ are binary variables.

$$Q_s^{ST,IN,min} \leq Q_{s,t}^{ST,IN} \leq Q_s^{ST,IN,max} x_{s,t}^{ST,IN}, \quad \forall s \in \Omega^{ST}, \forall t \in T \qquad (2.56)$$

$$Q_s^{ST,OUT,min} \leq Q_{s,t}^{ST,OUT} \leq Q_s^{ST,OUT,max} x_{s,t}^{ST,OUT}, \quad \forall s \in \Omega^{ST}, \forall t \in T \qquad (2.57)$$

$$x_{s,t}^{ST,IN} + x_{s,t}^{ST,OUT} \leq 1, \quad \forall s \in \Omega^{ST}, \forall t \in T \qquad (2.58)$$

$$x_{s,t}^{ST,IN} \in \{0, 1\}, \ \forall s \in \Omega^{ST}, \forall t \in T \qquad (2.59)$$

$$x_{s,t}^{ST,OUT} \in \{0, 1\}, \ \forall s \in \Omega^{ST}, \ \forall t \in T \qquad (2.60)$$

2.5.2.6 Linepack modeling

Linepack is considered as an additional storage of natural gas. The gas can be stored in the pipeline, as the amount of gas flowing into the node differs from the gas going out of the node. The amount of linepack is the volume of the gas in the pipeline dependent on the physical volume of the pipeline. By considering linepack, the flexibility of the natural gas system is increased [37]. The linepack volume at the base conditions is calculated as follows [32]:

$$V_{nm}^{LP} = C^{LP} \frac{T_b}{p_b} \frac{1}{Z_a T_a} D_{nm}^2 L_{nm} p_{ave,nm}^{LP}, \quad \forall (nm) \in \Omega^{LP} \qquad (2.61)$$

where

V_{nm}^{LP} = linepack volume in each pipeline (m^3),
C^{LP} = constant dependent on units (7.855×10^{-4}),
T_b = temperature based on normal cubic meter conditions (K),
p_b = pressure based on normal cubic meter conditions (MPa),
$p_{ave,nm}^{LP}$ = average absolute gas pressure in the pipeline (MPa),
T_a = average absolute gas temperature (K),
Z_a = average compressibility factor,
D_{nm} = pipeline diameter (mm), and
L_{nm} = pipeline length (km).

The majority of the symbols have been explained and defined earlier. The only symbol not defined yet is the average gas pressure in the pipeline. The most accurate calculation for the average gas pressure in the pipeline is as follows [32].

$$p_{ave,nm}^{LP} = \frac{2}{3}\left(p_n + p_m - \frac{p_n p_m}{p_n + p_m}\right), \quad \forall(nm) \in \Omega^{LP} \tag{2.62}$$

Once the linepack volume is calculated, it can be used as the initial linepack and the mass conservation for the linepack storage can be written as follows.

$$V_{nm,t}^{LP} = V_{nm,0}^{LP0} + Q_{nm,t}^{LP,IN} - Q_{nm,t}^{LP,OUT}, \quad \forall(nm) \in \Omega^{LP}, t = 1 \tag{2.63}$$

$$V_{nm,t}^{LP} = V_{nm,t-1}^{LP} + Q_{nm,t}^{LP,IN} - Q_{nm,t}^{LP,OUT}, \quad \forall(nm) \in \Omega^{LP}, \forall t \in T, t > 1 \tag{2.64}$$

$Q_{nm,t}^{LP,IN}$ and $Q_{nm,t}^{LP,OUT}$ denote the inflow and outflow rate of the gas pipeline, respectively. Often, the linepack is limited at the end of optimization period to the initial linepack at the start of the optimization period. This is to ensure that enough flexibility can be provided the following optimization period (i.e., the following day). It can be found as follows [37, 38].

$$V_{nm,t}^{LP} = V_{nm,0}^{LP0}, \quad \forall(nm) \in \Omega^{LP}, \forall t \in T, t = \max(T) \tag{2.65}$$

$$V_{nm,t}^{LP} \geq V_{nm,0}^{LP0}, \quad \forall(nm) \in \Omega^{LP}, \forall t \in T, t = \max(T) \tag{2.66}$$

The linepack volume is limited in each pipeline as shown in Eq. (2.67). Due to additional storage, Eq. (2.40) expands to Eq. (2.68).

$$V_{nm}^{LP,min} \leq V_{nm,t}^{LP} \leq V_{nm}^{LP,max}, \quad \forall(nm) \in \Omega^{LP}, \forall t \in T \tag{2.67}$$

$$\sum_{g \in \Omega_n^{GS}} Q_g^{GS} + \sum_{ic \in \Omega_n^{IC}} \left(Q_{ic}^{IMP} - Q_{ic}^{EXP}\right)$$
$$+ \sum_{s \in \Omega_n^{ST}} \left(Q_s^{ST,OUT} - Q_s^{ST,IN}\right) +$$
$$+ \sum_{p \in \Omega_n^{P2G}} Q_p^{P2G} + \sum_{nm \in \Omega_n^{LP}} \left(Q_{nm,t}^{LP,OUT} - Q_{nm,t}^{LP,IN}\right)$$
$$- \sum_{d \in \Omega_n^{GD}} Q_d^{GD} = \sum_{m \in \Omega_n^{NGS}} Q_{nm}, \quad \forall n, m \in \Omega^{NGS} \tag{2.68}$$

2.5.3 Conversion to a single-unit system

Based on the equations presented, a mixture of several units can be noticed. To express the natural gas flow in a more acceptable way as the payments are done for the amount of energy delivered as discussed earlier, the natural gas system is converted to power units. Furthermore, the entire system is converted in pu system, as such conversion ensures reduction of the computational burden.

First, the parameters should be converted to the power unit, and the conversion is shown as follows.

$$Q_{MW} = \frac{UCV}{3600} Q_{m^3/h} \tag{2.69}$$

UCV is upper calorific value often called Wobbe index and the unit is MJ/m^3 [33]. The UCV can also be found in kWh/m^3, in which case 3600 can be omitted and only conversion to MWh/m^3 is required to obtain the gas flow in MW units. The upper Wobbe index is from 14.1 to 15.5 kWh/m^3 [26]. In terms of MJ/m^3, it ranges from 50.76 to 55.8 MJ/m^3.

Second, to obtain the natural gas system in pu, the base power and base pressure should be chosen. As example, in Eqs. (2.32) and (2.33), the parameter C_{mn} is in $(m^3/h)/MPa$. As a first step, a conversion from $(m^3/h)/MPa$ to $(MW)/MPa$ is performed by using Eq. (2.69). Moreover, the base power and base pressure are used to convert the parameter to the pu system. Furthermore, all parameters from Eqs. (2.41) through (2.49) and Eqs. (2.54) through (2.67) are converted from m^3/h to the pu system in the same manner.

The remaining equations are Eqs. (2.50) through (2.53) which are related to the gas compressor. The gas compressor consumption should be converted to the pu system as well. Hence, new parameters are introduced to Eq. (2.53) as follows:

$$D_{gc}^{DGC,pu} = C^{GC} Z_a \frac{T_i}{E^{GC}\eta_{gc}^{GC}} \frac{c_k}{c_k-1} \left(CR^{\frac{c_k-1}{c_k}} - 1 \right) Q_{gc}^{GC,pu} f_{C1}^{GC} f_{C2}^{GC}, \ \forall gc \in \Omega^{GC} \tag{2.70}$$

where $Q_g^{GC,pu}$ and $D_{gc}^{DGC,pu}$ are in the pu system. f_{C1}^{GC} is a constant describing conversion from MW to m^3/h similar to Eq. (2.69) as $(3600/UCV)Q_{MW}$. f_{C2}^{GC} describes conversion from m^3/h to Mm^3/day and equals to 24×10^{-6}.

2.5.4 Linearization procedure

In order to reach the global optimum and feasbile solution, the natural gas system is often linearized. Due to the nonlinearity of the natural gas system, it is not guaranteed that the global optimum can be obtained. As a result of nonlinear problem formulation, a local optimum can be reached. Different approaches can be used to linearize and relax the natural gas system.

The nonlinear terms originate from Eq. (2.31). Several approaches will be discussed further on to linearize the gas flow equation.

The first approach is the iterative method as explained by Chen et al. [39]. In each timestep, a new updated value of the flow is calculated as a multiplication of the initial flow value and the new obtained value. However, the process is time consuming. Further on, Taylor first-order approximation can be used. In that case, as the flow is the function of two variables, the set of distributed points to split the range of node pressures has to be introduced. Taylor series expansion of the pipeline flow equation is explained in the work of Tomasgard et al. [40] and Bai [41]. Furthermore, the piecewise linear approximation of Weymouth equation can be found in [42, 43]. Another formulation for piecewise linearization is incremental formulation. In the work of Correa-Posada and Sánchez-Martín [37], the nonlinear pipeline flow equation is linearized using an incremental method and a simpler approach for linearization can be seen in another work [44] from the same authors based on piecewise approximation. Correa-Posada and Sánchez-Martín [37] explain the advantages of using incremental formulation for piecewise linearization over simple piecewise linearization. Using incremental formulation for piecewise linearization in regard to computational speed has better performance. The extended incremental method is used by Sirvent et al. [45] and is based on incremental formulation for piecewise linearization. The additional advantage of the extended incremental method is introducing error for maximum interpolation. The maximum interpolation error can be calculated as the maximum deviation between the original function and the approximated linear line. Moreover, the error decreases with an increase of the chosen data points. However, by including more chosen data points, the number of constraints increases, and consequently the computational time might increase. The review of piecewise linear approximations for the pipeline gas flow can be found in another work by Correa-Posada and Sánchez-Martín [46]. Along with that, the benefits and drawback of each of the methods are discussed. As a conclusion, the incremental method for piecewise linearization shows the best performance. The linearization process by the incremental piecewise function is shown in the following.

By assuming one directional flow in the pipelines, the variables of quadratic terms for pressure can be defined as follows:

$$(\pi_n - \pi_m)C_{nm}^2 = (Q_{nm})^2, \quad \forall(nm) \in \Omega^{NGS} \tag{2.71}$$

where

$$\pi_n = \left(p_n\right)^2, \quad \forall n \in \Omega^{NGS} \tag{2.72}$$

In Eq. (2.71), the nonlinear terms on the left-hand side are eliminated. The unit of π_n is MPa^2. The right-hand side can be linearized by a piecewise linear function as follows:

$$h(Q_{nm}) = (Q_{nm})^2, \quad \forall(nm) \in \Omega^{NGS}, \quad Q_{nm}^{min} \leq Q_{nm} \leq Q_{nm}^{max} \tag{2.73}$$

$$h(Q_{nm}) \approx h(Q_1) + \sum_{i=1}^{p} (h(Q_{i+1}) - h(Q_i))\delta_i, \quad \forall i \in P, \quad \forall(nm) \in \Omega^{NGS} \tag{2.74}$$

$$Q_{nm} = Q_1 + \sum_{i=1}^{p} (Q_{i+1} - Q_i)\delta_i, \quad \forall i \in P, \quad \forall(nm) \in \Omega^{NGS} \tag{2.75}$$

$$0 \leq \delta_i \leq 1, \quad \forall i \in P \tag{2.76}$$

$$x_i \leq \delta_i, \quad \forall i \in P - 1 \tag{2.77}$$

$$x_i \geq \delta_{i+1}, \quad \forall i \in P - 1 \tag{2.78}$$

$$\delta_i \in [0, 1], \quad \forall i \in P \tag{2.79}$$

$$x_i \in \{0, 1\}, \quad \forall i \in P - 1 \tag{2.80}$$

$$Q_i \in \left[Q_{nm}^{min}, Q_{nm}^{max}\right], \quad \forall i \in P \tag{2.81}$$

where
  $P =$ the set of segments,
  $S_i =$ points chosen for the linearization,
  $x_i =$ the binary variable, and
  $\delta_i =$ portion of the segment.

Finally, Eq. (2.31) can be rewritten as Eq. (2.71) and the nonlinear term is replaced by Eqs. (2.73) through (2.81). Moreover, the nonlinear terms in Eqs. (2.47) and (2.49) through (2.50) can be replaced as demonstrated in Eq. (2.72).

Eq. (2.62) can be replaced by arithmetic average for the average pressure calculation. In a case of presenting bi-directional flow, the reader is advised to follow works elsewhere [25, 37, 38].

2.5.5 Natural gas system integration

In recent years, installed capacity of RESs has been increasing at a high rate pushing towards the decrease of CO_2 emissions. However, not only the EPS can provide a clean energy that a future energy system is in need of. The natural gas system is a fossil fuel gas as long as it is produced in a conventional manner. The International Council on Clean Transportation [47] defines *fossil fuel–based natural gas* as follows: "In contrast to renewable sources of gas, fossil gas is a natural gas or an energy-carrying gas produced in a process that utilizes hydrocarbons originating from beneath the earth's surface, such as coal, natural gas, or petroleum, as a chemical feedstock or a primary energy source."

Hence, a high focus has been put on new technologies, such as hydrogen storages, biogas, and biomethane. The alternative to fossil fuel–based natural gas is renewable gas. Renewable gas includes biomethane, hydrogen, and synthetic methane, among others. The International Council on Clean Transportation [47] defines *renewable gas* as follows: "A gas source is renewable if it is naturally regenerating on a human timescale. Renewable energy includes wind, solar, geothermal, tidal, and ambient energy, hydropower, biomass, and biomethane."

Both biomass and biogas are biofuels obtained from sources such as wood and manure. Biomass is the primary form of biofuel, whereas biogas is derived from organic matter. Biofuels can be used to produce heat and electricity. However, biomass is a scarce resource and must be used carefully [47]. Furthermore, biogas can be upgraded to biomethane. Biomethane is a gas with composition similar to natural gas. Therefore, the upgrading plants can be connected to a gas transmission and distribution system where biomethane can be injected. Compared to fossil fuel–based natural gas, biomethane can have zero or negative life cycle GHG emissons. As an example, converting dairy manure into biomethane avoids the emission of methane into the atmosphere compared to natural decompostion of manure at the farms [47]. To summarize, by upgrading the biogas to biomethane, a clean and green gas can be injected to the natural gas system.

Another clean and green gas can be obtained by consuming the excess production from RESs. The electricity from the renewables can be used in the electrolysis process to produce hydrogen. However, hydrogen injection to the natural gas grid is limited. Therefore, a methanation process is required to obtain methane that can be injected into the natural gas system. In the recent years, hydrogen storages and biogas storages have increased the attention of many researchers. However, hydrogen storage facilities must

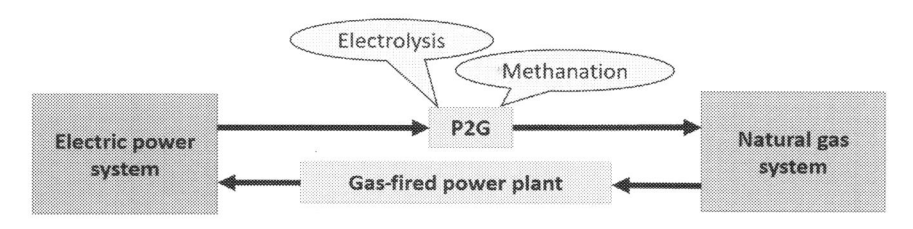

Figure 2.15 Linkages between the natural gas system and EPS.

be designed properly, as hydrogen is more explosive and aggressive toward material such as steel. A biogas usually has an input higher amount of CO_2. Hence, before biogas is injected in the storage, it is important to remove CO_2 as the mixture of water in the reservoirs and CO_2 creates acids that create problems for storage facilities.

To conclude, it is critical to ensure that electricity used for electrolysis comes from renewable-based sources. Using renewable energy in the natural gas system is leading toward neutral GHG emission pathways. Therefore, great importance is put on linking several subsystems to achieve flexible and renewable integrated energy systems. The linking units can be CHP plants and P2G. The linkages between the natural gas system and EPS are shown in Fig. 2.15.

CHP plants can be biomass-fired, coal-fired, or gas-fired power plants. The gas-fired plant (GFP) consumes the natural gas and provides electricity and heat. Electricity production provided by GFP depends on the gas consumption and generating electricity efficiency of GFP as shown in Eq. (2.82). To show the CHP conversion from natural gas to heat, Eq. (2.83) is presented. P2G converts excess electricity generated from the RES to hydrogen through a process called *electrolysis*. As mentioned earlier, only a part of hydrogen can be injected to the natural gas system, and thus a methanation process is applied to obtain methane that can then be injected into the natural gas system. All linkage constraints are illustrated as presented in Eqs. (2.82) through (2.85):

$$P_j^{GFP} = \eta^{GFP,e} Q_j^{D\_GFP}, \ \forall j \in \Omega^{GFP} \tag{2.82}$$

$$H_j^{GFP} = \eta^{GFP,h} Q_j^{D\_GFP}, \ \forall j \in \Omega^{GFP} \tag{2.83}$$

$$Q_p^{P2G} = \eta^{P2G,Q} D_p^{P2G}, \ \forall p \in \Omega^{P2G} \tag{2.84}$$

$$H_p^{P2G} = \eta^{P2G,H} D_p^{P2G}, \ \forall p \in \Omega^{P2G} \tag{2.85}$$

where

P_j^{GFP} = electricity generated from GFP,

H_j^{GFP} = heat generated from GFP,

$Q_j^{D\_GFP}$ = consumed natural gas of GFP,

$\eta^{GFP,e}$ = generated electricity efficiency of GFP,

$\eta^{GFP,h}$ = generated heat efficiency of GFP,

Q_p^{P2G} = gas production of P2G,

D_p^{P2G} = electricity consumption of P2G,

H_p^{P2G} = heat production of P2G,

$\eta^{P2G,Q}$ = P2G conversion efficiency from electricity to gas, and

$\eta^{P2G,H}$ = P2G conversion efficiency from electricity to heat.

2.5.6 Conclusion

In this section, an overview of the natural gas system was presented. The composition of natural gas and its important quality parameters were discussed. Two of the most commonly used equations for pipeline flow were illustrated. The most important parameters related to the pipeline flow equation were considered, and the gas flow dependency on pressure and pipeline parameters were explained. Natural gas modeling has been illustrated through several equations. First, nodal gas balance equations showed the conservation of gas flow in the natural gas system. Further on, a brief introduction to gas storages and gas compressors was given. Furthermore, the purpose and a model of the linepack were explained. Last, to decrease the computational burden and obtain gas flow in terms of energy, the conversions were demonstrated. Finally, the integration of a natural gas system with an EPS was presented and linking components were introduced.

References

[1] International Energy Agency, World energy outlook 2019, 2019. https://www.iea.org/reports/world-energy-outlook-2019. (Accessed 16 April 2021).

[2] International Energy Agency, Combined heat and power, evaluating the benefits of greater global investment, 2008, 2008. https://www.osti.gov/etdeweb/servlets/purl/21589339. (Accessed 16 April 2021).

[3] International Energy Agency, Market Report Series: Renewables 2019, 2019. https://www.iea.org/reports/renewables-2019/heat. (Accessed 16 April 2021).

[4] Environmental Protection Agency, Combined Heat and Power (CHP) Partnership: catalog of CHP technologies. https://www.epa.gov/chp/catalog-chp-technologies. (Accessed 16 April 2021).

[5] D Fischer, M Hatef, On heat pumps in smart grids: a review, Renew Sustain Energy Rev 70 (2017) 342–357.

[6] X Chen, C Kang, M O'Malley, Q Xia, J Bai, C Liu, et al., Increasing the flexibility of combined heat and power for wind power integration in china: modeling and implications, IEEE Trans Power Syst 30 (2017) 1848–1857.

[7] T Ommen, WB Markussen, B Elmegaard, Lowering district heating temperatures— impact to system performance in current and future Danish energy scenarios, Energy 94 (2016) 273–291.

[8] LH Hvidtfeldt, SP Leif, DTU International Energy Report 2015: Energy systems integration for the transition to non-fossil energy systems, 2015. https://backend.orbit. dtu.dk/ws/portalfiles/portal/119583507/ DTU_International_Energy_Report_2015_rev.pdf. (Accessed 16 April 2021).

[9] Y Dai, L Chen, Y Min, Q Chen, J Hao, K Hu, et al., Dispatch model for CHP with pipeline and building thermal energy storage considering heat transfer process, IEEE Trans Sustain Energy 10 (2019) 192–203.

[10] Z Pan, Q Guo, H Sun, Interactions of district electricity and heating systems considering time-scale characteristics based on quasi-steady multi-energy flow, Appl Energy 167 (2016) 230–243.

[11] J Wang, S You, Y Zong, C Træholt, ZY Dong, Y Zhou, Flexibility of combined heat and power plants: a review of technologies and operation strategies, Appl Energy 252 (2019) 113445.

[12] X Chen, MB McElroy, C Kang, Integrated energy systems for higher wind penetration in China: formulation, implementation, and impacts, IEEE Trans Power Syst 33 (2018) 1309–1319.

[13] P Byrne, R Ghoubali, A Diaby, Heat pumps for simultaneous heating and cooling, 2018. https://hal.archives-ouvertes.fr/hal-01990466/document. (Accessed 16 April 2021).

[14] K Holzapfel, Heat pumps—basic principles, in: AT De Almeida, AH Rosenfeld (Eds.), Demand-Side Management and Electricity End-Use Efficiency, in: NATO ASI Series (Series E: Applied Sciences), 149, Springer, Dordrecht, 1988, pp. 399–406.

[15] H Pieper, T Ommen, JK Jensen, B Elmegaard, WB Markussen, Comparison of COP estimation methods for large-scale heat pumps in energy planning tools., Energy 205 (2020) 117994.

[16] H Cai, C Ziras, S You, R Li, K Honoré, HW Bindner, Demand side management in urban district heating networks, Appl Energy 230 (2018) 506–518.

[17] J Yang, N Zhang, A Botterud, C Kang, On an equivalent representation of the dynamics in district heating networks for combined electricity-heat operation, IEEE Trans Power Syst 35 (1) (2020) 560–570.

[18] X Liu, J Wua, N Jenkins, A Bagdanavicius, Combined analysis of electricity and heat networks, Appl Energy 162 (2016) 1238–1250.

[19] M Pirouti, A Bagdanavicius, J Ekanayake, J Wu, N Jenkins, Energy consumption and economic analyses of a district heating network, Energy 57 (2013) 149–159.

[20] B.E. Larock, R.W. Jeppson, G.Z. Watters, Hydraulics of Pipeline Systems, CRC Press, Boca Raton, FL.

[21] S Huang, W Tang, Q Wu, C Li, Network constrained economic dispatch of integrated heat and electricity systems through mixed integer conic programming, Energy 179 (2019) 464–474.

[22] H Lund, S Werner, R Wiltshire, S Svendsen, J Eric, F Hvelplund, et al., 4th Generation District Heating (4GDH): integrating smart thermal grids into future sustainable energy systems, Energy 68 (2014) 1–11.

[23] B Liu, K Meng, ZY Dong, W Wei, Optimal dispatch of coupled electricity and heat system with independent thermal energy storage, IEEE Trans Power Syst 34 (2019) 3250–3263.

[24] A Kargarian, J Mohammadi, J Guo, S Chakrabarti, M Barati, G Hug, et al., Toward distributed/decentralized DC optimal power flow implementation in future electric power systems, IEEE Trans Smart Grid 9 (2018) 2574–2594.

[25] J Munoz, N Jimenez-Redondo, J Perez-Ruiz, J Barquin, Natural gas network modeling for power systems reliability studies, in, in: Proc. 2003 IEEE Bologna Power Tech Conference Proceedings, 2003.

[26] EnerginetSecurity of Gas Supply Report 2018, Energinet, Fredericia, Syddanmark, Denmark, 2018.

[27] W Wei, J Wang, Modeling and Optimization of Interdependent Energy Infrastructures, Springer, Cham, Switzerland, 2020.

[28] EnerginetHow to Measure Gas Quality, Energinet, Fredericia, Syddanmark, Denmark, 2018.

[29] S Mokhatab, WA Poe, JG Speight, Handbook of Natural Gas Transmission and Processing, Elsevier, 2006.

[30] EnerginetFuture Natural Gas Qualities—Fact Sheet, Energinet, Fredericia, Syddanmark, Denmark, 2017.

[31] EnerginetRules for Gas Transport (Version 16.1), Energinet, Fredericia, Syddanmark, Denmark, 2017.

[32] E Shashi Menon, Gas Pipeline Hydraulics, CRC Press, Boca Raton, FL, 2005.

[33] Q Zeng, J Fang, J Li, Z Chen, Steady-state analysis of the integrated natural gas and electric power system with bi-directional energy conversion, Appl Energy 184 (2016) 1483–1492.

[34] EnerginetSystem Plan 2018—Electricity and Gas in Denmark, Energinet, Fredericia, Syddanmark, Denmark, 2018.

[35] JM Leroy, Underground natural gas storage: ensuring a secure and flexible gas supply, 2020. https://gie.eu/index.php/gie-publications/presentations. (Accessed 16 April 2021).

[36] Energinet; Danish Energy Agency, Technology data: energy storages, technology descriptions and projections for long-term energy system planning, Energinet, Danish Energy Agency, Fredericia, Copenhagen (2018).

[37] CM Correa-Posada, P Sánchez-Martín, Integrated power and natural gas model for energy adequacy in short-term operation, IEEE Trans Power Syst 30 (6) 3347–3355.

[38] A Schwele, C Ordoudis, J Kazempour, P Pinson, Coordination of power and natural gas systems: convexification approaches for linepack modeling, in: Proc. 2019 IEEE Milan PowerTech, 2019.

[39] Z Chen, J Fang, Q Zeng, S Nielsen, W Hu, C Su, Harmonized Integration of Gas, District Heating and Electric Systems, Aalborg University, Aalborg, Denmark, 2017.

[40] A Tomasgard, F Rømo, M Fodstad, K Midthun, Optimization models for the natural gas value chain, in: G Hasle, KA Lie, E Quak (Eds.), Geometric Modelling, Numerical Simulation, and Optimization, Springer, Berlin, Heidelberg, 2007, pp. 521–558.

[41] L Bai, Doctoral Dissertations: "Co-Optimization of Gas-Electricity Integrated", University of Tennessee, Knoxville, 2017.

[42] J Ye, R Yuan, Integrated natural gas, heat, and power dispatch considering wind power and power-to-gas, Sustainability 9 (4) (2017) 602.

[43] G Li, R Zhang, T Jiang, H Chen, L Bai, H Cui, X Li, Optimal dispatch strategy for integrated energy systems with CCHP and wind power, Appl Energy 192 (2017) 408–419.

[44] C.M Correa-Posada, P Sánchez-Martín, Stochastic contingency analysis for the unit commitment with natural gas constraints, in: Proc. 2013 IEEE Grenoble Conference, 2013.

[45] M Sirvent, N Kanelakis, B Geisler, P Biskas, Linearized model for optimization of coupled electricity and natural gas systems, J Mod Power Syst Clean Energy 5 (3) (2017) 364–374.

[46] CM Correa-Posada, P Sánchez-Martín, Gas network optimization: a comparison of piecewise linear models, Chem Eng Sci (2014) submitted.

[47] International Council on Clean Transportation, Gas definitions for the European Union, 2019. https://theicct.org/publications/gas-definitions-european-union. (Accessed 16 April 2021).

CHAPTER 3

Uncertainty modeling

3.1 Introduction

In multi–energy system optimization problems, the system operators make decisions based on predictions of renewables, electricity demand, heat load, gas load, and so forth. Among these prediction parameters, the renewable energy sources, led by wind power, have larger prediction errors and hence have a more obvious impact on the decision making of multi–energy systems. With the fast development and wide deployment of wind power, the related uncertainty is becoming more and more obvious. In such a situation, the system operators must make decisions with the uncertainty taken into account.

Stochastic optimization (SO) and robust optimization (RO) are the two most popular techniques to cope with uncertainty in multi-energy system optimization problems. The former represents the uncertainty with plausible scenarios sampled from a known probability distribution and focuses on the expected operational cost associated with each scenario [1]. The latter represents the uncertainty with an uncertainty set that is based on the known boundary information of uncertainty and pursues the feasibility and minimum operational cost under the worst case [2].

Representation of uncertainty is critical in SO and RO, which directly impacts the decision results. Depending on the specific uncertainty optimization techniques used, the representation and modeling of uncertainty can be quite different [3].

In SO, scenario generation is one of the most commonly used techniques for uncertainty representation. The basic idea of scenario representation is to generate a set of scenarios, wherein each scenario represents a possible realization of uncertainty factors. This is an approximate representation of actual probability distribution of uncertainty factors. A scenario set reflecting the actual probability distribution is necessary to guarantee a reliable and economic decision because it is closely related to the real-time regulation capacity of the multi-energy systems.

A number of scenario generation methods have been developed in the literature. Traditional methods such as Monte Carlo samping and Latin

Optimal operation of integrated multi-energy systems under uncertainty.   Copyright © 2022 Elsevier Inc.
DOI: https://doi.org/10.1016/B978-0-12-824114-1.00010-X   All rights reserved.

hypercube sampling usually neglect the spatial and temporal correlation of stochastic process. Since multi-energy system optimization problems usually involve multiple time periods and multiple locations of renewables or loads, it is essential to account for the spatial and temporal correlations of these uncertainty factors to improve the efficacy of the decision making.

Several scenario generation methods accounting for the spatial and temporal correlations have been proposed. Pinson and Girard [4] utilize an exponential function to construct the temporal correlation of wind power outputs. On the basis of the work of Pinson and Girard [4], Ma et al. [5] consider the optimization of the control parameter in the exponential function to represent the correlation coefficient more accurately. Zhang et al. [6] propose evaluation indices for the scenario set based on the exponential function. However, the exponential function method in the mentioned works [4–6] is not suitable for modeling spatial correlation. Other authors [7]-[9] consider the spatial correlation of actual wind power production and forecast errors. Zhang et al. [7] utilize the Gaussian Copula to formulate the conditional distribution of forecast errors of multiple wind farms. Wang et al. [8, 9] utilize the Gaussian mixture model (GMM) to represent the joint distribution of actual wind power production and forecast errors with respect to different forecast values of multiple wind farms, respectively. Although both the Copula and GMM methods can incorporate the spatial and temporal correlations simultaneously, they suffer from the curse of dimensionality. Tang et al. [10] construct the temporal and spatial correlations simultaneously by combining the exponential function and Copula method. In addition, Gibbs sampling is utilized to reduce the complexity of sampling. However, the conditional distribution in Gibbs sampling needs to select the Copula function and calculate the marginal distribution of each random variable. Tang et al. [10] select the Gaussian Copula and cumulative empirical distribution to calculate the conditional distribution but may influence the accuracy and computational efficiency of the scenario generation.

In RO, the uncertainty set is one of the most commonly used techniques for uncertainty representation. The basic idea of the uncertainty set is to utilize the boundary information of each dimension of the uncertainty factors for uncertainty representation. This is quite different from the probability distribution in SO. Conventionally, the box-like uncertainty set with budget constraints are the most widely adopted for uncertainty representation in RO.

One of the main focuses for uncertainty sets in RO is the conservativeness adjustment. The conservativeness of the box-like uncertainty set mainly originates from that it includes some small chance or rare events. For example, all dimensions of an uncertainty factor locate on their boundaries simultaneously. The budget constraints adjust the consavativeness of the original box-like uncertainty set by shrinking its size [11–13]. However, the size adjustment is not based on historical data. If the uncertainty set is shrunk too much, the robustness cannot be guaranteed; otherwise, conservative decision may be made. Dai et al. [14] and An and Zeng [15] propose the multiband uncertainty set and uncertainty set variant, respectively, which are still based on budget constraints. Therefore, the dilemma for robustness and conservativeness remains. Lorca and Sun [16] utilize the union of several separated regions instead of a single convex set to reduce the conservativeness, which is only applicable for cases where the range of isolated regions are all known and fixed.

Some other works construct data-driven uncertainty sets. Guan and Wang [17] and Li et al. [18] utilize historical data to approach the probability distribution and determine the feasible dispatch range of uncertainty, respectively. But these two works do not belong to RO.

Generally, the data-driven uncertainty sets in the RO field can be classified into two categories. The first group models the correlation constraints of uncertainty sets to reduce the conservativeness, but the estimated parameters have a great influence on the description accuracy of the uncertainty [19, 20]. The second group utilizes the convex hull of historical data to construct the uncertainty set. Ding et al. [21] and Zhang et al. [22] utilize the ellipse and diamond convex hulls, respectively, to cover the historical data in each period. But these two works only adapt to a specifically shaped dataset. Velloso et al. [23] construct a high-dimensional convex hull that considers temporal and spatial correlation simultaneously. But a large number of vertices are needed for the precise description of uncertainty.

This chapter proposes new uncertainty modeling methods for both SO and RO. First, Gibbs sampling and the GMM are combined to generate spatial-temporal-correlated scenarios for uncertainties in SO. Second, a partition-combine uncertainty set in RO is proposed to reduce the conservativeness, which can describe the irregular shape of historical data with the smallest area.

3.2 Scenario generation with spatial-temporal correlations in SO

To provide reasonable inputs for the multi-energy optimization problems, a scenario set that can reflect the probability distribution of uncertainty factors needs to be generated. In this section, a scenario generation method considering both spatial and temporal correlation is detailed.

The temporal correlation and spatial correlation are formulated separately using different methods, and they are linked by inverse sampling. The temporal correlation is constructed by the exponential covariance function. By the spatial correlation, the GMM method is adopted to model the conditional joint distribution of actual wind power outputs with respect to forecast values. Based on the conditional joint distribution modeled by the GMM, Gibbs sampling is used to improve the sampling efficiency.

3.2.1 Temporal correlation modeling based on exponential function

According to Appendix A, the stochastic process obeying non–Gaussian distribution can be transformed into a random vector obeying Gaussian distribution by using Nataf inverse transformation.

It is assumed the joint distribution of a random vector X follows the multivariate Gaussian distribution, $X \sim N(\mu, \Sigma)$, where μ is an N_T-dimensional mean vector, which is a zero vector, and Σ is the covariance matrix, satisfying

$$\Sigma = \begin{bmatrix} \sigma_{1,1}\sigma_{1,2}...\sigma_{1,N_T} \\ \sigma_{2,1}\sigma_{2,2}...\sigma_{2,N_T} \\ \vdots \ddots \vdots \\ \sigma_{N_T,1}\sigma_{N_T,2}...\sigma_{N_T,N_T} \end{bmatrix} \tag{3.1}$$

$$\sigma_{t_1,t_2} = \text{cov}(X_{t_1}, X_{t_2}) = \exp\left(-\frac{|t_1 - t_2|}{\varepsilon}\right), \quad 1 \le t_1, t_2 \le N_T \tag{3.2}$$

where σ_{t_1,t_2} is the covariance of X_{t_1} and X_{t_2}, and ε is the parameter to control the correlation, which is obtained by optimization [5].

In Eq. (3.1), Σ is a positive definite matrix with all diagonal elements equal to 1. In Eq. (3.2), the exponential function method [4] is adopted, which proved realistic in view of the empirical correlations observed. The multivariate normal distribution of X is uniquely determined by the covariance matrix Σ.

3.2.2 Spatial correlation modeling based on GMM

Let $w^a = [w_1^a, w_2^a, ..., w_n^a]$ denote the actually realized wind power outputs of n wind farms, and let $w^f = [w_1^f, w_2^f, ..., w_n^f]$ denote the corresponding forecast values. The conditional probability distribution can be computed as

$$f\left(\boldsymbol{w}^a | \boldsymbol{w}^f\right) = \frac{f_1\left(\boldsymbol{w}^a, \boldsymbol{w}^f\right)}{f_2\left(\boldsymbol{w}^f\right)} = \frac{f_1\left(w_1^a, w_2^a, ..., w_n^a, w_1^f, w_2^f, ..., w_n^f\right)}{f_2\left(w_1^f, w_2^f, ..., w_n^f\right)} \tag{3.3}$$

where f is the conditional probability distribution of $\boldsymbol{w}^a$ given $\boldsymbol{w}^f$, f_1 is the joint PDF of actual outputs and forecast values of all wind farms, and f_2 is the joint PDF of forecast values of all wind farms.

In Eq. (3.3), computing the joint distribution function f_1 and f_2 is required to determine the conditional joint distribution f. The GMM method is adopted to build the joint distributions f_1 and f_2 due to its capability in fitting the non-Gaussian correlated random variables.

Mathematically, a GMM for the PDF of a random vector $[\boldsymbol{w}^a, \boldsymbol{w}^f]^T$ is a convex combination of multivariate Gaussian distribution functions with an adjustable parameter set $\gamma = \left\{\mu_m^{\text{spa}}, \Sigma_m^{\text{spa}}, \omega_m\right\}_{m=1}^M$. Taking f_1 as an example, the GMM method can be expressed as [24] follows:

$$f_1\left(\boldsymbol{w}^a, \boldsymbol{w}^f\right) = \sum_{m=1}^M \omega_m N_m\left(\boldsymbol{w}^a, \boldsymbol{w}^f; \mu_m^{\text{spa}}, \Sigma_m^{\text{spa}}\right) \tag{3.4}$$

$$\sum_{m=1}^M \omega_m = 1, \omega_m > 0 \tag{3.5}$$

$$N_m\left(\boldsymbol{w}^a, \boldsymbol{w}^f; \mu_m^{\text{spa}}, \Sigma_m^{\text{spa}}\right) = \frac{e^{-\frac{1}{2}\left(\left[\boldsymbol{w}^a, \boldsymbol{w}^f\right]^T - \mu_m^{\text{spa}}\right)^T \left(\Sigma_m^{\text{spa}}\right)^{-1} \left(\left[\boldsymbol{w}^a, \boldsymbol{w}^f\right]^T - \mu_m^{\text{spa}}\right)}}{(2\pi)^{n/2} \det\left(\Sigma_m^{\text{spa}}\right)^{1/2}} \tag{3.6}$$

where $N_m(\cdot)$ denotes the $2n$-dimensional multivariate Gaussian distribution function and each Gaussian distribution function is called a *Gaussian component*, M is the number of mixture components, ω_m is the weight of the mth component, $\boldsymbol{\mu}_m$ is the $2n$-dimensional mean vector, and Σ_m is the $2n \times 2n$ covariance matrix of the mth component.

An important attribute of the GMM is its ability to approach arbitrarily shaped densities by adjusting its parameter set γ [25]. The GMM has been verified to be able to model different uncertainties well. The parameter set γ of the GMM can be determined by the parameter estimation function *gmdistribution* in Matlab. Similarly, the parameter set of the GMM for f_2 and other joint distribution can be determined by this method.

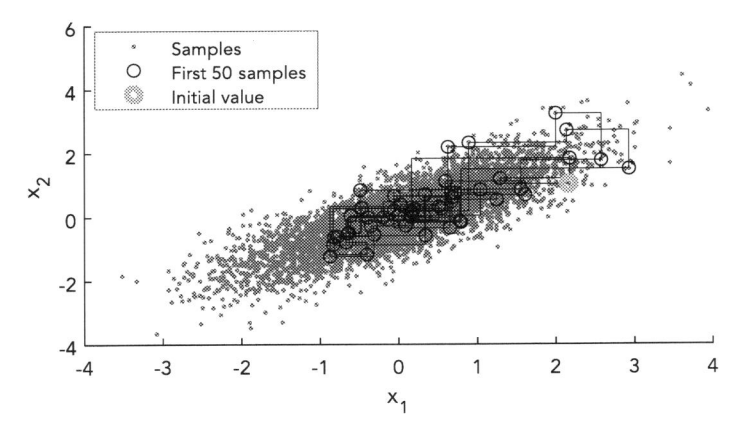

Figure 3.1 The process of Gibbs sampling with two random variables.

3.2.3 Gibbs sampling

In statistics, Gibbs sampling, also called *alternating conditional sampling*, is a Markov chain Monte Carlo algorithm for high-dimensional data. It is called *Monte Carlo* because it draws samples from specified probability distributions; the Markov chain comes from the fact that each sample is dependent on the previous sample and the values of the constructed Markov Chain converge toward the target distribution [26].

The idea in Gibbs sampling is to generate posterior samples by sweeping through each random variable of the random vector to sample from its conditional distribution with the remaining random variables fixed to their current values. The method is based on each random variable's posterior density, conditional on the other random variables. This gives an accurate representation of marginal posterior densities.

In other words, Gibbs sampling is an efficient way of reducing a multidimensional sampling problem to a lower-dimensional sampling problem when the direct sampling from the multidimensional joint distribution is difficult. The entire random vector is subdivided into smaller subvectors (e.g., a single random variable). For one iteration of the algorithm, each subvector samples from its posterior density, conditional on the other subvector's current values [27].

Fig. 3.1 shows the Gibbs sampling process with two random variables: x_1 and x_2. Starting from the initial sampling value, the Gibbs sampling first takes a step only along the x_1 direction with the fixed value of x_2, then only along the x_2 direction with the fixed value of x_1. This process shows

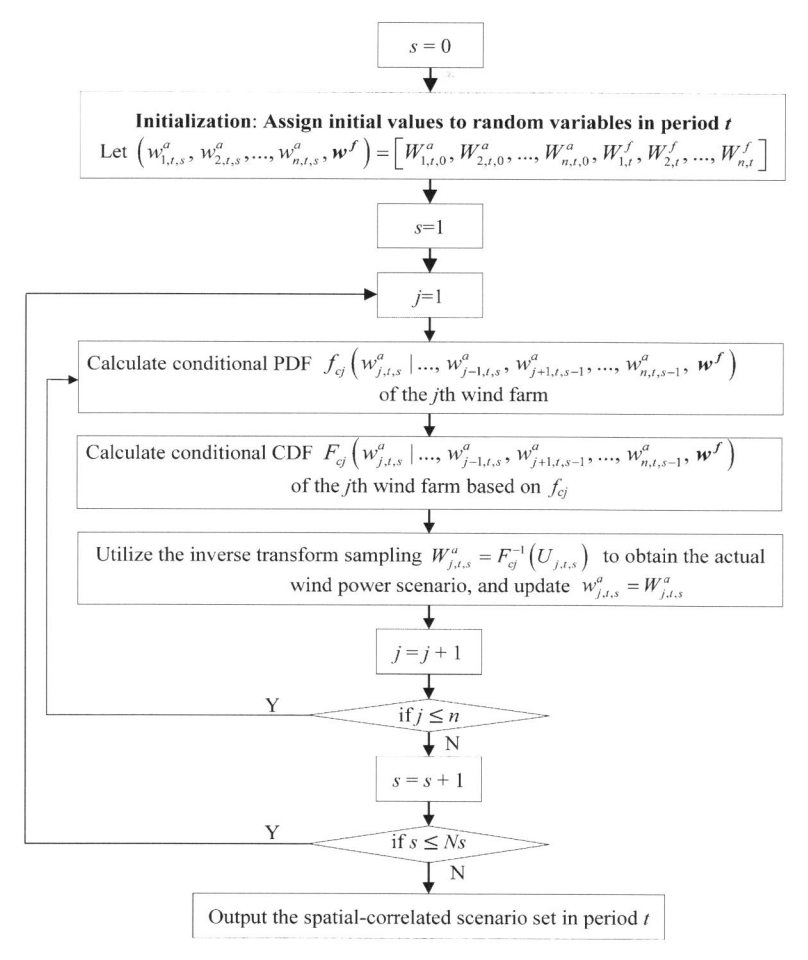

Figure 3.2 Gibbs sampling in period t.

how Gibbs sampling sequentially samples the value of each random variable separately, conditional on the values of other random variables [28]. This sampling procedure is repeated until the predefined iterations are satisfied.

By utilizing the Gibbs sampling technique, the joint distribution in Eq. (3.3) can be approximately simulated by sampling from the conditional distribution of each variable. Fig. 3.2 details the process of using Gibbs sampling to generate a spatial-correlated scenario set with N_s scenarios. As Gibbs sampling in each period t is independent of the other periods, Fig. 3.2 only shows the sampling in a single period.

In Fig. 3.2, in the initialization, the random variables $\boldsymbol{w}^f =$ $(w_{1,t}^f, w_{2,t}^f, ..., w_{n,t}^f)$ are known parameters, fixed as the forecast values $[W_{1,t}^f, W_{2,t}^f, ..., W_{n,t}^f]$; the initial values $[W_{1,t,0}^a, W_{2,t,0}^a, W_{3,t,0}^a, ..., W_{n,t,0}^a]$ of random variables $(w_{1,t,s}^a, w_{2,t,s}^a, ..., w_{n,t,s}^a)$ are also set as the forecast values. The calculation of conditional PDF f_{cj} for the jth wind farm in scenario s is similar to Eqs. (3.3) through (3.6). The calculation of conditional CDF F_{cj} for the jth wind farm in scenario s is based on the accumulation of sample points on f_{cj}. Given the probability value $U_{j,t,s}$ of sampling points, a scenario value $W_{j,t,s}^a$ for random variable $w_{j,t,s}^a$ can be obtained through the inverse transformation F_{cj}^{-1}. Implementing the process for all periods, a scenario set with spatial correlation can be derived.

It should be noted that samples from early iterations may not represent the actual joint distribution—that is, Gibbs sampling needs a burn-in period to converge to the true distribution [29]. Therefore, it is common to discard samples in early iterations.

3.2.4 Procedure for scenario generation with spatial-temporal correlations

The procedure of the spatial-temporal correlated scenario generation method is detailed as follows. Step 1 is used for the construction of temporal correlation, and Step 2 is used to construct the spatial correlation and generate spatial-temporal correlated scenarios:

Step 1 Construct the temporal correlation.

Step 1.1 For the jth ($j = 1, 2, ..., n$) wind farm, construct the temporal covariance matrix Σ_j^{tem} of actual wind power outputs in different time periods, and each element in Σ_j^{tem} can be calculated by Eq. (3.2).

Step 1.2 Generate N_S scenarios of Gaussian random series$(X_{j,1,s}, X_{j,2,s}, ..., X_{j,t,s}, ..., X_{j,T,s}), j = 1, 2, ..., n; s = 1, 2, ..., N_S$ for each wind farm, which follows the joint distribution $N_j\left(\mu_j^{\text{tem}}, \Sigma_j^{\text{tem}}\right)$. μ_j^{tem} is an N_T-dimensional zero vector.

Step 1.3 Calculate the cumulative distribution function (CDF) value of the Gaussian random variable $U_{j,t,s} = \Phi(X_{j,t,s})$. $\Phi(\cdot)$ is a Gaussian CDF; $U_{j,t,s}$ is the random data that follow the uniform distribution in $[0, 1]$.

Step 2 Construct spatial correlation and generate spatial-temporal correlated scenarios.

Step 2.1 Calculate the conditional CDF $F_{cj}(j = 1, 2, ..., n)$ of actual wind power output for the jth wind farm based on the corresponding conditional PDF $f_{cj}(j = 1, 2, ..., n)$.

Step 2.2 Utilize the inverse transform sampling $W_{j,t,s}^a = F_{cj}^{-1}(U_{j,t,s})$ to obtain the actual wind power scenario.

Step 2.3 Implement Step 2.1 and Step 2.2 for each time period t and each scenario s.

Step 2.4 Discard the first N_b scenarios in the burn-in period, and take the remaining scenarios as the final spatial-temporal correlated scenario set of actual wind power outputs.

3.3 Partition-combine uncertainty set modeling in RO

A partition-combine method is proposed to build the uncertainty set in RO. It reduces the conservativeness by partitioning the original box-like uncertainty set and cutting off its areas without historical data while maintaining the same level of robustness of the decision. The enveloped space of the uncertainty set can properly adapt to the data provided.

The construction of the partition-combine uncertainty set can be divided into three steps. First, the box-like uncertainty set enclosing all of the historical data is partitioned into subsets with the same size, with the removal of empty subsets followed. Second, the remaining subsets with new derived boundaries are combined to reformulate the new uncertainty set, namely the partition-combine uncertainty set. Last, the inner subsets are identified to reduce the computational burden.

3.3.1 Set partition and combination

In the modeling of the partition-combine uncertainty set, only the spatial correlation is considered. Therefore, the uncertainty in different periods can be treated independently. Take the uncertainty set for two wind farms as an example. In period t, the output of the two wind farms is represented by the two-dimensional vector $\tilde{w}_t$. The traditional box-like uncertainty set is represented by the region surrounded by the blue rectangle in Fig. 3.3.

The mathematical formulation of the box-like uncertainty set can be described as follows:

$$U = \left\{ \tilde{w}_t | \underline{\mathbf{W}}_t \leq \tilde{w}_t \leq \bar{\mathbf{W}}_t \right\} \tag{3.7}$$

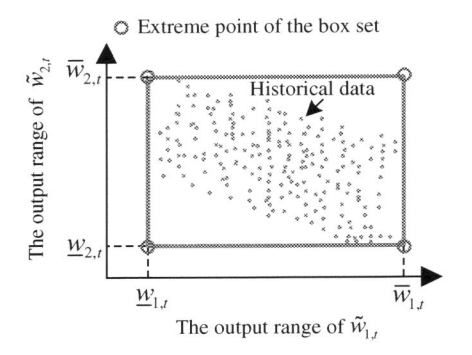

Figure 3.3 Box-like uncertainty set.

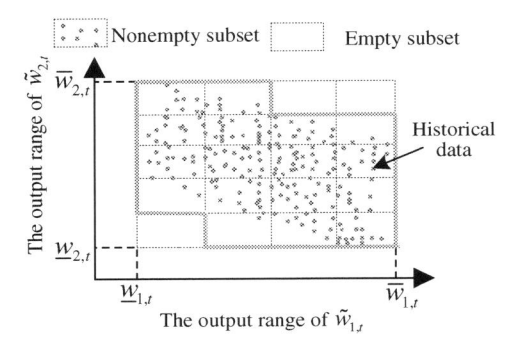

Figure 3.4 Partition the box set into equal subsets, and remove empty subsets.

where $\bar{\mathbf{W}}_t$ and $\underline{\mathbf{W}}_t$ are the upper and lower boundaries of two wind farms, respectively.

As the worst-case scenarios of RO always locate at the extreme points of the uncertainty set, the uncertainty parameter $\tilde{w}_t$ is often represented by the extreme points on the boundaries of the set, which is more tractable for computation. Then the uncertainty set in Eq. (3.7) can be replaced by

$$U = \{\tilde{\boldsymbol{w}}_t^{\mathrm{ex}} | \tilde{\boldsymbol{w}}_t^{\mathrm{ex}} = \underline{\mathbf{W}}_t \boldsymbol{Z}_t^- + \bar{\mathbf{W}}_t \boldsymbol{Z}_t^+, \ \boldsymbol{Z}_t^+ + \boldsymbol{Z}_t^- = 1, \ \boldsymbol{Z}_t^+, \boldsymbol{Z}_t^- \in \{0,1\}\} \quad (3.8)$$

where $\tilde{\boldsymbol{w}}_t^{\mathrm{ex}}$ are the extreme points of the set, and for the box-like set, its extreme points are just the vertices; $\boldsymbol{Z}_t^+$ and $\boldsymbol{Z}_t^-$ are binary variables to decide which extreme point is reached.

In Fig. 3.3, the box-like uncertainty set contains some blank region without historical data, which may result in a conservative decision. On the basis of Fig. 3.3, Fig. 3.4 and Fig. 3.5 show the basic principle of the partition–combine uncertainty set.

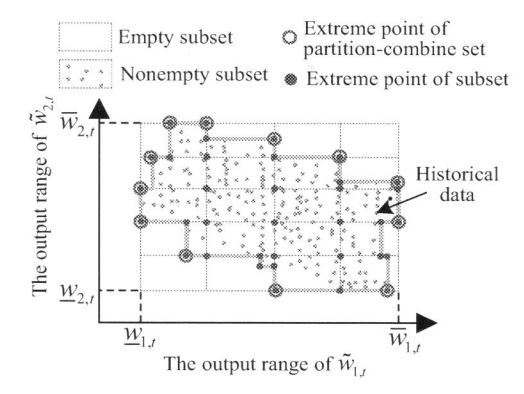

Figure 3.5 Combine the nonempty subsets under new obtained boundaries.

First, the original box-like uncertainty set is partitioned into subsets with the same size according to the predefined number of divisions in each dimension, as shown in Fig. 3.4. All of the subsets can be divided into two categories (i.e., the empty and nonempty subsets) depending on whether they contain historical data. As the empty subsets may result in a conservative solution, they are identified and removed to reduce the conservativeness.

Second, the new boundaries of each of the nonempty subsets are redefined according to the range of its covered historical data. Such a measure can further reduce the size of the uncertainty set to decrease the conservativeness. After that, these remaining subsets with new boundaries are combined. Then, the partition-combine uncertainty set is obtained as shown in Fig. 3.5.

Similar to Eq. (3.8), the extreme points of each subset are utilized to formulate the partition-combine uncertainty set U_{new}, which can be expressed as

$$
U_{new} = \left\{ \tilde{\boldsymbol{w}}_{t,new} \left|
\begin{array}{ll}
\tilde{\boldsymbol{w}}_{t,new} = \sum_{i=1}^{N} \left(\underline{\mathbf{w}}_{t,i} \boldsymbol{Z}_{t,i}^{-} + \bar{\mathbf{w}}_{t,i} \boldsymbol{Z}_{t,i}^{+} \right) X_{t,i} & (a) \\
\boldsymbol{Z}_{t,i}^{+} + \boldsymbol{Z}_{t,i}^{-} \leq 1, \boldsymbol{Z}_{t,i}^{+}, \boldsymbol{Z}_{t,i}^{-} \in \{\mathbf{0}, \mathbf{1}\}, \forall i & (b) \\
\sum_{i=1}^{N} X_{t,i} = 1, X_{t,i} \in \{0, 1\} & (c)
\end{array}
\right. \right\}
\tag{3.9}
$$

where $\tilde{\boldsymbol{w}}_{t,new}$ is the uncertainty parameter of the partition-combine uncertainty set in period t, and it can be any purple point in Fig. 3.5; i is the index for subsets; N is the total number of subsets; $\bar{\mathbf{w}}_{t,i}/\underline{\mathbf{w}}_{t,i}$ are the upper/lower boundaries of the ith subset; and $X_{t,i}$ is a binary variable, with $X_{t,i} = 1$ representing that $\tilde{w}_{t,new}$ is taken at one of the extreme points of the ith subset.

To make the solving procedure tractable, two auxiliary variables $Y_{t,i}^-$ and $Y_{t,i}^+$ are introduced to reformulate the two nonlinear terms $Z_{t,i}^- X_{t,i}$ and $Z_{t,i}^+ X_{t,i}$ in Eq. (3.9). Set $Y_{t,i}^- = Z_{t,i}^- X_{t,i}$ and $Y_{t,i}^+ = Z_{t,i}^+ X_{t,i}$, then introduce constraints concerning $Y_{t,i}^-$ and $Y_{t,i}^+$ to make an equivalent transformation as follows:

$$
U_{new} = \left\{ \tilde{w}_{t,new} \left|
\begin{array}{ll}
\tilde{w}_{t,new} = \sum\limits_{i=1}^{N} \left(\underline{w}_{t,i} Y_{t,i}^- + \bar{w}_{t,i} Y_{t,i}^+ \right) & (a) \\
Y_{t,i}^+ + Y_{t,i}^- \leq 1, Y_{t,i}^+, Y_{t,i}^- \in \{0, 1\}, \forall i & (b) \\
\sum\limits_{i=1}^{N} X_{t,i} = 1, X_{t,i} \in \{0, 1\} & (c) \\
Y_{t,i}^+, Y_{t,i}^- \leq X_{t,i} & (d)
\end{array}
\right. \right\}
\quad (3.10)
$$

After the equivalent transformation, the new uncertainty set in Eq. (3.10) is a mixed–integer linear model, which is similar to Eq. (3.8).

It should be noted that Eq. (3.10) utilizes the collection of subsets instead of the extreme points of the smallest convex hull to represent the uncertainty set. This is because finding extreme points of the smallest convex hull is a challenging task, which is well known in the field of computational geometry, especially when the dimension of uncertainty is high.

Even though the collection of subsets contains nonextreme points that may increase the computational burden, a method to reduce the size of variables of the uncertainty set is proposed in the following section.

3.3.2 Identification of inner subsets

Fig. 3.6 divides all of the subsets into two categories: the periphery subsets and the inner subsets. The periphery subsets are those of which at least one edge is on the blue boundary of the partition-combine set. On the contrary, for the inner sets, there is no edge on the boundaries.

As can be seen from Fig. 3.6, only the vertices of the periphery subsets can become the extreme points of the partition-combine set. Therefore, identifying the inner subsets can simplify the description of the uncertainty set and reduce the size of variables describing the uncertainty set.

Fig. 3.7 shows how to identify inner subsets. As can be seen from this figure, an inner subset always has two adjacent subsets in each dimension,

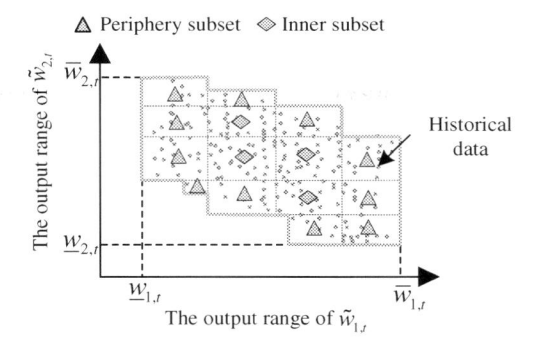

Figure 3.6 Periphery and inner subsets.

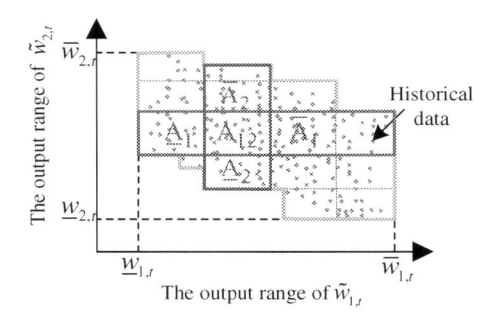

Figure 3.7 Inner subsets identification.

which make its four edges not fall on the blue boundary of the partition–combine uncertainty set. Here, take the identification of subset A_{12} as an example. For the first dimension of A_{12}, which corresponds to the first wind farm, we lock the region surrounded by the red rectangle. In this region, there exist two adjacent subsets: $\underline{A}_1$ and $\bar{A}_2$. Similarly, in the second dimension of A_{12}, which corresponds to the second wind farm, we lock the region surrounded by the purple rectangle, and there also exist two adjacent subsets: $\underline{A}_2$ and $\bar{A}_2$. Then, according to the adjacent subsets, we can judge A_{12} as an inner subset.

After finding out all of the inner subsets, the following constraint is added to Eq. (3.10).

$$X_{t,i} = 0, \ i \in innersubsets \tag{3.11}$$

Then, the partition-combine uncertainty set will be only the combination of periphery subsets, and the number of uncertainty variables to be solved can be reduced.

3.4 Case study

3.4.1 Performance of the spatial-temporal correlated scenario set

Four wind farms with spatial and temporal correlations were used to validate the performance of the proposed scenario generation method. The data of the four wind farms are from the western Denmark dataset [30–31]. The historical wind power measurements have an hourly resolution, and each wind farm has 4300 historical observations. Each wind farm has a capacity of 100 MW, and the wind power data has been scaled to match the wind farm capacity. As the dataset does not give the corresponding forecast values, the forecast values were generated by using the moving average method.

Before generating the spatial-temporal correlated scenario set, the correlation features of actual wind power outputs are first analyzed. Fig. 3.8 shows the scattered data between each pair of wind farms.

One of the obvious features in Fig. 3.8 is that the actual wind power outputs for each pair of wind farms are positively correlated. The bottom-left and top–right areas have a larger data density, which means that the PDF of actual wind power outputs for each wind farm is a multimodal function and it is hard for the traditional Gaussian distribution to fit this feature.

The key to conditional joint PDF in Eq. (3.3) lies in the calculation of the joint distribution function f_1 and f_2. The fitting performance of GMM on f_1 and f_2 is validated by each wind farm and aggregation of all wind farms. Fig. 3.9 shows the performance of the GMM method on fitting the PDF of both forecast and actual values. From this figure, it can be seen that the PDFs of forecast and actual values of each wind farm and aggregation of all wind farms are characterized by a multimodal feature, wherein the GMM method can fit the probability distribution histogram well, indicating that the GMM has a satisfactory level of accuracy.

Fig. 3.10 shows the joint distribution function of forecast and actual values for single wind farm and aggregation of all wind farms, which are fitted by the GMM method. From this figure, it can be seen that the GMM method with five Gaussian mixture components can fit the multimodal feature of the PDF accurately.

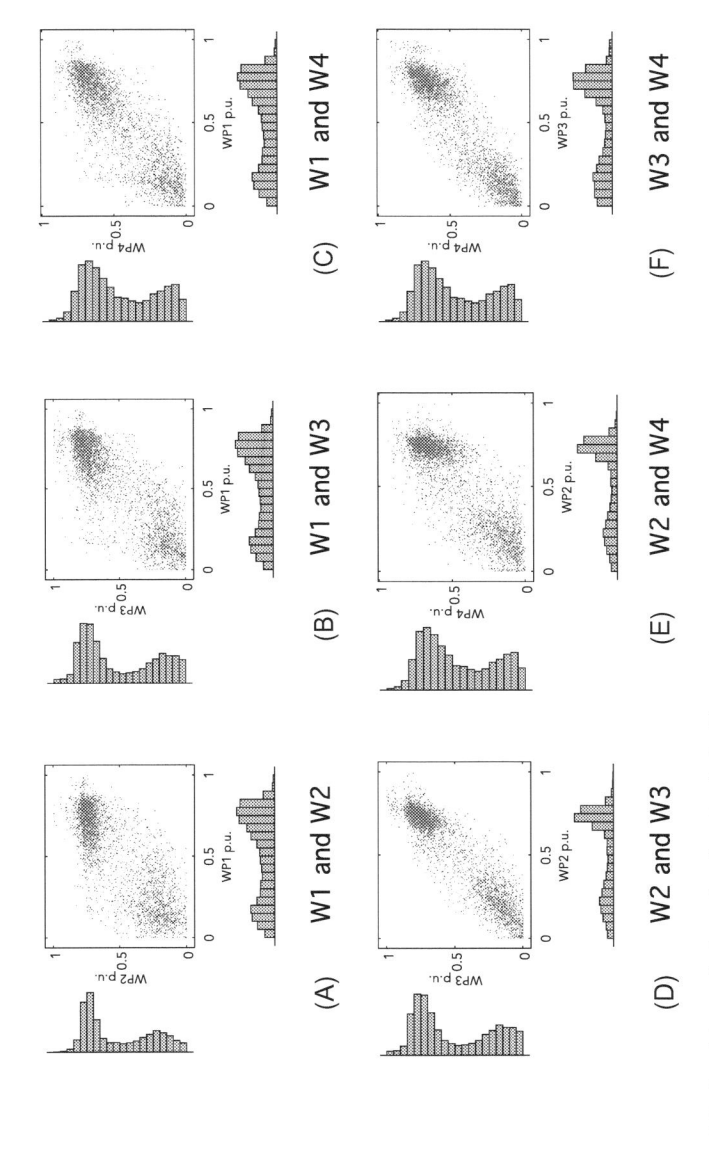

Figure 3.8 Spatial correlation of each pair of wind farms.

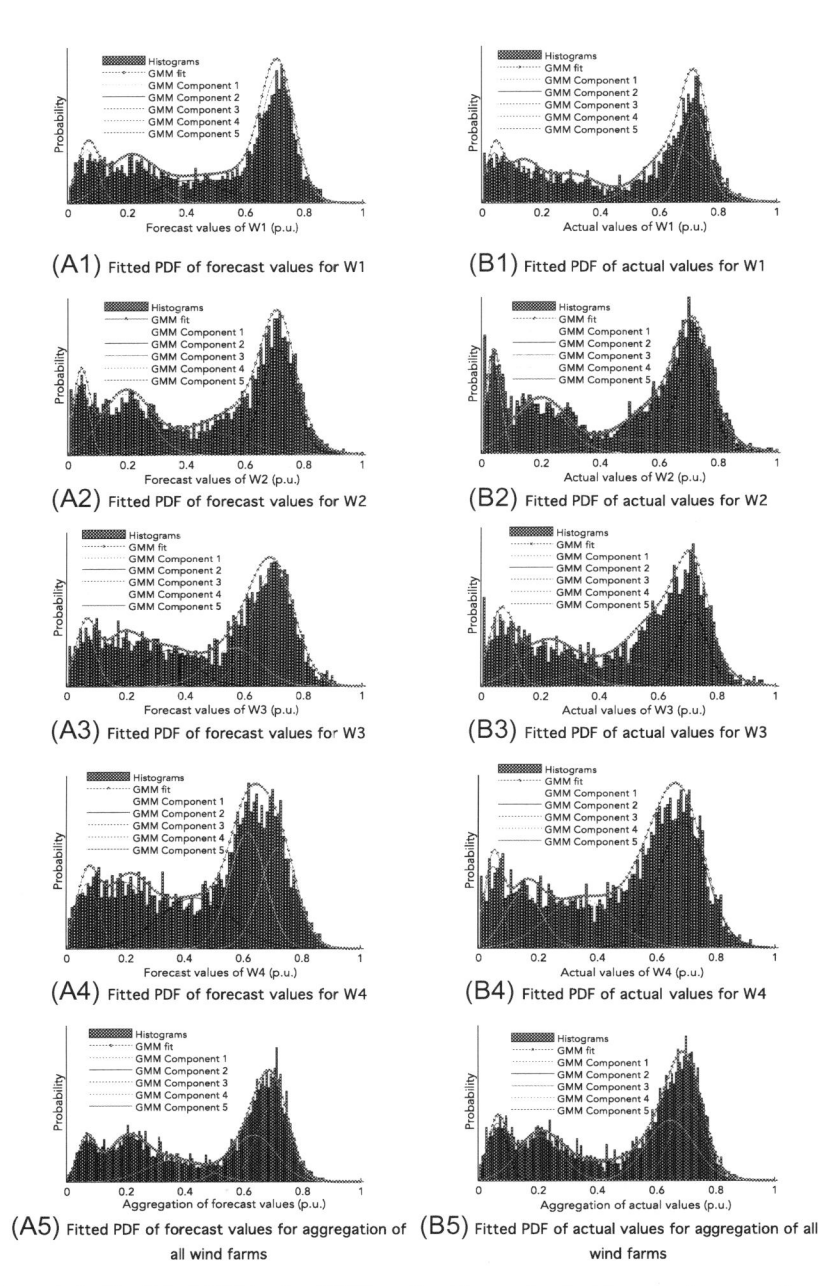

Figure 3.9 The fitted PDF by the GMM method for each wind farm and aggregation of all wind farms.

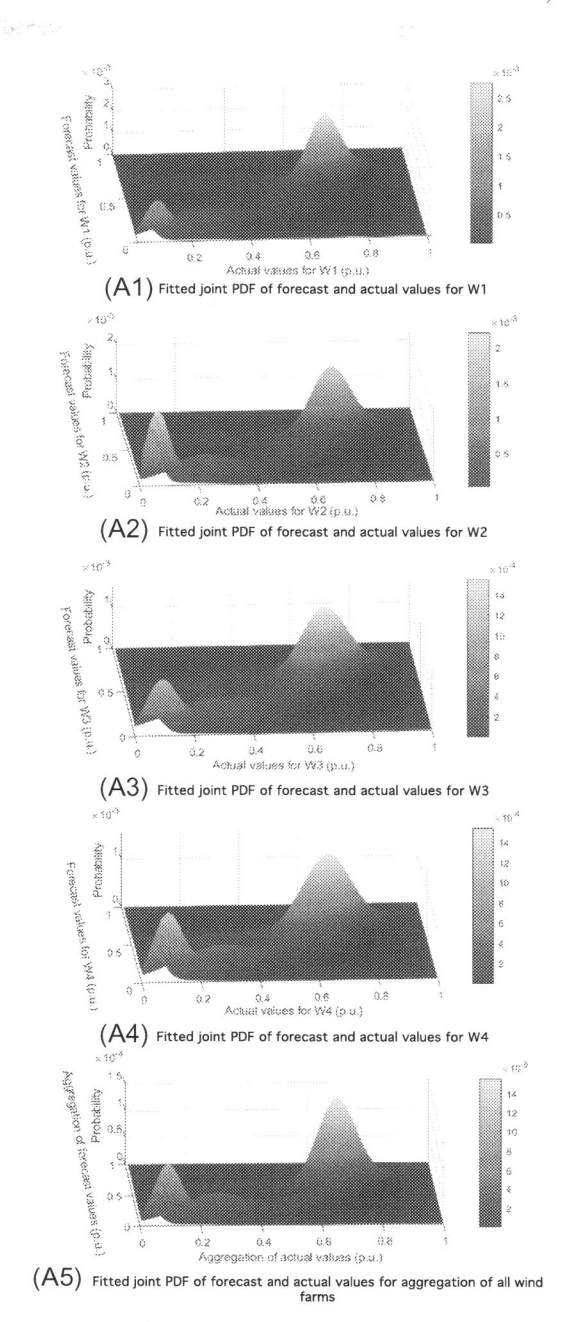

(A1) Fitted joint PDF of forecast and actual values for W1

(A2) Fitted joint PDF of forecast and actual values for W2

(A3) Fitted joint PDF of forecast and actual values for W3

(A4) Fitted joint PDF of forecast and actual values for W4

(A5) Fitted joint PDF of forecast and actual values for aggregation of all wind farms

Figure 3.10 The fitted joint PDF of forecast and actual outputs of each wind farm and aggregation of all wind farms by the GMM method.

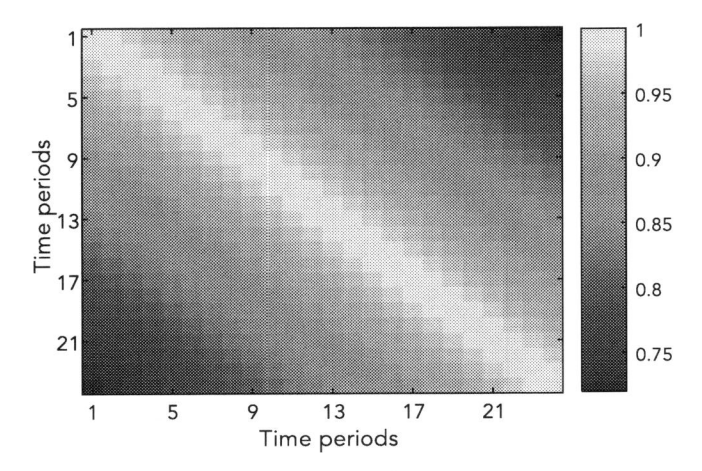

Figure 3.11 The temporal correlation between different time periods.

Fig. 3.11 shows the temporal correlation between different time periods based on the exponential function. As can be seen from this figure, the temporal correlation is closely related to the length of two time periods. The temporal correlation decreases with the length between two time periods.

For comparison, the scenario generation method in the work of Tang et al. [10] was also conducted to generate scenarios. Their method [10] combines the Gaussian Copula and exponential function to construct spatial and temporal correlation. For simplicity, the method of Tang et al. [10] is abbreviated as *Copula-exponential*, whereas the method proposed in this chapter is called the *GMM-exponential method*. The scenario set for each wind farm and aggregate outputs of all wind farms are given in Fig. 3.12. During the process of scenario generation, the total number of scenarios N_{sc} is set as 2000, and N_b is set as 1500 for the burn-in period. Therefore, 500 scenarios remain in Fig. 3.12.

To qualify the performance of these two methods, two indices are used to evaluate the quality of the scenario set [6]. One is the coverage rate, which aims to evaluate whether the actual wind power outputs can be covered by the scenario set at each period. The other is the envelope area of the scenario set, which is used to test whether the scenario set can cover the actual wind power outputs with the smallest possible area. They are expressed as follows.

$$index\ 1 = \frac{1}{T}\sum_{t=1}^{T} B_t, \quad B_t = \begin{cases} 1 & \text{if } w_t^{obs} \in \left[\min_{s\in[1,N_S-N_b]} (W_{t,s}^a), \max_{s\in[1,N_S-N_b]} (W_{t,s}^a)\right] \\ 0 & \text{otherwise} \end{cases}$$

$$(3.12)$$

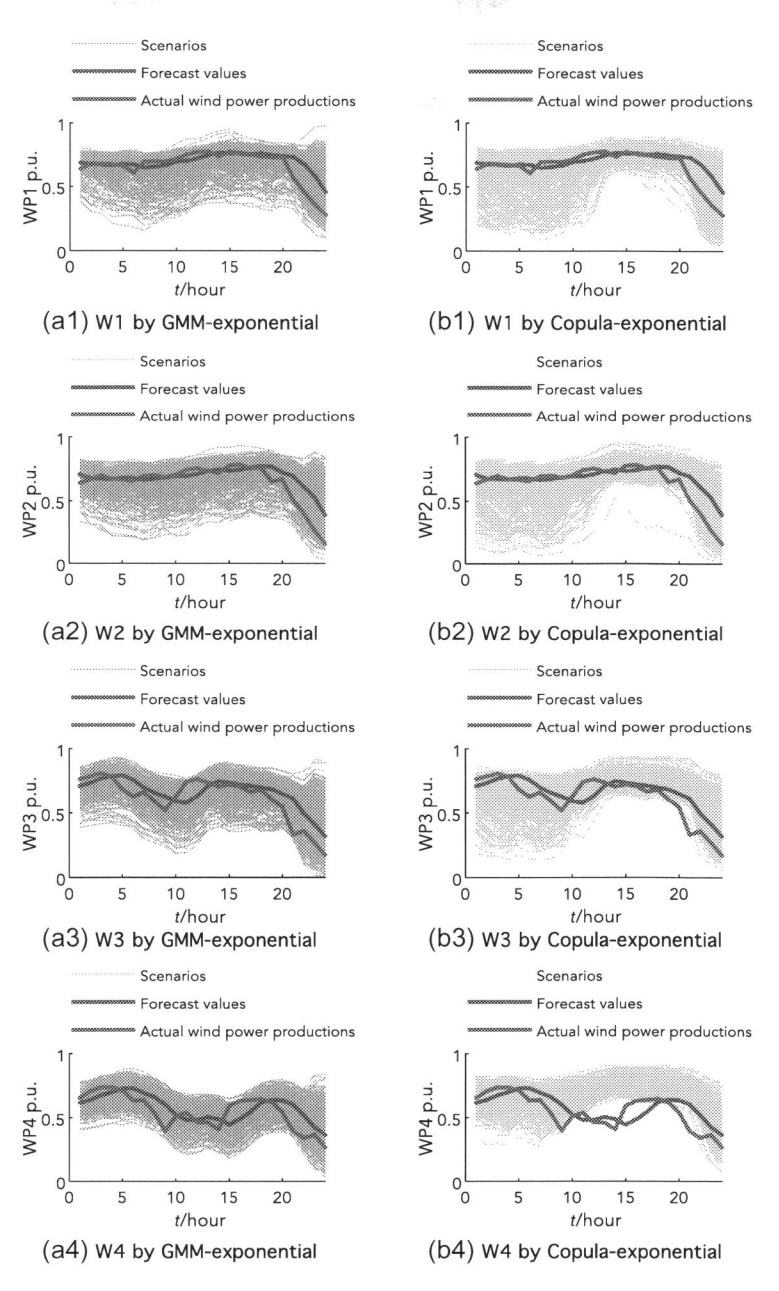

Figure 3.12 Scenario sets generated by the GMM-exponential and Copula-exponential methods.

Table 3.1 Comparison of scenario quality and calculation time between the GMM-exponential and Copula-exponential methods.

Scenario Generation Method	Copula-exponential	GMM-exponential
Coverage rate	0.9653	0.9875
Envelope area	0.5213	0.4422
Calculation time (s)	923	45

$$index\,2 = \frac{1}{T}\sum_{t=1}^{T}\left[\min_{s\in[1,N_S-N_b]}(W_{t,s}^{a}) - \max_{s\in[1,N_S-N_b]}(W_{t,s}^{a})\right] \qquad (3.13)$$

Table 3.1 compares the two scenario generation methods in scenario quality and calculation time. To guarantee the robustness of test results, the data of 30 days are selected to generate scenarios. This table gives the average value of the 30 days. The results show that the GMM-exponential method not only saves calculation time by 95% but also guarantees a larger coverage rate and a smaller envelope area than the Copula-exponential method.

3.4.2 Performance of the partition-combine uncertainty set

To validate the effectiveness of the partition–combine uncertainty set in reducing conservativeness, two cases are set for the comparison of the data space covered by the box-like set and partition–combine set. The wind power data are from western Denmark [29, 30], as shown in Table 3.2. In Case 1, the uncertainty sets of two wind farms with spatial correlations are modeled based on the box-like set and partition–combine set, respectively. In Case 2, the number of wind farms is set as 3.

In Table 3.2, the number of subsets is equal to the product of partitions of each wind farm. Indices i, j, k represent the partitions of three wind farms, respectively. And when $i \times j = 1$ in Case 1 and $i \times j \times k = 1$ in Case 2, the uncertainty set is a box-like set. The ratio in the last column of Table 3.2 is defined by the ratio of data space covered by the partition–combine uncertainty set to the original data space covered by the box-like uncertainty set. As can be seen from Table 3.2, when the number of subsets increases from 1 to 81, the ratio decreases gradually in both Case 1 and Case 2. This is because when the number of subsets increases, more regions without historical data will be removed, and the partition–combine uncertainty set has a smaller size to enclose the historical data.

Table 3.2 Comparative results between the box-like set and partition-combine uncertainty set.

Case	Wind Farms (#)	Partition Scheme $(i \times j)/(i \times j \times k)$	Subsets (#)	Ratio of the Covered Data Space by Uncertainty Sets (%)
Case 1	2	1×1	1	100
		2×2	4	67.72
		3×3	9	63.01
		4×4	16	59.19
		5×5	25	53.70
		2×18	36	47.85
		3×12		50.87
		4×9		48.92
		6×6		50.43
		7×7	49	46.57
		8×8	64	47.18
		9×9	81	44.31
Case 2	3	$1 \times 1 \times 1$	1	100
		$2 \times 1 \times 2$	4	53.12
		$3 \times 1 \times 3$	9	41.14
		$2 \times 4 \times 2$	16	41.14
		$5 \times 5 \times 1$	25	30.94
		$6 \times 1 \times 6$	36	25.17
		$6 \times 2 \times 3$		28.37
		$3 \times 2 \times 6$		29.89
		$2 \times 1 \times 18$		25.79
		$7 \times 1 \times 7$	49	22.63
		$8 \times 2 \times 4$	64	24.65

It should be noted that there are different partition schemes for the same total number of subsets. Take Case 1 as an example. When the total number of subsets is 36, the partition scheme for two wind farms can be 2×18, 3×12, 4×9, and 6×6. As can be seen from Table 3.2, a different partition scheme usually corresponds to a different result in reducing conservativeness. At present, it can rely on the simulation results to judge which partition scheme is better when the number of subsets is fixed.

In addition, Fig. 3.13 depicts the number of identified inner subsets of Case 1. By the identification of inner subsets, the number of binary variables that are used to describe the partition–combine uncertainty set will be reduced and thus reduce the computational burden for the energy optimization problems.

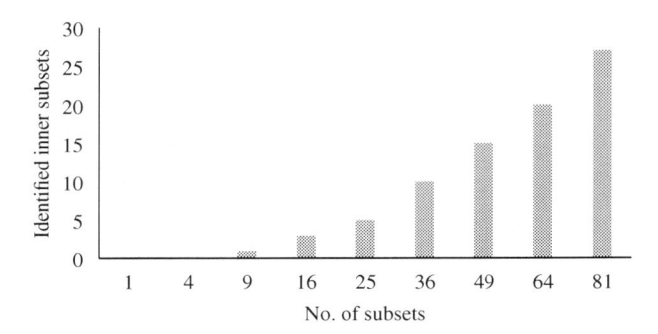

Figure 3.13 The number of identified inner subsets under different partition schemes for Case 1.

3.5 Conclusion

This chapter proposed new uncertainty modeling methods for both SO and RO, respectively.

For SO, a scenario generation method considering the spatial and temporal correlations is presented. The GMM based on conditional probability distribution is used to model the spatial correlation, and the exponential function is used to model the temporal correlation. In addition, Gibbs sampling reduces the high-dimensional sampling problem to a single-dimension problem. The simulation results show that the proposed scenario generation method can accurately fit the joint probability distribution and generate scenarios with higher quality.

For RO, a data-driven partition-combine uncertainty set is modeled, which can properly describe the irregular shape of historical data by the partition and combination of subsets. The simulation results show that the partition-combine uncertainty set can cover the historical data with smaller data space and the inner subsets identification method can reduce the scale of variables describing the uncertainty set.

References

[1] JR Birge, F Louveaux, Introduction to Stochastic Programming, Springer Science & Business Media, New York, 2011.
[2] A Ben-Tal, L El Ghaoui, A Nemirovski, Robust Optimization, Princeton University Press, Princeton, NJ, 2009.
[3] QP Zheng, J Wang, AL Liu, Stochastic optimization for unit commitment—a review,, IEEE Trans. Power Syst. 30 (4) (2014) 1913–1924.
[4] P Pinson, R Girard, Evaluating the quality of scenarios of short-term wind power generation, Appl Energy 96 (2012) 12–20.

[5] X Ma, Y Sun, H Fang, Scenario generation of wind power based on statistical uncertainty and variability, IEEE Trans Sustain Energy 4 (4) (2013) 894–904.

[6] M Zhang, X Ai, J Fang, W Yao, W Zuo, Z Chen, et al., A systematic approach for the joint dispatch of energy and reserve incorporating demand response, Appl Energy 230 (2018) 1279–1291.

[7] N Zhang, C Kang, Q Xia, J Liang, Modeling conditional forecast error for wind power in generation scheduling,, IEEE Trans Power Syst 29 (3) (2014) 1316–1324.

[8] Z Wang, C Shen, F Liu, X Wu, C Liu, F Gao, Chance-constrained economic dispatch with non-Gaussian correlated wind power uncertainty, IEEE Trans Power Syst 32 (6) (2017) 4880–4893.

[9] Z Wang, C Shen, F Liu, A conditional model of wind power forecast errors and its application in scenario generation, Appl Energy 212 (2018) 771–785.

[10] C Tang, Y Wang, J Xu, Y Sun, B Zhang, Efficient scenario generation of multiple renewable power plants considering spatial and temporal correlations, Appl Energy 221 (2018) 348–357.

[11] R Jiang, J Wang, Y Guan, Robust unit commitment with wind power and pumped storage hydro, IEEE Trans Power Syst 27 (2) (2012) 800–810.

[12] D Bertsimas, E Litvinov, X.A Sun, J Zhao, T Zheng, Adaptive robust optimization for the security constrained unit commitment problem, IEEE Trans Power Syst 28 (1) (2013) 52–63.

[13] C Zhao, Y Guan, Unified stochastic and robust unit commitment, IEEE Trans Power Syst 28 (3) (2013) 3353–3361.

[14] C Dai, L Wu, H Wu, A multi-band uncertainty set based robust SCUC with spatial and temporal budget constraints, IEEE Trans Power Syst 31 (6) (2016) 4988–5000.

[15] Y An, B Zeng, Exploring the modeling capacity of two-stage robust optimization: variants of robust unit commitment model, IEEE Trans Power Syst 30 (1) (2015) 109–122.

[16] A Lorca, X.A Sun, Multistage robust unit commitment with dynamic uncertainty sets and energy storage, IEEE Trans Power Syst 32 (3) (2017) 1678–1688.

[17] Y Guan, J Wang, Uncertainty sets for robust unit commitment, IEEE Trans Power Syst 29 (3) (2014) 1439–1440.

[18] C Li, J Zhao, T Zheng, E Litvinov, Data-driven uncertainty sets: robust optimization with temporally and spatially correlated data, in: Proc. 2016 IEEE Power and Energy Society General Meeting, 2016.

[19] C Zhao, Y Guan, Data-driven stochastic unit commitment for integrating wind generation, IEEE Trans Power Syst 31 (4) (2016) 2587–2596.

[20] F Qiu, Z Li, J Wang, A data-driven approach to improve wind dispatchability, IEEE Trans Power Syst. 32 (1) (2017) 421–429.

[21] T Ding, J Lv, R Bo, Z Bie, F Li, Lift-and-project MVEE based convex hull for robust SCED with wind power integration using historical data-driven modeling approach, Renew Energy 92 (2016) 415–427.

[22] Y Zhang, X Ai, J Wen, J Fang, H He, Data-adaptive robust optimization method for the economic dispatch of active distribution networks, IEEE Trans Smart Grid 10 (4) (2018) 3791–3800.

[23] A Velloso, A Street, D Pozo, JM Arroyo, NG Cobos, Two-stage robust unit commitment for co-optimized electricity markets: an adaptive data-driven approach for scenario-based uncertainty sets, IEEE Trans Sustain Energy 11 (2) (2019) 958–969.

[24] Z Wang, C Shen, F Liu, A conditional model of wind power forecast errors and its application in scenario generation, Appl Energy 212 (2018) 771–785.

[25] Z Wang, C Shen, F Liu, X Wu, C-C Liu, F Gao, Chance-constrained economic dispatch with non-Gaussian correlated wind power uncertainty, IEEE Trans Power Syst 32 (6) (2017) 4880–4893.

[26] S Glen, Gibbs sampling: definition & overview. https://www.statisticshowto.com/gibbs-sampling/. (Accessed 17 April 2021).

[27] I Yildirim, Bayesian Inference: Gibbs Sampling, Technical Note, University of Rochester, Rochester, NY, 2012.

[28] V Fang, MCMC: The Gibbs Sampler, simple example w/Matlab code, 2014. https://victorfang.wordpress.com/2014/04/29/mcmc-the-gibbs-sampler-simple-example-w-matlab-code/. (Accessed 17 April 2021).

[29] L Martino, H Yang, D Luengo, J Kanniainen, J Corander, A fast universal self-tuned sampler within Gibbs sampling, Digit Signal Process 47 (2015) 68–83.

[30] P Pinson, Wind energy: forecasting challenges for its operational management, Stat Sci 28 (4) (2013) 564–585.

[31] P Pinson, Wind scenarios. https://sites.google.com/site/datasmopf/wind-scenarios. (Accessed 17 April 2021).

CHAPTER 4

Optimal operation of the multi-energy building complex

4.1 Introduction

The conflict between the limited urban land resources and increasing construction land demand is getting more and more obvious due to the rapid progress of urbanization [1]. This is a major limiting factor for sustainable urban development, especially for developing countries. Meanwhile, the traditional method for urban construction tends to build the urban strictly in accordance with the functional zoning, which induces the unbalanced condition of electric loads among different regions of the urban. In that case, the secure and efficient operation of the urban power grid would be challenging. To address these issues, the building complex (BC) is employed in both urban construction and urban transformation around the world, such as La Défense in Paris, France, Roppongi in Tokyo, Japan, and 100 new BCs in Hangzhou, China [2]. The BC is defined as an organic combination of buildings with multiple functions in a limited property group [3]. The BC has two main advantages: (1) the conflict between the limited urban land resources and increasing construction land demand can be alleviated, and (2) people can enjoy more convenience for their fast-paced daily life [4]. Therefore, the BC has attracted more and more attention due to its centrality of living facilities and convenience for daily life [4].

However, the BC has a large occupied area and massive buildings integration with high population density, which leads to several operational problems for both the urban power grid and the buildings. First, a large amount of electricity is consumed to maintain the normal operation of the heating, ventilation, and air conditioning (HVAC) and other electrical appliances in buildings [5]. Second, the HVAC loads have special characteristics and working rules, which results in a high peak–valley load difference [6]. The high peak–valley load difference will also pose significant challenges on the urban power grid [7]. Third, the BC will also be charged for the reserve services to compensate for the imbalance resulting from the high

Optimal operation of integrated multi-energy systems under uncertainty.
DOI: https://doi.org/10.1016/B978-0-12-824114-1.00007-X Copyright © 2022 Elsevier Inc.
All rights reserved.

peak-valley load difference [8]. Therefore, reducing the peak-valley load difference will benefit both the urban power grid and the BC. Therefore, energy management methods are required to maintain optimal operation of the BC and urban power grid. Meanwhile, both minimizing the operating cost and reducing the peak-valley load difference should be considered in the energy management of the BC.

A great deal of progress has been made in modeling the thermal behavior of the building envelope and its HVAC system. In the research [9–12], building thermal models containing HVAC systems were proposed to simulate the heat transfer through a building envelope, such as the lumped thermal resistance and capacitance network model [9], the statistical regression model [10], the neural network model [11], and support vector machine [12]. However, the parameters used in these models are not thermal parameters of the physical materials of the building envelope. This means that the thermal parameters used in these models are not available and estimation of the thermal parameters based on a large scale of historical measurements is needed. Commonly used estimation methods include maximum likelihood estimation [13] and finite difference approximation [14]. Parameter estimation is difficult, and its process is time consuming because the building envelope consists of multiple layers of construction material and each layer has different thermal characteristics. Furthermore, accurate estimation would be a challenging task due to the changing outdoor environment and indoor human behaviors. Therefore, detailed physical models without parameter estimation are used to simulate the thermal behaviors of the building envelope in this chapter according to design data of the building, the ASHRAE handbook [15], and the site survey. Then, the physical thermal performance of the BC can also be described clearly by using detailed physical models in this chapter.

Great progress has been achieved for HVAC control in an individual building. Over the past few decades, several methods to design the PID controller for HVAC control have been proposed [16]. However, the tuning of PID controller parameters is a time-consuming, expensive, and difficult task in a time-varying environment [17], such as an individual building with thermal dynamics. The nonlinear control method [18] is effective for HVAC control but requires identifying stable states and complex mathematical analysis for the controller design. The rule-based ON/OFF control method is widely used for HVAC control [19] due to its simplicity and ease of implementation. However, this method cannot control the HVAC in an optimal way. In that case, neither the operating cost minimization of the individual building nor the operational coordination with the urban power grid can be realized. Compared with those methods for HVAC control, optimization-based control uses the model of the HVAC, building thermal model, inputs for these models, and disturbances to predict future states of the building [20]. Then, the optimal control action for the HVAC can be obtained by optimizing multiple objectives with defined constraints. Therefore, the optimization-based controller is suitable for controlling the HVAC using weather and occupancy predictions and other available information in an optimal way [20].

A single building with HVAC has limited flexibility, so an aggregator is required to aggregate a large number of buildings. Several studies have explored the use of aggregated buildings to provide ancillary services, such as the frequency reserves for transmission system operators and the congestion alleviation for distribution systems operators. A control strategy for aggregated buildings was proposed by Conte et al. [21] to contribute to the power system frequency control under both normal and emergency conditions. Two methods were presented in the work of Hanif et al. [22] to use aggregated building loads to resolve congestion for a distribution network.

These existing research works have made good contributions to the scheduling and provision of ancillary services from aggregated buildings. However, very limited work has focused on the optimal energy management for a BC with different types of buildings integrated. Meanwhile, the urban power grid operator is more concerned about loss minimization and peak–valley load difference reduction, whereas cost minimization and comfort maximization are of primary interest to the BC. Therefore, both the energy management of the buildings of the BC and the operational objectives of the urban power grid should be considered. To address these problems, an integrated optimal scheduling and predictive control method in a hierarchical structure for energy management of a BC is proposed in this chapter.

4.2 Configuration of a BC

The structure of a BC is shown in Fig. 4.1, mainly including multiple multi-energy buildings (MEBs), the photovoltaic (PV) station, and the energy management system of the BC (BCEMS) and building energy management systems (BEMSs) installed in smart buildings. The energy devices of a MEB include an HVAC system, a combined heating and power (CHP) unit, a heat

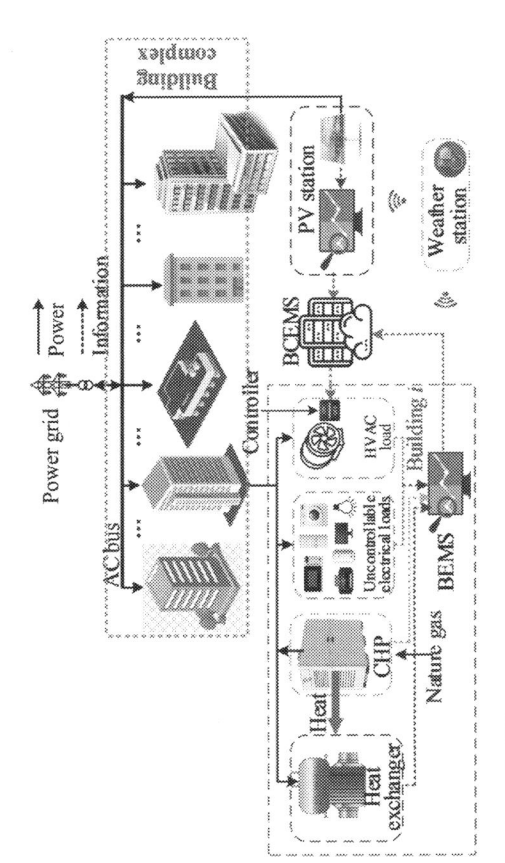

Figure 4.1 The physical structure of a BC.

exchanger, and other electric appliances. The proposed energy management method is based on information exchange between the BCEMS and the BEMSs using the communication infrastructure of the BC.

BCEMS. The BCEMS monitors the energy consumption of buildings and interacts with each BEMS to optimize energy usage. With the uploaded predictive data from the BEMSs, the optimal control schedules for each HVAC and CHP unit are obtained and issued to the corresponding HVAC controller and CHP controller in the individual building.

BEMS. Prediction data of solar radiation and outdoor temperature are obtained by the BEMSs from the weather station. Meanwhile, the measured data of each building—that is, the indoor temperature and uncontrollable electric load—are used as input data to the BEMSs. The energy consumption of each building is predicted based on the input data and the PMIB by its BEMS. Then, the measured data and predicted results from BEMSs are uploaded to the BCEMS.

Communication infrastructure. A bidirectional communication infrastructure is required between the BEMSs and BCEMS, as shown in Fig. 4.1. In addition, unidirectional communication links are required between the BEMSs and the market operator, BEMSs, and local weather stations.

In this chapter, four types of loads in each MEB of the BC are considered: (1) controllable electric load (i.e., HVAC loads), (2) uncontrollable electric load (e.g., loads of lighting and other electric appliances), (3) internal heat gain (e.g., heat dissipation of human bodies and electric appliances), and (4) heat gain due to solar radiation. The energy performance of different type of MEBs is mainly determined by the following factors: (1) number of stories, floor height, floor space, and the materials and structure of the building envelope; (2) HVAC schedules, and loads of lighting and electric appliances; and (3) occupied hours and hourly human occupancy rate. The MEBs of the BC can be divided into four categories [23]: office buildings, shopping malls, residential buildings, and other types of buildings.

4.3 PMIB with the HVAC system

The BEMSs of the MEBs not only consider local information for control purposes but also exchange the information of MEBs with the BCEMS for scheduling purposes. Therefore, the PMIB is developed to facilitate the BEMSs and BCEMS for optimal energy management of the BC.

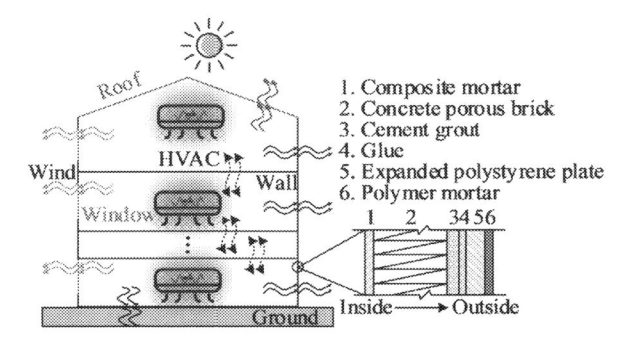

Figure 4.2 The thermal dynamic model of an individual MEB.

4.3.1 Thermal dynamics of a building

The thermal environment of an individual MEB is mainly determined by three factors: (1) the heating/cooling energy generated by the heating/cooling equipment (HVAC or other equipment); (2) heat gain due to solar radiation; and (3) the internal heat gains from human bodies, lighting, and electric appliances. The heat exchange with outdoor air mainly consists of thermal convection, thermal conduction, and heat storage in thermal mass. The thermal dynamic model of an individual MEB is shown in Fig. 4.2, which can be described by using detailed physical structures, according to the design data of the buildings [15].

The proposed model consists of several transient energy balance equations for external walls and internal air according to Fourier's thermal conducting equation approach [24]. The convective heat transfer represents a heat transfer through convection between the air and interior/exterior surface by means of heat fluid motion. It includes forced convection from the indoor air to the interior surface (as shown in Eq. (4.1)) and convection from the exterior surface to outdoor air (as shown in Eq. (4.2)):

$$Q_{c\_in} = k_{c\_in}A(T_{room} - T_A) \tag{4.1}$$

$$Q_{c\_ex} = k_{c\_ex}A(T_B - T_{out}) \tag{4.2}$$

where Q_{c_in}/Q_{c_ex} are the heat transfer through convection from indoor air to the interior surface/exterior surface to outdoor air (kW), k_{c_in}/k_{c_ex} are the convective heat transfer coefficient of the internal/external surface of the building envelope [kW/(m$^2 \cdot$ °C)], T_{room} is the temperature of indoor

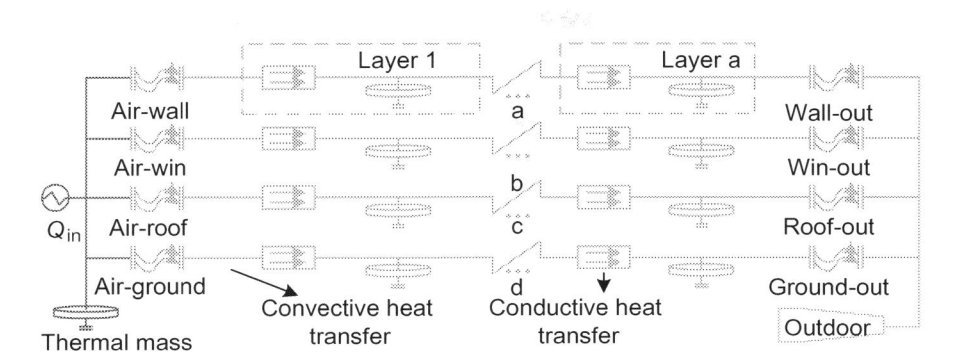

Figure 4.3 The thermal dynamic model of an individual building on the MAT-LAB/Simulink platform.

air (°C), T_A/T_B are the surface temperature of the internal layer/external layer of the building envelope (°C), and T_{out} is the outdoor temperature (°C).

The conductive heat transfer represents the heat transfer by conduction between two layers of the building envelope (i.e., walls, windows, and roof). Heating/cooling is transferred from the high-temperature side to the low-temperature side. In the proposed model, the building envelope consists of multiple layers of construction material, and each layer has different thermal characteristics, as shown in Fig. 4.2. The conductive heat transfer in each layer of the building envelope is governed by the Fourier law and is described by Eq. (4.3):

$$Q_k = k_k(A/D)(T_A - T_B) \qquad (4.3)$$

where Q_k is the heat transfer by conduction between two layers of the building envelope (kW) and k_k is the material thermal conductivity coefficient [kW/(m·°C)].

The material or the combination of materials of the building envelope can serve as thermal mass to store heat energy, which forms the heat storage capability of a building. The heat storage in thermal mass is shown in Eq. (4.4):

$$Q_m = cm_m(dT_B/dt) \qquad (4.4)$$

where Q_m is the heat storage in thermal mass of the building envelope (kW), c is the specific heat capacity of the thermal mass [J/(kg·°C)], and m_m is the mass of material of the building envelope (kg).

The proposed model is implemented on the MATLAB/Simulink platform, as shown in Fig. 4.3. The heating/cooling energy is generated by the heating or cooling system. Indoor air absorbs the heating/cooling energy and transfers it to the floor, walls, roof, and windows by convection, as shown in Eqs. (4.1) and (4.2). When the heating/cooling energy moves from air to the building envelope, it begins the heat transfer through the multiple layers of the building envelope by conduction (i.e., walls, windows, and roof), as shown in Eq. (4.3). During the heat conduction process, each layer can store energy and discharge energy, as shown in Eq. (4.4), which leads to the temperature difference between the interior and exterior surface of each layer. Then, the building envelope conducts the heat from the interior surface and transfers it to the outdoor air through the exterior surface. Finally, heating/cooling energy is transferred from the exterior surface of the building envelope to the outdoor air by convection.

4.3.2 The prediction model of loads of an individual building

(1) Controllable loads
  * Space heating load

For each individual MEB, its space heating load is supplied by the HVAC. For individual MEB i ($i = 1,2,\ldots,N$), the heat exchange with the outdoor air through ventilation is shown in Eq. (4.5) and the heat generated by HVAC i is shown in Eq. (4.6):

$$Q_{\text{flow}}^{(i)}(t) = \begin{cases} c_{\text{air}} m_{\text{flow}}^{(i)} \left(T_{\text{out}}^{(i)}(t) - T_{\text{room}}^{(i)}(t) \right) & control^{(i)}(t) = 1 \\ 0 & control^{(i)}(t) = 0 \end{cases} \qquad (4.5)$$

$$Q_{\text{heat}}^{(i)}(t) = \begin{cases} c_{\text{air}} m_{\text{flow}}^{(i)} \left(T_{\text{HVAC}}^{(i)}(t) - T_{\text{room}}^{(i)}(t) \right) & control^{(i)}(t) = 1 \\ 0 & control^{(i)}(t) = 0 \end{cases} \qquad (4.6)$$

where $Q(i)$ flow is the heat exchange with the outdoor air through ventilation in building i (kW); c_{air} is the specific heat capacity of the air [J/(kg·°C)]; $m(i)$ flow is the mass flow rate of air exchange between the outdoor air and indoor air (kg/h), which is calculated by Eq. (4.7); $Q(i)$ heat is the heat generated by HVAC i (kW); $T(i)$ HVAC is the supply air temperature of HVAC i (°C); and $control^{(i)}$ is the control variable of HVAC i ("1" represents the ON state of HVAC i, whereas "0" represents the OFF state).

$$m_{\text{flow}}^{(i)} = \rho_{\text{air}} V_{\text{air}}^{(i)} = \rho_{\text{air}} \times Flow^{(i)} \times S_v^{(i)} \qquad (4.7)$$

Then, the electric power consumption of HVAC i is calculated in Eq. (4.8):

$$P_{\text{HVAC}}^{(i)}(t) = Q_{\text{heat}}^{(i)}(t)/COP \qquad (4.8)$$

where $P(i)$ HVAC is the electric power consumption of HVAC i (kW) and COP is the coefficient of performance of the HVAC.

The electricity cost of HVAC i over h hours is shown in Eq. (4.9):

$$C_{HVAC}^{(i)} = \int_0^h c^{(i)} P_{HVAC}^{(i)}(t) dt \tag{4.9}$$

where $C(i)$ HVAC is the energy cost of the HVAC and $c^{(i)}$ is the electricity purchasing price (RMB/kWh).

- Hot water load

For each individual MEB, its hot water load is supplied by the CHP unit and the heat exchanger. A combined CHP consumes natural gas and generates electricity and heating simultaneously. The relationship between the nature gas consumption and power generation is shown in Eq. (4.10):

$$P_{CHP}^{(i)}(t) = \eta_e \times Q_{gas}^{(i)} \tag{4.10}$$

where $P(i)$ CHP is the electric power generation of CHP unit i (kW), η_e is the coefficient of the combined CHP unit, and $Q(i)$ gas is the nature gas consumption (kW).

The heating generation of the CHP unit i is illustrated in Eq. (4.11):

$$Q_y^{(i)}(t) = c_y \times q_y \times \left(T_{y1}^{(i)}(t) - T_{y2}^{(i)}(t) \right) \tag{4.11}$$

where $Q_y^{(i)}$ is the heating generation of CHP unit i (kW); c_y is the flow rate of the exhaust gas of CHP i (kg/s); q_y is the specific heat capacity of the exhaust gas of CHP i [J/(kg·°C)]; and $T_{y1}^{(i)}$ and $T_{y2}^{(i)}$ are the temperature of the exhaust gas into the heat exchanger and the temperature of the exhaust gas into the environment, respectively (°C).

The hot water load is mainly satisfied by the heating generated by CHP i ($Q_y^{(i)}$). If the heating generated by CHP i is not enough to cover the hot water load, the heat exchanger will use electricity to supply the remaining hot water load. Therefore, the electricity consumption by the heat exchanger is shown in Eq. (4.12):

$$P_{water}^{(i)}(t) = \begin{cases} \dfrac{Q_{water}^{(i)}(t) - Q_y^{(i)}(t)}{N_{water}} & Q_{water}^{(i)}(t) > Q_y^{(i)}(t) \\ 0 & other \end{cases} \tag{4.12}$$

where $P_{water}^{(i)}$ is the electricity consumption by heat exchanger i (kW), $Q_{water}^{(i)}$ is the total hot water load of MEB i (kW) and can be calculated by Eq. (4.13), and N_{water} is the coefficient of the heat exchanger:

$$Q_{water}^{(i)}(t) = \rho_r \times C_{water} \times (t_r - t_l) \times percent_b^{(i)}(t) \tag{4.13}$$

where ρ_r is the water density (kg/m^3), C_{water} is the specific heat capacity of water [J/(kg·°C)], and t_r and t_l are the supply and return temperatures of hot water (°C).

The cost of CHP unit i over h hours is shown in Eq. (4.14):

$$C_{CHP}^{(i)} = \int_0^h c_{gas}^{(i)} P_{CHP}^{(i)}(t) dt \tag{4.14}$$

where $C(i)$ CHP is the energy cost of the CHP unit and $c_{gas}^{(i)}$ is the nature gas purchasing price (RMB/kWh).

The electricity cost of heat exchanger i over h hours is shown in Eq. (4.15).

$$C_{water}^{(i)} = \int_0^h c^{(i)} P_{water}^{(i)}(t) dt \tag{4.15}$$

(2) Uncontrollable loads

* Uncontrollable electric loads

Uncontrollable electric loads include lighting and electric appliance loads, as shown in Eq. (4.16):

$$P_{unload}^{(i)}(t) = percent_e^{(i)}(t) \times \left(Light_{per}^{(i)} + Appliance_{per}^{(i)} \right) \times S_f^{(i)} \tag{4.16}$$

where $P_{unload}^{(i)}$ is the uncontrollable electric load (kW), $percent_{b(t)}^{(i)}$ is the hourly human occupancy rate in building i, $Light_{per}^{(i)}$ is the power density of lighting in building i (W/m^2), $Appliance_{per}^{(i)}$ is the power density of electric appliance in building i (W/m^2), and $S_f^{(i)}$ is the floor space of building i (m^2).

Electricity cost of uncontrollable electric load over h hours is shown in Eq. (4.17):

$$C_{unload}^{(i)} = \int_0^h c^{(i)} P_{unload}^{(i)}(t) dt \tag{4.17}$$

where $C_{unload}^{(i)}$ is the power cost of the uncontrollable electric load (RMB).

- Internal heat gains

The internal heat gains from human bodies, lights, and electric appliances are closely associated with hourly occupancy rate, hourly utilization rate of lighting and appliances, and heat density of the electric appliances [23]. Therefore, the internal heat gains are described as Eqs. (4.18) through (4.20):

$$Q_{body}^{(i)}(t) = percent_b^{(i)}(t) \times Body\gamma_{per}^{(i)} \times \frac{S_f^{(i)}}{Body\gamma_S^{(i)}} \tag{4.18}$$

$$Q_{app}^{(i)}(t) = percent_e^{(i)}(t) \times \left(Light_{per}^{(i)} + Appliance_{per}^{(i)}\right) \times S_f^{(i)} \times \varepsilon^{(i)} \tag{4.19}$$

$$Q_{rand}^{(i)}(t) = Q_{body}^{(i)}(t) + Q_{app}^{(i)}(t) \tag{4.20}$$

where $Q_{body}^{(i)}$ is the internal heat gains from human bodies of building i (kW), $percent_{b(t)}^{(i)}$ is the hourly human occupancy rate in building i, $Body_{per}^{(i)}$ is the heat emission of human body in average of building i (W/capita), $Body_s^{(i)}$ is the per capita area in building i (m²), $\varepsilon^{(i)}$ is the ratio of heat dissipation of the electric appliances in building i, and $Q_{app}^{(i)}$ is the internal heat gains from appliances and lighting of building i (kW).

- Heat gain due to solar radiation

Heat gain due to solar radiation is contributed by solar radiation transmitted across the windows and solar radiation on the opaque surface of the external walls. Solar radiation transmitted across the windows is calculated by Eq. (4.21), and it is assumed that the surfaces of the total window are distributed in the south, west, north, and east orientations of the walls in a building uniformly [25]:

$$Q_{sr}^{(i)}(t) = \sum_{j \in J} \tau_{win} \times SC \times A_{win,j}^{(i)} \times I_{T,j}(t) \tag{4.21}$$

where $Q_{sr}^{(i)}$ is the heat gain due to solar radiation transmitted across the windows in building i (kW), τ_{win} is the glass transmission coefficient of the windows, SC is the shading coefficient of the windows, $A_{win,j}^{(i)}$ is the area of the total window surface at the jth wall orientation (m²), and $I_{T,j}$ is the total solar radiation on the walls/windows surface at the j-wall orientation (kW/m²).

Solar radiation on the opaque surface of the external walls is calculated by summing the heat contribution due to solar radiation on each external wall (south, west, north, and east orientations) according to ISO 13790 [26], as shown in Eq. (4.22). In addition, the external surface heat resistance for convection and radiation of the external wall j, $R_{se,j}$, is considered in Eq. (4.22), and a typical method to calculate the $R_{se,j}$ is given in work by the ISO [27], which takes both radiation and convection terms into account:

$$Q_{sw}^{(i)}(t) = \sum_{j \in J} \alpha_w \times R_{se,j} \times A_{wall,j}^{(i)} \times I_{T,j}(t) \tag{4.22}$$

where $Q_{sw}^{(i)}$ is the heat gain due to solar radiation on the opaque surface of the external walls in building i (kW), α_w is the absorbance coefficient of the external surface of the wall, $R_{se,j}$ is the external surface heat resistance for convection and radiation of external wall j (m²·K/W), and $A_{wall,j}^{(i)}$ is the area of the total wall surface at the jth wall orientation (m²).

$I_{T,j}$ in Eqs. (4.21) and (4.22) is determined according to the method presented by Duffie and Beckman [28], which is a commonly used method to calculate the total solar radiation on a tilted surface [29]. It can be calculated as the sum of various types of solar radiation (i.e., beam, diffuse, and reflected radiation), as shown in Eq. (4.23):

$$I_T = I_b \times R_b + I_d \times \left(\frac{1 + \cos\beta}{2}\right) + I \times \rho_g \times \left(\frac{1 - \cos\beta}{2}\right) \tag{4.23}$$

where I_b, I_d, and I represent beam, diffuse, and total radiation on a horizontal surface, respectively (kW/m²); ρ_g is the ground reflectance and is taken as 0.2 [29, 30]; and R_b is the geometric factor which is defined as the ratio of beam radiation on a tilted surface to that on a horizontal surface. R_b is expressed in Eq. (4.24):

$$R_b = \frac{\cos\theta}{\cos\theta_z} \tag{4.24}$$

where θ and θ_z are incidence and zenith angles.

(3) Output of the PMIB

The operating cost of MEB i over h hours is calculated as follows:

$$C_{Building}^{(i)} = C_{HVAC}^{(i)} + C_{unload}^{(i)} + C_{water}^{(i)} + C_{CHP}^{(i)} \tag{4.25}$$

where $C_{Building}^{(i)}$ is the energy cost of building i (RMB).

The electric power consumption of MEB i over h hours is described as follows:

$$P_{Building}^{(i)}(t) = P_{HVAC}^{(i)}(t) + P_{unload}^{(i)}(t) + P_{water}^{(i)}(t) - P_{CHP}^{(i)}(t) \tag{4.26}$$

The heating/cooling energy absorbed by indoor air in MEB i is described as follows:

$$Q_{in}^{(i)}(t) = Q_{flow}^{(i)}(t) + Q_{heat}^{(i)}(t) + Q_{rand}^{(i)}(t) + Q_{sr}^{(i)}(t) + Q_{sw}^{(i)}(t) \tag{4.27}$$

4.4 Formulation of the hierarchical method

4.4.1 Formulation of the control layer

Two predictive control schedules of the HVAC in MEB i over the kth prediction horizon are set up first, as shown in Eqs. (4.28) and (4.29):

$$controlIMP_k^{on,(i)}(t) = \begin{cases} 1 \times H_k^{(i)}(t) & t \in [t_k, t_{k+1}] \\ controlTRA_k^{(i)}(t) & t \in [t_{k+1}, t_{k+p}] \end{cases} \tag{4.28}$$

$$controlIMP_k^{off,(i)}(t) = \begin{cases} 0 \times H_k^{(i)}(t) & t \in [t_k, t_{k+1}] \\ controlTRA_k^{(i)}(t) & t \in [t_{k+1}, t_{k+p}] \end{cases} \tag{4.29}$$

where $H_k^{(i)}$ is the state index function of HVAC i ("1" means HVAC i is enabled, and "0" represents HVAC i is disabled), as shown in Eq. (4.30); $controlTRA_k^{(i)}$ is the rule-based ON/OFF control actions[1] with the indoor temperature being constrained within a defined comfort range, as shown in Eqs. (4.31) and (4.32):

$$H_k^{(i)}(t) = \begin{cases} 1 & t \in [t_k, t_{k+p}] \& t \in t_{work}^{(i)} \\ 0 & t \in [t_k, t_{k+p}] \& t \notin t_{work}^{(i)} \end{cases} \tag{4.30}$$

where $t_{work}^{(i)}$ is the occupied hours of MEB i.

$$controlTRA_k^{(i)}(t) =$$
$$\begin{cases} 1 \times H_k^{(i)}(t) & \begin{aligned} & T_{room}^{(i)}(t) < T_{low} || \\ & \left(T_{room}^{(i)}(t) \in [T_{low}, T_{high}] \& controlTRA_k^{(i)}(t-1) = 1 \right) \end{aligned} \\ 0 \times H_k^{(i)}(t) & \begin{aligned} & T_{room}^{(i)}(t) > T_{high} || \\ & \left(T_{room}^{(i)}(t) \in [T_{low}, T_{high}] \& controlTRA_k^{(i)}(t-1) = 0 \right) \end{aligned} \end{cases}$$
$$\tag{4.31}$$

$$T_{low} < T_{room}^{(i)}(t) < T_{high} \tag{4.32}$$

Then, the indoor temperature $T_{room}^{(i)}(t_{k+1})$ at t_{k+1} and the operating cost of MEB i (shown in Eq. (4.25)) with the predictive control schedules over

---

[1] The rule-based ON/OFF control is a two-position control method that turns off the HVAC when the indoor temperature reaches the upper limit and turns on the HVAC when it reaches the lower limit [19].

Table 4.1 The predicted results of buildings and PV.

Predictive Control Schedule	Predicted Results of Building i		PV
	Operating Cost with Penalty Factor	Electric Power Consumption	
$controlIMP$on,(i) $k(t)$	$CrIMP$on,(i) Building,k	$PIMP$on,(i) Building,$k(t)$	$P_{PV,k}(t)$
$controlIMP$off,(i) $k(t)$	$CrIMP$off,(i) Building,k	$PIMP$off,(i) Building,$k(t)$	
$control TRA(i)$ $k(t)$	$CrTRA(i)$ Building,k	$PTRA(i)$ Building,$k(t)$	

$t_k\tilde{}t_{k+p}$ can be predicted based on the PMIB. To constrain the indoor temperature at t_{k+1} within the comfort range after implementing the predictive control schedules at time t_k, a penalty factor is introduced, as shown in Eq. (4.33), then the predicted results of MEB i over $t_k\tilde{}t_{k+p}$ (i.e., the operating cost with penalty) are obtained, as shown in Eq. (4.34):

$$r_k^{(i)} = \begin{cases} 1 & T_{\text{room}}^{(i)}(t_{k+1}) \in \left[T_{\text{low}}, T_{\text{high}}\right] \\ e^{T_{\text{room}}^{(i)}(t_{k+1}) - T_{\text{high}}} & T_{\text{room}}^{(i)}(t_{k+1}) \in \left(T_{\text{high}}, +\infty\right) \\ e^{T_{\text{low}} - T_{\text{room}}^{(i)}(t_{k+1})} & T_{\text{room}}^{(i)}(t_{k+1}) \in (-\infty, T_{\text{low}}) \end{cases} \tag{4.33}$$

$$Cr_{Building,k}^{(i)} = C_{HVAC,k}^{(i)} \times r_k^{(i)} + C_{unload,k}^{(i)} + C_{water,k}^{(i)} + C_{gas,k}^{(i)} \tag{4.34}$$

where $r^{(i)}{}_k$ is the penalty factor and $Cr^{(i)}{}_{Building,k}$ is the predicted results of MEB i with penalty. The operating cost with penalty factor ($CrIMP^{on,(i)}{}_{Building,k}$, $CrIMP^{off,(i)}{}_{Building,k}$ and $CrTRA^{(i)}{}_{Building,k}$) and electric power consumption ($PIMP^{on,(i)}{}_{Building,k(t)}$, $PIMP^{off,(i)}{}_{Building,k(t)}$ and $PTRA^{(i)}{}_{Building,k(t)}$) of MEB i over $t_k\tilde{}t_{k+p}$ can be predicted under three control schedules of the HVAC in MEB i (as shown in Table 4.1). The power output of PV ($P_{PV,k}(t)$) over the kth prediction horizon can be forecasted. Then, the predicted results are uploaded to BCEMS from BEMSs, as shown in Table 4.1.

4.4.2 Formulation of the scheduling layer

A multiobjective optimization problem is formulated in the scheduling layer. The objectives are reducing the peak–valley load difference and minimizing the operating cost of the BC.

(1) Reducing the peak–valley load difference

The peak–valley load difference can be represented by the mean square error (MSE) of the electric tie-line power of the BC over the kth prediction horizon [31]. The matrix $\mathbf{x_k}$ is considered as the control variables at t_k for

MEB i, as shown in Eq. (4.35).

$$x_k(i) = \begin{cases} 1 & controlIMP_k^{\text{on},(i)}(t) \\ 0 & controlIMP_k^{\text{off},(i)}(t) \end{cases} \tag{4.35}$$

Then, the first objective function of the BC is formulated as follows:

$$f_1(x) = \sqrt{\frac{1}{p+1} \sum_{j=0}^{p} \left(PIMP_{\text{line},k}(t_{k+j}) - PIMP_{\text{ave},k} \right)^2} \tag{4.36}$$

$$PIMP_{\text{line},k}(t) = \sum_{i=1}^{N} PIMP_{\text{Building},k}^{(i)}(t) - P_{\text{PV},k}(t) \tag{4.37}$$

$$PIMP_{\text{Building},k}^{(i)}(t) = \begin{cases} PIMP_{\text{Building},k}^{\text{on},(i)}(t) & x_k(i) = 1 \\ PIMP_{\text{Building},k}^{\text{off},(i)}(t) & x_k(i) = 0 \end{cases} \tag{4.38}$$

$$PIMP_{\text{ave},k} = \frac{\sum_{j=0}^{p} PIMP_{\text{line},k}(t_{k+j})}{p+1} \tag{4.39}$$

where $PIMP_{\text{line},k}(t)$ is the tie–line power of the BC under the hierarchical method and $PIMP^{(i)}{}_{\text{Building},k(t)}$ is the electric power consumption of MEB i under the hierarchical method.

(2) Minimizing the operating costs

The electricity cost with the penalty factor is utilized to describe the electricity consumption of the BC. The second objective function is shown in Eq. (4.40).

$$f_2(x_k) = \sum_{i=1}^{N} CrIMP_{\text{Building},k}^{(i)} \tag{4.40}$$

$$CrIMP_{\text{Building},k}^{(i)} = \begin{cases} CrIMP_{\text{Building},k}^{\text{on},(i)} & x_k(i) = 1 \\ CrIMP_{\text{Building},k}^{\text{off},(i)} & x_k(i) = 0 \end{cases} \tag{4.41}$$

The multiobjective optimization problem is converted into a single-objective optimization problem using the weighted sum method [32]. The two objective functions (i.e., $f_1(x)$ and $f_2(x)$) are first normalized separately, as shown in Eq. (4.42):

$$\begin{cases} f_1(x_k)' = \frac{f_1(x_k)}{f_{1,\text{TRA}}(x_k)} \\ f_2(x_k)' = \frac{f_2(x_k)}{f_{2,\text{TRA}}(x_k)} \end{cases} \tag{4.42}$$

$$f_{1,\text{TRA}}(x_k) = \sqrt{\frac{1}{p+1} \sum_{j=0}^{p} \left(PTRA_{\text{line},k}(t_{k+j}) - PTRA_{\text{ave},k}\right)^2} \tag{4.43}$$

$$PTRA_{\text{line},k}(t) = \sum_{i=1}^{N} PTRA_{\text{Building},k}^{(i)}(t) - P_{\text{PV},k}(t) \tag{4.44}$$

$$PTRA_{\text{ave},k} = \frac{\sum_{j=0}^{p} PTRA_{\text{line},k}(t_{k+j})}{p+1} \tag{4.45}$$

$$f_{2,\text{TRA}}(x_k) = \sum_{i=1}^{N} CrTRA_{\text{Building},k}^{(i)} \tag{4.46}$$

where $f_{1,\text{TRA}}(x_k)$ and $f_{2,\text{TRA}}(x_k)$ are the MSE of tie-line power and the electricity cost with the penalty factor of the BC with the rule-based ON/OFF control method of the HVACs, respectively; $PTRA_{\text{line},k}(t)$ is the tie-line power under the rule-based ON/OFF control method of the HVACs; and $PTRA(i)$ Building,$k(t)$ is the electric power consumption of MEB i under the rule-based ON/OFF control method of the HVACs.

Then, the normalized objective functions (i.e., $f_1(x_k)'$ and $f_2(x_k)'$) are aggregated with the weights (i.e., ω_1 and ω_2) of each objective function, as shown in Eq. (4.47):

$$\min f(x_k) = \omega_1 f_1(x_k)' + \omega_2 f_2(x_k)' \tag{4.47}$$

where ω_1 and ω_2 are the weight coefficient of objective functions, which meet Eq. (4.48).

$$\omega_1 + \omega_2 = 1 \tag{4.48}$$

4.4.3 Simulation setup

In the hierarchical method, the whole scheduling timeline is divided into several time slots and each time slot is allocated with a duration for each control horizon. The hierarchical algorithm for the BC is illustrated in Fig. 4.4.

Control layer. At current time t_k and $k = 1$, the BEMSs get the current forecasting data over the time horizon from t_k to t_{k+p} (i.e., the kth

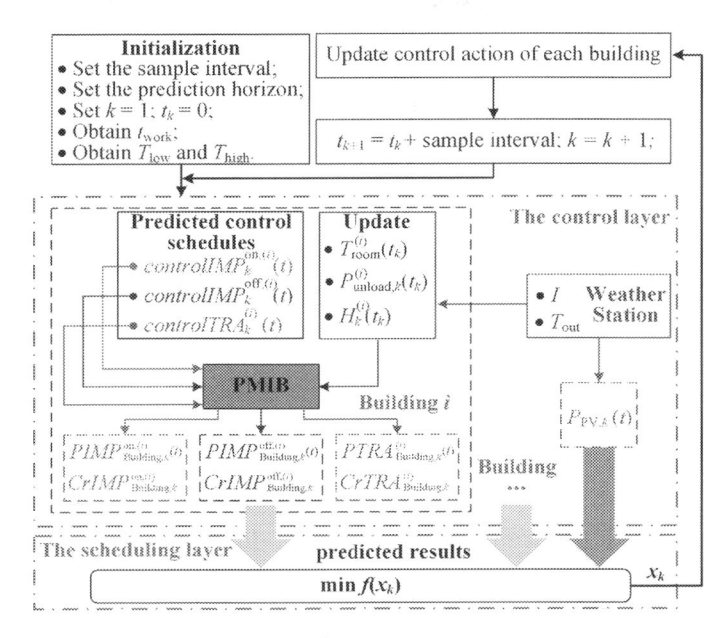

Figure 4.4 The hierarchical algorithm for the BC.

prediction horizon). In addition, the predictive control schedules of the HVAC in each individual MEB i at t_k (i.e., ON/OFF states of the HVAC) are set up. Finally, the predicted results of all MEBs are uploaded to the BCEMS.

Scheduling layer. The BCEMS solves a multiobjective optimization problem. It minimizes the operating costs and reduce the peak-valley load difference based on the predicted results from the BEMSs. The optimized results are a sequence of control schedules of the HVAC and the control schedules of the CHP unit in individual MEB i at t_k (i.e., $\boldsymbol{x_k}$). Then $\boldsymbol{x_k}$ is transferred to the control layer to be used as inputs of PMIB in Simulink.

At the next time slot $t_k + 1$ and $k = 2$, the BEMSs, and BCEMS get updated forecasting data for the next prediction horizon and the optimization problem is solved again. The time horizon moves forward by a one-time slot for the new optimization until all control schedules of the HVAC in each individual MEB are determined during the whole timeline.

Table 4.2 Parameters of the BC.

ID	Type	Amount	Floor Space (m²/per)	Stories
A	Office building	3	1242	14
B	Residential building	92	600	6
C	Shopping mall	2	4000	5
D	Other	3	1200	6

Table 4.3 Operational information of the BC.

ID	Occupied Hours	Electricity Price (RMB/kWh)	Ventilation Rate [m³/(h•m²)]	Light(i) per(W/m²)	Appliance(i) per(W/m²)
A	07:00–18:00	0.924	8–10	9	15
B	00:00–24:00	0.4883	8–10	7	15
C	08:00–21:00	0.924	18–20	10	13
D	07:00–18:00	0.924	17–19	9	5

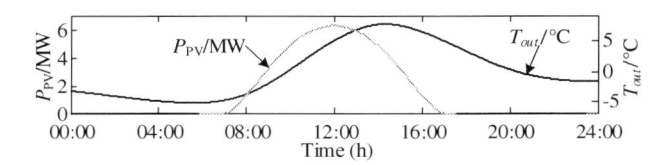

Figure 4.5 Outdoor temperature and active power output of PV.

4.5 Results and discussions

4.5.1 Case setting

Parameters and the operational information of the BC are shown in Table 4.2 and Table 4.3 [23, 33]. Material and its thermal parameters of the building envelope are shown in Table 4.4 [15].

The comfort range of each MEB is set to be 22 to 24°C during occupied hours. The initial T_{room} of the MEBs are set by random values within the comfort range. The initial ON/OFF status of each HVAC is set to be ON. COP of the HVAC is set as 4. The heat emission of the human body in average $Body^{(i)}_{per}$ is set as 43.84 W/capita [34], and the percentage of electrical appliance heat dissipation ε is set as 0.2 [35]. The equal weights method [36] is used to determine the weight coefficient in this chapter. Therefore, ω_1 and ω_2 are set as 0.5 equally. The outdoor temperature and active power output of PV on a typical winter day in China are shown in Fig. 4.5. The uncontrollable electric loads for each type of building are shown

Table 4.4 Material and its thermal parameters of the building envelope.

	Material	Thickness (mm)	Density (kg/m³)	Specific Heat Capacity (J/kg/°C)	Thermal Conductivity Coefficient (W/m/°C)	Convective Heat Transfer Coefficient (W/m²/°C)
Window	Glass	12	2000	940	0.76	25
	Glass	12	2000	940	0.76	32
Wall	Composite mortar	20	1800	1050	0.93	24
	Concrete porous brick	240	1450	1050	0.738	—
	Cement grout	20	1800	1050	0.93	—
	Expanded polystyrene board	30	20	1380	0.08	—
	Expanded polystyrene board	40	20	1380	0.08	—
	Polymer mortar	3	1800	1050	0.93	34
Roof	Cast-in-place reinforced concrete roof	120	920	2500	1.74	12
	Lightweight aggregate concrete	80	1050	1600	0.89	—
	Cement grout	20	1050	1800	0.93	—
	Extruded polystyrene board	40	1380	28	0.08	—
	Extruded polystyrene board	30	1380	28	0.08	—
	Cement grout	20	1050	1800	0.93	38
Floor	Solid wood flooring	12	700	2510	0.17	12
	Blockboard	15	300	1890	0.093	—
	Mineral wool board/Glass wool board	30	100	1220	0.14	—
	Cement grout	20	1800	1050	0.93	—
	Cast-in-place reinforced concrete floor	100	2500	920	1.74	40

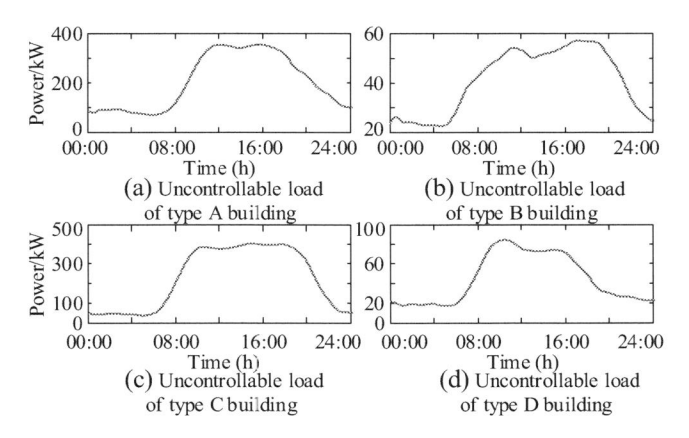

Figure 4.6 Uncontrollable electric load of four types of MEBs.

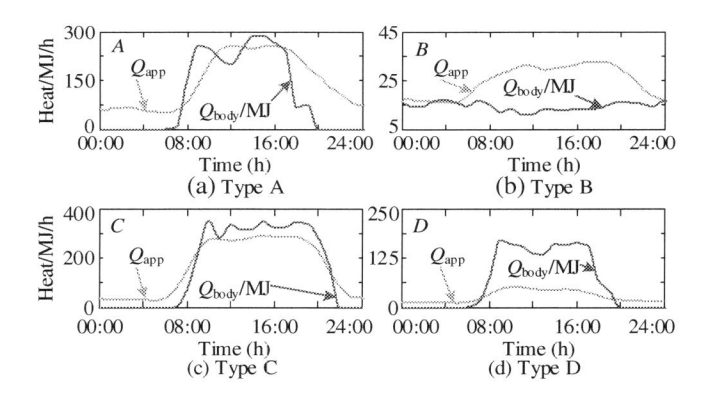

Figure 4.7 Internal heat gains of four types of MEBs (1 MJ/h = 0.2778 kWh).

in Fig. 4.6. The internal heat gains of each type of building are shown in Fig. 4.7. The nature gas price is 2.28 RMB/m^3.

HVACs of the MEBs are divided into five regions according to their indoor temperatures, as shown in Table 4.5. The initial operational schedules of the CHP are shown in Fig. 4.8. The hot water loads of MEBs are shown in Fig. 4.9.

4.5.2 Results and discussions

(1) Simulation results: the control layer

Two comparative scenarios are employed to verify the effectiveness of the proposed optimal energy management method:

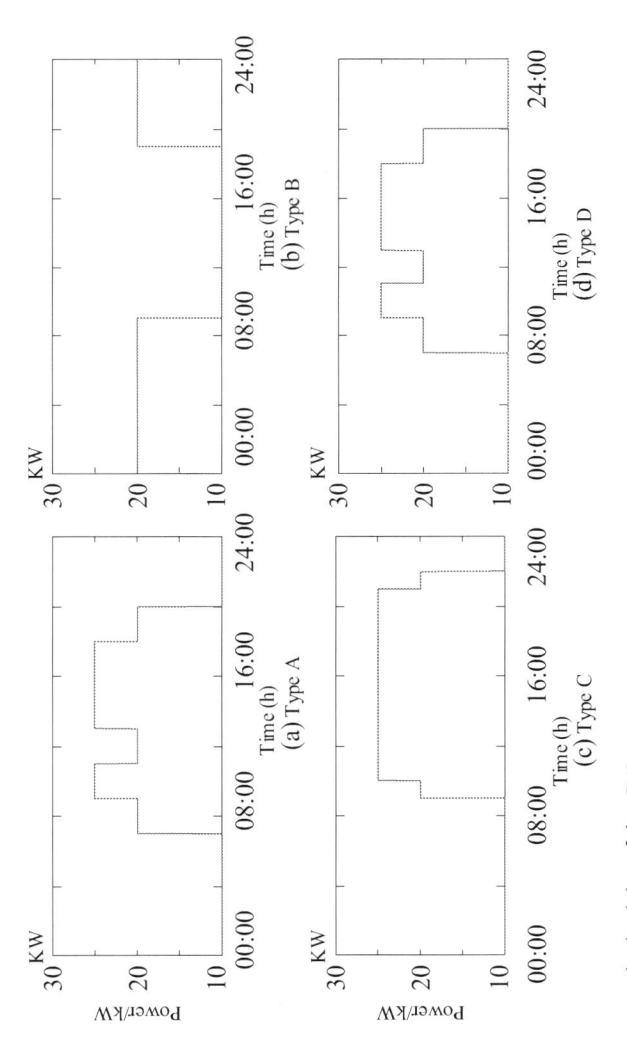

Figure 4.8 Operational schedules of the CHP.

Table 4.5 Region of the HVACs according to their indoor temperatures.

Region	Indoor Temperature and Operational Status
I	$23°C < T_{room} \leq 24°C$ & HVAC OFF
II	$23°C < T_{room} \leq 24°C$ & HVAC ON
III	$22°C \leq T_{room} \leq 23°C$ & HVAC ON
IV	$22°C \leq T_{room} \leq 23°C$ & HVAC OFF
V	Out of range[a]

[a] Since HVACs are turned off during the unoccupied hours, the indoor temperatures would drop out of the comfort range in this case.

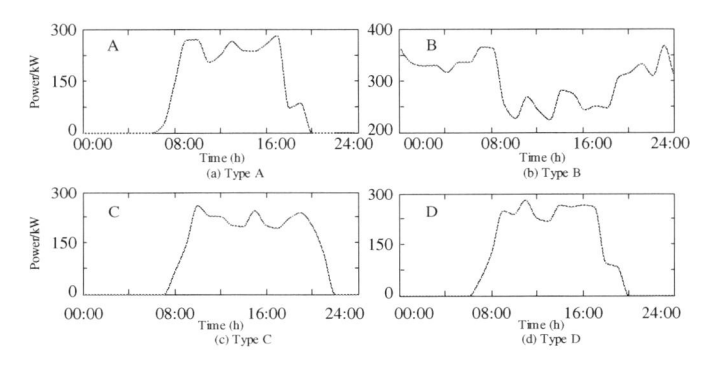

Figure 4.9 Hot water loads of buildings.

Scenario I (reference case): The HVAC of each individual MEB is operated with a rule-based ON/OFF control algorithm without coordinating with other HVACs. The CHP units are not optimized, and the predicted schedules of CHP units (as shown in Fig. 4.8) are used in this scenario. The power consumption of the BC is not optimized in this scenario.

Scenario II: The hierarchical algorithm with coordination between the BCEMS and the BEMSs of each MEB is implemented. The power consumption of the BC is optimized in the scheduling layer, and the HVAC and CHP unit of each MEB is controlled in the control layer.

As shown in the enlarged area at 16:00 in Fig. 4.10(A), the MEBs with the indoor temperatures close to the lower limit are large in quantity in Scenario I. This means that the operational states of HVACs are switched from OFF to ON intensively at 15:51, resulting in percussive loads to the power grid in Scenario I. Compared with the results in Scenario I, the MEBs are uniformly distributed at 16:00 and MEBs with the indoor temperatures close to the lower limit is small in quantity in Scenario II, as shown in the enlarge area

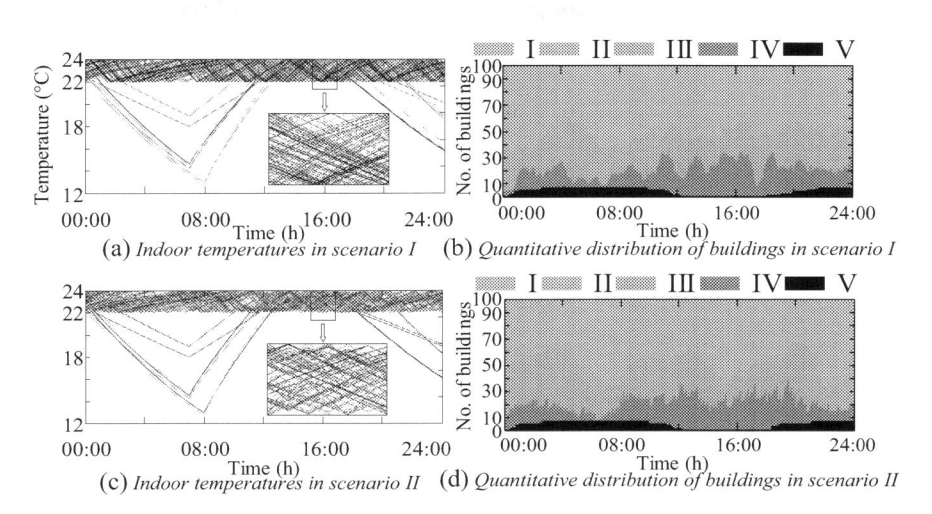

Figure 4.10 Indoor temperatures and quantitative distribution of MEBs in both scenarios.

in Fig. 4.10(C). In this case, the HVACs are not switched from OFF to ON intensively. The percussive loads to the power grid are also avoided.

Fig. 4.10(B) reveals that the number of MEBs in regions I through IV are all frequently fluctuating over time in Scenario I. From 8:00 to 20:00, the areas of region II and III, corresponding to the ON state of HVACs, experienced peaks and valleys frequently. This means that the electric load of the BC fluctuates, which causes a high peak–valley load difference. In contrast to the results in Scenario I, the fluctuations of the number of MEBs in regions I through IV are smoothed and the quantitative distribution of the buildings is more uniform in Scenario II, as shown in Fig. 4.10(D). Since HVACs of types A, C, and D MEBs are turned off during their unoccupied hours, their indoor temperatures drop out of the comfort range gradually, as shown in the dark areas of Fig. 4.10(B) and (D).

The number of HVACs in operation in Scenario I is shown in Fig. 4.11. From 07:00 and 08:00, the number of HVACs of types A, D, and C MEBs put into operation are 3, 3, and 2, respectively, which increases the electric load of the BC. The number of HVACs in the ON state of type B MEBs drops from 80 at 07:00 to the minimum level (i.e., 33 at 15:15), as a consequence of gradually increasing outdoor temperatures during this period (as shown in Fig. 4.5), which decreases the electric load of the BC. The increased electric load from 07:00 to 08:00 and gradually decreasing electric load from 07:00 to 15:15 result in the high peak–valley demand difference.

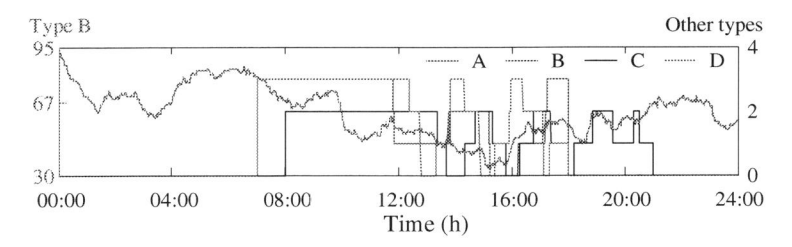

Figure 4.11   The number of HVACs that are in operation in Scenario I.

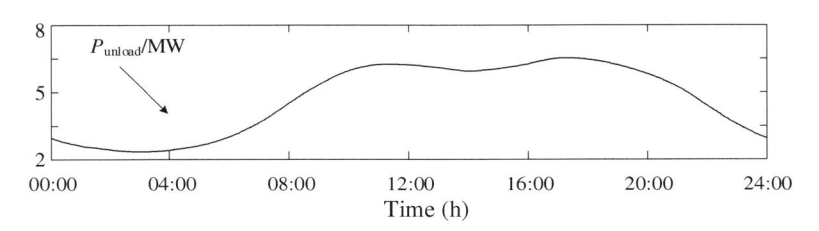

Figure 4.12   Uncontrollable load of MEBs in Scenario I.

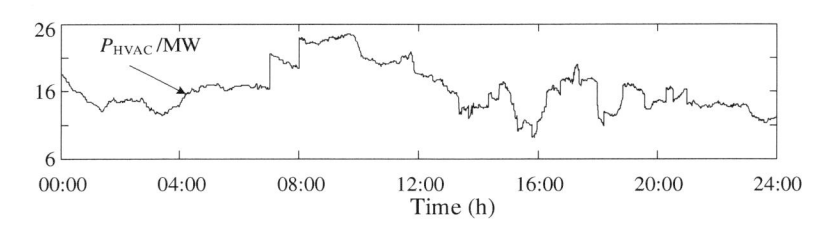

Figure 4.13   Power consumptions of HVACs in MEBs in Scenario I.

The energy schedules of MEBs in Scenario I are shown in Fig. 4.12 through Fig. 4.16. As can be observed, the power consumption of HVACs takes a large part of the toal power consumption of MEBs. Meanwhile, due to the fluctuations of the power consumption curve of HVACs, there are fluctuations of the total power consumption curve of MEBs. From 00:00 to 07:00, only HVACs of type B are in use, resulting in low HVAC load, as shown in Fig. 4.13. With HVACs of other types of MEBs put into use during occupied hours, two peaks of P_{HVAC} appear at 07:00 and 08:00. Then, the P_{HVAC} gradually decreases with gradually increasing outdoor temperature.

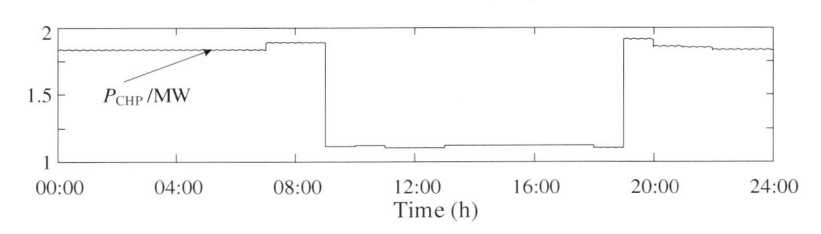

Figure 4.14 Power generation of CHPs in MEBs in Scenario I.

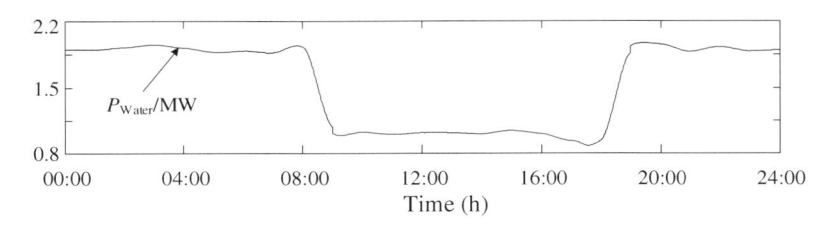

Figure 4.15 Power consumption of water heaters in MEBs in Scenario I.

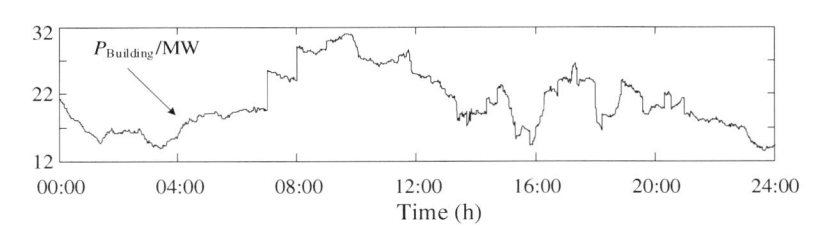

Figure 4.16 Total power consumptions of MEBs in Scenario I.

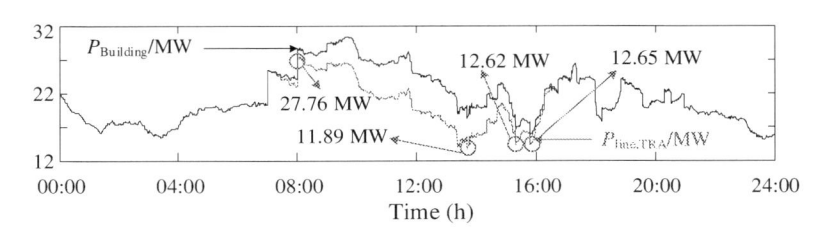

Figure 4.17 Tie-line power of the BC in Scenario I.

After 18:00, it maintains a lower level because HVACs of types A, C, and D are not dispatched during unoccupied hours.

The electric tie-line power of the BC in Scenario I is shown in Fig. 4.17. From 00:00 to 07:00, only HVACs of type B are in use, resulting in low electric tie-line power (P_{line}), as shown in Fig. 4.17. With HVACs of

Table 4.6 Power consumption of the buildings in Scenario I.

ID	Average Power Consumption/kW	Total Power Consumption/kW
A	1012.72	3038.15
B	199.00	18,308.01
C	2337.72	4675.44
D	817.39	2452.18

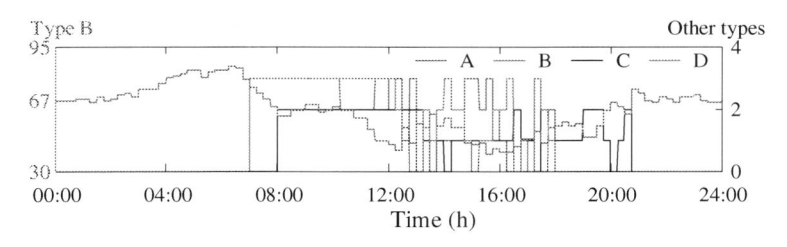

Figure 4.18 The number of HVACs that are in operation in Scenario II.

other types of MEBs put into use during occupied hours, two peaks of P_{line} appear at 07:00 and 08:00. Then, the P_{line} gradually decreases with gradually increasing outdoor temperature. After 18:00, P_{line} maintains a lower level due to out of operation of HVACs of types A, C, and D MEBs during unoccupied hours.

The power consumption of the MEBs in Scenario I is calculated based on the control schedules of the HVACs (see Fig. 4.11 and Fig. 4.13) and the PMIB, as shown in Table 4.6.

To reduce the peak demand from 7:00 to 8:00 in Scenario II with the hierarchical method, HVACs of type B MEBs are switched from OFF to ON gradually from 04:00 to 07:00 to preheat the type B MEBs and storage heating energy before the coming of the instantaneous loads of types A, C, and D MEBs from 07:00 to 08:00, as shown in Fig. 4.18. Then, HVACs of type B MEBs can be switched off from 7:00 to 8:00 without disturbing the indoor temperature comfort levels, which further reduces the peak–valley load difference from 7:00 to 16:00. The same results can also be observed in Fig. 4.10(D), in which the area of regions II and III widens gradually from 04:00 to 07:00 and narrows from 7:00 to 8:00 accordingly due to the preheating control schedules of HVACs of type B MEBs from 04:00 to 07:00 and the turning off schedules of them from 7:00 to 8:00.

As shown in Fig. 4.17, from 12:00 to 16:00, the electric load of the BC decreases due to the decreasing heating demand with high outdoor

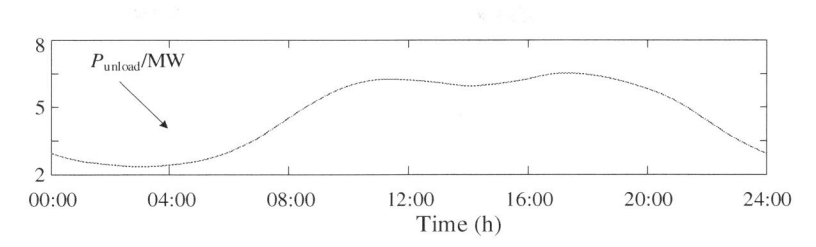

Figure 4.19 Uncontrollable load of MEBs in Scenario II.

temperature. Meanwhile, the power output of PV increases dramatically and reaches the peak value at 14:18, which causes valley demand of the BC at 13:42 with 11.89 MW and 15:48 with 12.62 MW in Scenario I. It further leads to the increase of the peak–valley load difference, as represented by the blue solid line in Fig. 4.17. Simulation results show that the peak–valley load difference is 15.87 MW and the MSE of tie-line power is 3670.5 MW in Scenario I. To cope with that in Scenario II, HVACs of type B MEBs are switched from ON to OFF from 10:30 to 12:30 and from 13:45 to 15:45 to lower the indoor temperatures of the buildings in advance of the times of the valley demands (i.e., 13:42 and 15:48), as shown in Fig. 4.18. In this case, HVACs of type B MEBs can be switched on during the times of the valley demands without disturbing the indoor temperature comfort levels, which increases the valley demand and further reduces the peak–valley load difference.

Comparing the schedules of HVACs in Scenario II (see Fig. 4.18) with those in Scenario I (see Fig. 4.11), it can be concluded that the HVACs can be scheduled according to the energy management of the BC under the hierarchical method.

(2) Simulation results: the scheduling layer

The BCEMS solves the multiobjective optimization problem in Eq. (4.47) based on the predicted results from the BEMSs. The optimized power consumption of the buildings in Scenario II is shown in Fig. 4.19 through Fig. 4.24. To compare the differences between Scenario I and II, we also added the results in Scenario I in Fig. 4.19 through Fig. 4.24. It can be observed from Fig. 4.20 that the hierarchical method can mitigate the peak–valley load difference by adjusting P_{HVAC}. As shown in Fig. 4.24, the peak of tie-line power is reduced from 27.76 MW in Scenario I to 26.65 MW in Scenario II, which is reduced by 3.999%. The peak–valley load differences are 11.43 MW in Scenario II and 15.87 MW in

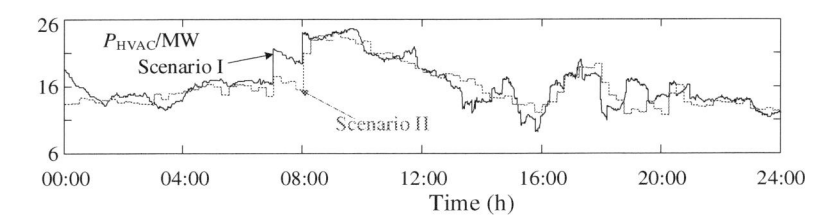

Figure 4.20 Power consumptions of HVACs in MEBs in Scenario II.

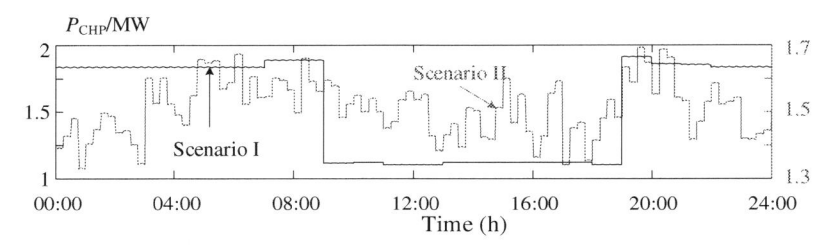

Figure 4.21 Power generation of CHPs in MEBs in Scenario II.

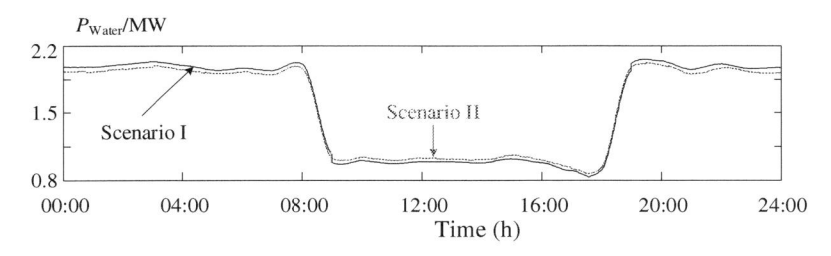

Figure 4.22 Power consumption of water heaters in MEBs in Scenario II.

Scenario I, respectively, which contributes 27.977% reduction. The MSE of tie-line power in the two scenarios are 2809.13 MW and 3670.5 MW, which is reduced by 23.47%. As shown in Fig. 4.21, the schedules of CHP in Scenario II are optimized, whereas those in Scenario I are not optimized.

The comparison results of the tie-line power and the operating costs in two scenarios are shown in Fig. 4.24 and Table 4.7. In summary, the operating cost minimization and peak-valley load difference reduction can be realized by the proposed hierarchical method.

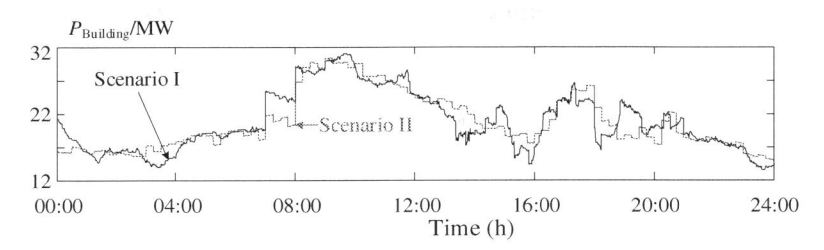

Figure 4.23 Total power consumption of MEBs in Scenario II.

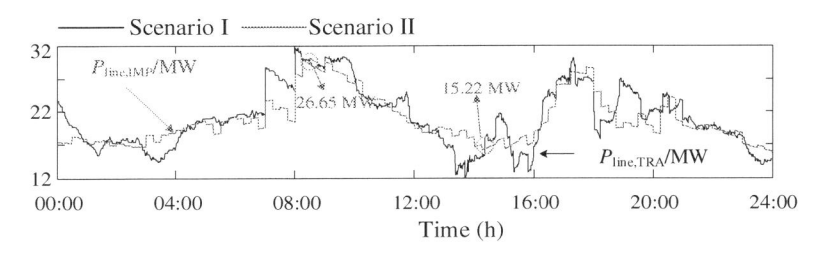

Figure 4.24 Tie-line power in both scenarios.

Table 4.7 Operating cost of the BC in both scenarios.

Load	Operating Costs in Scenario I/ × 10⁴RMB	Operating Costs in Scenario II/ × 10⁴RMB	Reduction/%
P_{unload}	6.9128	6.9128	—
P_{HVAC}	22.4896	22.4168	0.3237
$P(i)\ water$	1.8726	1.8563	0.8704
$P(i)\ CHP$	4.5117	4.4621	1.0994
$P_{Building}$	35.7867	35.6480	0.3876

4.6 Conclusion

This chapter proposed an optimal energy management method for the BC to optimally schedule the energy usage of the BC and control the HVACs and CHP units of the MEBs. In the control layer, each BEMS predicts the energy consumption of the individual building and controls its HVAC and CHP based on the PMIB. In the scheduling layer, the BCEMS solves the multiobjective optimization problem based on the predicted results from the BEMSs to reduce its peak–valley load difference and minimize its operating cost. The energy management is done with information exchange between the two layers and the interactions between the BCEMS and BEMSs.

Compared with the traditional rule-based ON/OFF control method of the HVAC, the proposed optimization method can control the HVAC and CHP unit in an optimal way based on updated forecasts of weather, occupancy, and other available information of the MEB. The peak–valley load difference and the operating cost of the BC can be reduced by implementing the proposed optimization method.

References

[1] L Zuo, Z Zhang, KM Carlson, GK MacDonald, KA Brauman, Y Liu, et al., Progress towards sustainable intensification in China challenged by land-use change, Nat Sustain 1 (2018) 304–313, doi:10.1038/s41893-018-0076-2.

[2] J Zhang, Y Wang, C Wang, R Wang, F Li, Quantifying the emergy flow of an urban complex and the ecological services of a satellite town: a case study of Zengcheng, China, J Clean Prod 163 (2017) S267–S276, doi:10.1016/j.jclepro.2016.02.059.

[3] J Lou, J Xu, K Wang, Study on construction quality control of urban complex project based on BIM, Procedia Eng 174 (2017) 668–676, doi:10.1016/j.proeng.2017.01.215.

[4] J Wang, P Xu, J Li, X Sun, W Tian, Towards an efficient cyber-physical system for first-mile taxi transit in urban complex, IEEE 19th International Symposium on High Assurance Systems Engineering, 2019 (2019), doi:10.1109/HASE.2019.00012.

[5] X Jin, T Jiang, Y Mu, C Long, X Li, H Jia, et al., Scheduling distributed energy resources and smart buildings of a microgrid via multi-time scale and model predictive control method, IET Renew Power Gener 13 (2019) 816–833, doi:10.1049/iet-rpg.2018.5567.

[6] T Jiang, Z Li, X Jin, H Chen, X Li, Y Mu, Flexible operation of active distribution network using integrated smart buildings with heating, ventilation and air-conditioning systems, Appl Energy 226 (2018) 181–196, doi:10.1016/j.apenergy.2018.05.091.

[7] H Yang, H Pan, F Luo, J Qiu, Y Deng, M Lai, et al., Operational planning of electric vehicles for balancing wind power and load fluctuations in a microgrid, IEEE Trans Sustain Energy 8 (2017) 592–604, doi:10.1109/TSTE.2016.2613941.

[8] J Villar, R Bessa, M Matos, Flexibility products and markets: literature review, Electr Power Syst Res 154 (2018) 329–340, doi:10.1016/j.epsr.2017.09.005.

[9] M Maasoumy Haghighi, AL Sangiovanni-Vincentelli, Modeling and Optimal Control Algorithm Design for HVAC Systems in Energy Efficient Buildings, Technical Report No. UCB/EECS-2011-12 (2011) https://escholarship.org/uc/item/0475s253Accessed 17 April 2021.

[10] M Aydinalp-Koksal, VI Ugursal, Comparison of neural network, conditional demand analysis, and engineering approaches for modeling end-use energy consumption in the residential sector, Appl Energy 85 (2008) 271–296, doi:10.1016/j.apenergy.2006.09.012.

[11] AH Neto, FAS Fiorelli, Comparison between detailed model simulation and artificial neural network for forecasting building energy consumption, Energy Build 40 (2008) 2169–2176, doi:10.1016/j.enbuild.2008.06.013.

[12] Z Hou, Z Lian, An application of support vector machines in cooling load prediction,, Int Work Intell Syst Appl 2 (2009) 1–4, doi:10.1109/IWISA.2009.5072707.

[13] P Bacher, H Madsen, Identifying suitable models for the heat dynamics of buildings, Energy Build 43 (2011) 1511–1522, doi:10.1016/j.enbuild.2011.02.005.

[14] VSKV Harish, A Kumar, Reduced order modeling and parameter identification of a building energy system model through an optimization routine, Appl Energy 162 (2016) 1010–1023, doi:10.1016/j.apenergy.2015.10.137.

[15] American Society of Heating, Refrigerating, and Air-Conditioning Engineers, ASHRAE Handbook: Fundamentals, American Society of Heating, Peachtree Corners, GA, 2005.

[16] ZR Radakovic, VM Milosevic, SB Radakovic, Application of temperature fuzzy controller in an indirect resistance furnace, Appl Energy 73 (2002) 167–182, doi:10.1016/S0306-2619(02)00077-6.

[17] SM Attaran, R Yusof, H Selamat, A novel optimization algorithm based on epsilon constraint-RBF neural network for tuning PID controller in decoupled HVAC system, Appl Therm Eng 99 (2016) 613–624, doi:10.1016/j.applthermaleng.2016.01.025.

[18] GD Pasgianos, KG Arvanitis, P Polycarpou, N Sigrimis, A nonlinear feedback technique for greenhouse environmental control, Comput Electron Agric 40 (2003) 153–177, doi:10.1016/S0168-1699(03)00018-8.

[19] M Avci, M Erkoc, A Rahmani, S Asfour, Model predictive HVAC load control in buildings using real-time electricity pricing, Energy Build 60 (2013) 199–209, doi:10.1016/j.enbuild.2013.01.008.

[20] A Afram, F Janabi-Sharifi, Theory and applications of HVAC control systems— a review of model predictive control (MPC), Build Environ 72 (2014) 343–355, doi:10.1016/j.buildenv.2013.11.016.

[21] F Conte, S Massucco, F Silvestro, Frequency control services by a building cooling system aggregate, Electr Power Syst Res 141 (2016) 137–146, doi:10.1016/j.epsr.2016.07.003.

[22] S Hanif, T Massier, HB Gooi, T Hamacher, T Reindl, Cost optimal integration of flexible buildings in congested distribution grids, IEEE Trans Power Syst 32 (2017) 2254–2266, doi:10.1109/TPWRS.2016.2605921.

[23] China Academy of Building ResearchEnergy Efficiency Design Standards for Public Buildings: GB50189-2015, China Architecture & Building Press, Beijing Shi, China, 2015.

[24] B Tashtoush, M Molhim, M Al-Rousan, Dynamic model of an HVAC system for control analysis, Energy 30 (2005) 1729–1745, doi:10.1016/j.energy.2004.10.004.

[25] M Ozel, K Pihtili, Optimum location and distribution of insulation layers on building walls with various orientations, Build Environ 42 (2007) 3051–3059, doi:10.1016/j.buildenv.2006.07.025.

[26] ISOISO 13790:2008: Energy Performance of Buildings—Calculation of Energy Use for Space Heating and Cooling, ISO, Geneva, Switzerland, 2008.

[27] ISOISO 6946:2007: Building Components and Building Elements—Thermal Resistance and Thermal Transmittance—Calculation Method, ISO, Geneva, Switzerland, 2007.

[28] JA Duffie, WA Beckman, Solar Engineering of Thermal Process, Wiley, New York, 1991.

[29] O Adil Zainal, R Yumrutaş, Validation of periodic solution for computing CLTD (cooling load temperature difference) values for building walls and flat roofs, Energy 82 (2015) 758–768, doi:10.1016/j.energy.2015.01.088.

[30] X Jin, J Wu, Y Mu, M Wang, X Xu, H Jia, Hierarchical microgrid energy management in an office building, Appl Energy 208 (2017) 480–494, doi:10.1016/j.apenergy.2017.10.002.

[31] Z-K Feng, W-J Niu, C-T Cheng, X-Y Wu, Peak operation of hydropower system with parallel technique and progressive optimality algorithm, Int J Electr Power Energy Syst 94 (2018) 267–275, doi:10.1016/j.ijepes.2017.07.015.

[32] S Ganguly, NC Sahoo, D Das, Multi-objective planning of electrical distribution systems using dynamic programming, Int J Electr Power Energy Syst 46 (2013) 65–78, doi:10.1016/j.ijepes.2012.10.030.

[33] JC Lam, RYC Chan, CL Tsang, DHW Li, Electricity use characteristics of purpose-built office buildings in subtropical climates, Energy Convers Manag 45 (2004) 829–844, doi:10.1016/S0196-8904(03)00197-3.

[34] S Murakami, S Kato, J Zeng, Combined simulation of airflow, radiation and moisture transport for heat release from a human body, Build Environ 35 (2000) 489–500, doi:10.1016/S0360-1323(99)00033-5.

[35] F Qi, X Jin, Y Mu, H Jia, X Yu, X Xu, et al., Model predictive control based scheduling method for a building microgrid, Proc 2017 IEEE Power & Energy Society General Meeting (2017), doi:10.1109/PESGM.2017.8274239.a.

[36] F Fang, QH Wang, Y Shi, A novel optimal operational strategy for the CCHP system based on two operating modes, IEEE Trans Power Syst 27 (2012) 1032–1041, doi:10.1109/TPWRS.2011.2175490.

CHAPTER 5

MPC-based real-time dispatch of multi-energy building complex

5.1 Introduction

Recent years have experienced a rapid increase in power consumption in buildings worldwide due to the rapid process of urbanization of the world's population [1]. It has been shown that global use of electricity in buildings grew on average by 2.5% per year since 2010, and it increased by nearly 6% per year [2]. According to the International Energy Agency, buildings' share of the worldwide energy usage is approximately 40%, with almost half of it being used in their heating, ventilation, and air conditioning (HVAC) systems [3]. Therefore, issues on energy consumption reduction of buildings will become more prominent in China. A building complex (BC) is an organic combination of buildings with multiple functions in a limited property group, which forms highly intensive open public spaces. Therefore, issues on the BC's energy consumption reduction and efficiency improvement will become more prominent.

Power consumption related to thermal appliance operation for heating/cooling purposes in a building, such as HVAC and electric chillers, represents a very high portion in load demand. For most of the applications of heating/cooling devices, it is only required to control the temperature in a suitable zone without disturbing the temperature comfort level [4]. This provides an opportunity to effectively reduce the energy usage/cost and the peak demand of buildings by scheduling the heating/cooling devices in an optimal way. Hence, in principle, buildings and the BC have the potential to become a huge source of flexibility to reduce the load demand, facilitate integration of intermittent renewable generation, and provide ancillary services to power systems [4].

Furthermore, the BC offers an opportunity and a desirable infrastructure to dispatch smart buildings in an optimal way by utilizing advanced energy management technologies and intelligent communication technologies [5]. Several benefits and opportunities can be achieved by applying the energy

Optimal operation of integrated multi-energy systems under uncertainty. Copyright © 2022 Elsevier Inc.
DOI: https://doi.org/10.1016/B978-0-12-824114-1.00003-2
All rights reserved. **111**

management on the BC: (1) buildings can enjoy energy/cost savings [6]; (2) intermittent renewable generation can be more efficiently integrated [7]; (3) distributed energy resources (DERs) at the building side, such as controllable distributed generators (DGs), storage devices, and electric vehicles, are more efficiently managed; and (4) power imbalance of the BC can be balanced by optimal scheduling the DER and smart buildings without being charged for reserve service from the utility grid [8]. Therefore, energy management of the BC has attracted more and more attention in recent years.

Studies have been carried out to investigate the energy management methods for a smart building or a BC. Optimal day-ahead scheduling methods were investigated for a smart building or a BC in several works [7–11], achieving different operational objectives such as operating cost reduction and pollutants emission reduction. The main idea is that optimization technologies are used to decide day-ahead optimal schedules of the smart building or the BC based on day-ahead forecasting data, day-ahead electricity prices, and technical information of energy devices in the smart building or the BC. However, the inherent uncertainties of the day-ahead forecasting data are not considered, which leads to a discrepancy between the power really exchanged with the utility grid and the planned one. To further consider the uncertainties associated with the day-ahead forecasting data, stochastic day-ahead scheduling methods were proposed in the work of Wu et al. [12] and Thomas et al. [13] by using scenario-based methods and scenario reduction techniques. Nevertheless, the stochastic optimal scheduling methods assume that the day-ahead forecasting data follows certain probability density functions. The probability density functions require sufficient historical data, which limits the application of the stochastic method [14]. Meanwhile, the probability density functions may be inappropriate to describe the uncertainties in practical applications [15]. The robust day-ahead scheduling method is another way to consider the uncertainties for scheduling a smart building, as proposed by Wang et al. [16, 17]. However, concerns of its practical application are the conservativeness, as the robust optimal scheduling method is based on the worst-case analysis [18]. The preceding optimal day-ahead scheduling methods have made good contributions to the optimization of a BC. However, since the operational time scales of a BC are different between the day-ahead stage and actual operational stage, the day-ahead methods with one single time scale are difficult to capture the temporal dependencies of the schedules of the DERs and smart buildings between the two stages. In this context, the optimal schedules determined by the day-ahead methods leave questions on their feasibility and actual performances for practical applications on the BC during the real-time operational stage. Furthermore, although the uncertainties of the day-ahead forecasting data are considered at the day-ahead scheduling stage, their dynamic random fluctuations are not considered at the actual operational stage, which still causes errors between the schedules of two stages and further induces fluctuations of the electric tie-line power linking the BC and the utility grid. As stated by Bao et al. [19], balancing the errors and smoothing the fluctuations are important for the benefits of the utility grid and the BC.

To cope with the preceding problems, multi-time scale and multistage scheduling methods have gained much attention recently. A multi-time scale and coordinated scheduling method for a multi-energy microgrid was proposed in the work of Bao et al. [19]. A two-stage robust scheduling method for a smart building was proposed by Wang et al. [20] to minimize the operating cost. A multi-time scale stochastic scheduling method was proposed by Rahmani-Andebili [21] to schedule deferrable appliances and energy resources of a smart building in an optimal way considering the uncertainties of forecasting data. However, the flexibility of the building with heat inertia has not been fully explored in the preceding work. In the work of Jin et al. [22], a two-stage hierarchical microgrid energy management method in an office building by scheduling thermal mass of the building and plug-in electric vehicles was proposed. The daily operating cost can be reduced at the day-ahead stage and the fluctuations of the electric tie-line power can be smoothed to some extent at the real-time stage with the proposed method. Although the flexibility of the building was considered by Jin et al. [22], the optimal scheduling method used at the real-time stage was open-loop and single period based in nature. In other words, the optimal schedules at the actual operational stage are optimized using the single-optimization strategy based on the operating status of the smart building and forecasting data at the current operational period rather than the predictive operating status and forecasting data over a future time horizon [23]. In this case, concerns regarding the optimal adjustments of the controllable units of the smart building at the actual operational stage, such as insufficient adjustments, excessive adjustments, and untimely adjustments, would arise [24]. This leads to bad control performance of the single-optimization strategy in a time-varying context with uncertainties associated with forecasting data.

Different from single optimization, the model predictive control (MPC)-based scheduling method leads to better control performance against uncertainties because of its capability to handle the future behavior of the optimized system (i.e., the BC in this chapter), demand and renewable generation forecasts, and the constraints of the optimized system. MPC computes a sequence of decision variable adjustments over a future time horizon iteratively based on an underlying optimization model and forecasting data of uncertain variables [25]. In other words, MPC is a rolling process that runs the embedded optimization model repeatedly with updated forecasts, which has better control performance in a time-varying context, as verified by Guo et al. [26]. Motivated by the attractive features of the MPC method in a time-varying context, an MPC-based scheduling strategy for the BC is developed at the real-time dispatch stage in this chapter. The real-time dispatch stage can smooth the fluctuations of the electric tie-line power of the BC caused by the errors of the day-ahead forecasting data.

5.2 Configuration and modeling of the BC

5.2.1 Configuration of the BC

The physical configuration of a BC is shown in Fig. 5.1, mainly including multiple smart buildings, DERs (i.e., controllable DGs and battery), the energy management system of the BC (BCEMS), and building energy management systems (BEMSs) installed in each smart building. The energy systems of a smart building include an electric chiller for cooling purposes, the renewable generations, and other electric appliances. The proposed optimal scheduling method is based on information exchange between the BCEMS and the BEMSs thanks to the communication infrastructure of the BC. The functions of the BCEMS and BEMS are described as follows.

Energy management system of the BC. The BCEMS monitors the energy consumption of buildings and interacts with every BEMS to optimize their energy usage within the user's comfort range. With the proposed scheduling method and the forecasting data from the BEMSs, optimal schedules of the DERs and the smart buildings are obtained and issued to the corresponding DG controllers and the BEMSs. The optimal schedules of the DERs include operational schedules at the day–ahead stage (i.e., the unit commitment of the DGs and the charging/discharging schedules of the battery) and optimal dispatch results at the real-time stage (i.e., the optimal dispatches of power outputs of the DGs and the charging/discharging power of the battery). The optimal schedules of each smart building at both stages are the optimized

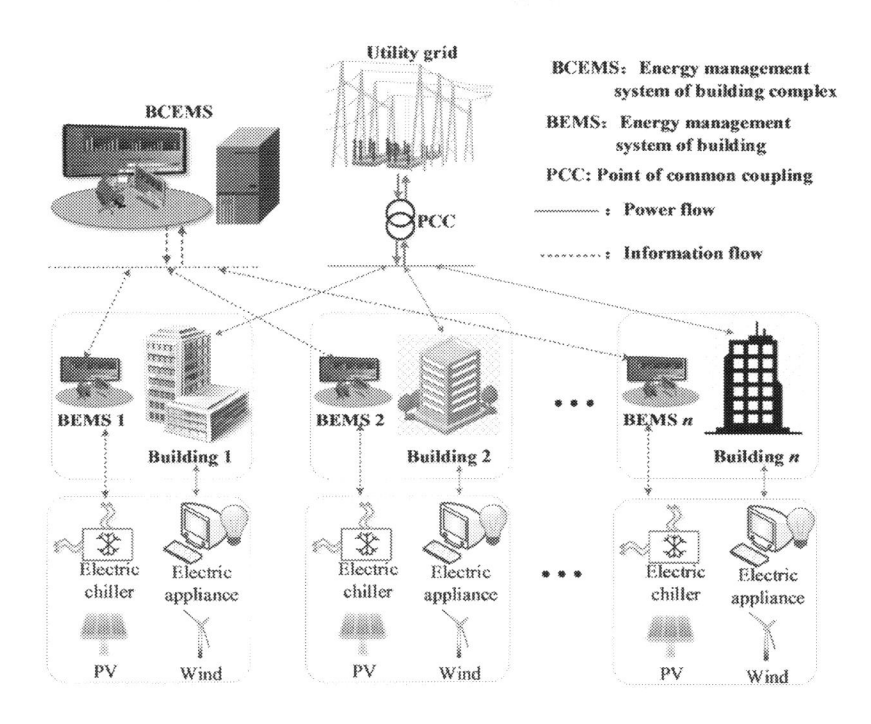

Figure 5.1 Schematic illustration of the multi-energy BC.

power consumption profiles of the electric chiller, which are used as control variables to fully explore the flexibility of the buildings.

Building energy management system. Forecasting data of solar radiation and outdoor temperature is obtained by BEMS from the weather station through the communication links. Meanwhile, the forecasting data of each building (i.e., electric power consumption of the electric appliance and internal heat gain) are obtained by BEMSs. All forecasting data from BEMSs are uploaded to the BCEMS. With the optimized load profile of each building issued by the BCEMS and thermal model of the building, the indoor temperature schedules for each building are calculated by the BEMS and issued to the indoor temperature controller.

Communication infrastructure. For the optimal scheduling method, a bi-directional communication infrastructure is required between the BEMSs and the BCEMS, as shown in Fig. 5.1. In addition, unidirectional communication links are required between the BEMSs and the market operator, BEMSs and the local weather station, and BCEMS and the DG controllers.

5.2.2 Building modeling

Considering a summer cooling scenario, a building is modeled as a single isothermal air volume [27]. The mathematical relationship among the indoor temperature, cooling demand, and outdoor temperature is formulated to investigate the thermal performance of a building by using the building thermal equilibrium equation, as shown in Eq. (5.1) [28]:

$$\rho \times C \times V \times \frac{dT_{in}}{dt} = \dot{Q}_{wall} + \dot{Q}_{win} + \dot{Q}_{in} + \dot{Q}_{sw} + \dot{Q}_{sg} - \dot{Q}_{EC} \tag{5.1}$$

where ρ is the density of the air in the building (kg/m^3); C is the specific heat capacity in the building [J/(kg·°C)]; V is the volume of the air in the building (m^3); T_{in} is the indoor temperature (°C); and $\dot{Q}_{wall}$ is the heat transfer through the external walls (kW), which is calculated by summing the contribution of each wall of a building, as shown in Eq. (5.2). The roof of a building is accounted for as part of the external walls [29]. $\dot{Q}_{win}$ is the heat transfer across the windows (kW). It is calculated by summing the contribution of each window of a building, as shown in Eq. (5.3). $\dot{Q}_{in}$ is the internal heat gains (kW). $\dot{Q}_{sw}$ is the heat contribution due to the solar radiation on the opaque surface of the external walls (kW). It is calculated by summing the heat contribution due to solar radiation on each external wall (south, west, north, and east orientations) according to ISO 13790 [30], as shown in Eq. (5.4). $\dot{Q}_{sg}$ is the whole solar radiation transmitted across the windows (kW). It is calculated according to Eq. (5.5). It is assumed that the total window surfaces are distributed in the south, west, north, and east orientations of the walls in a building uniformly [32]. $\dot{Q}_{EC}$ is the cooling power generated by the cooling equipment (kW). We have the following equations:

$$\dot{Q}_{wall} = \sum_{j \in J} U_{wall} \times F_{wall,j} \times (T_{out} - T_{in}) \tag{5.2}$$

$$\dot{Q}_{win} = \sum_{j \in J} U_{win} \times F_{win,j} \times (T_{out} - T_{in}) \tag{5.3}$$

$$\dot{Q}_{sw} = \sum_{j \in J} \alpha_w \times R_{se,j} \times U_{wall} \times F_{wall,j} \times I_{T,j} \tag{5.4}$$

$$\dot{Q}_{sg} = \sum_{j \in J} \tau_{win} \times SC \times F_{win,j} \times I_{T,j} \tag{5.5}$$

where U_{wall} and U_{win} are the heat transfer coefficient of the wall/window of the building [W/(m^2·K)]; $F_{wall,j}$ is the area of the total wall surface at the jth wall orientation (m^2); and $F_{win,j}$ is the area of the total window surface at the jth wall orientation (m^2), and it is assumed that the total window surfaces are distributed in the south, west, north, and east orientations of the walls in a building uniformly. T_{out} is the outdoor temperature (°C), α_w is the absorbance coefficient of the external surface of the wall, and $R_{se,j}$ is the external surface heat resistance for convection and radiation of the external wall j. A typical method to calculate $R_{se,j}$ is given by the ISO [31], which takes both radiation and convection terms into account. τ_{win} is the glass transmission coefficient of the windows and SC is the shading coefficient of the windows. $I_{T,j}$ is the total solar radiation on the walls/ windows surface at the j-wall orientation (kW/m²). It is determined according to the method presented in Duffie and Beckman [33], which is a commonly used method to calculate the total solar radiation on a tilted surface [34]. It can be calculated as the sum of various type of solar radiation, such as beam, diffuse, and reflected radiation, as shown in Eq. (5.6) :

$$I_T = I_b \times R_b + I_d \times \left(\frac{1 + \cos\beta}{2}\right) + I \times \rho_g \times \left(\frac{1 - \cos\beta}{2}\right) \tag{5.6}$$

where I_b, I_d, and I represent beam, diffuse, and total radiation on a horizontal surface, respectively (kW/m²); ρ_g is the ground reflectance and is taken as 0.2 in the present study [34]; and R_b is a geometric factor that is defined as the ratio of beam radiation on a tilted surface to that on a horizontal surface, which is shown in Eq. (5.7):

$$R_b = \frac{\cos\theta}{\cos\theta_z} \tag{5.7}$$

where θ and θ_z are incidence and zenith angles.

The thermal mass of a building can provide inertia. Like other technologies to store energy, this inherent property can be used to store energy at peak periods and preheat or precool the building without any additional investment cost. Therefore, the model of virtual energy storage system (VESS) is developed considering this inherent property of a building. The basic idea of the VESS is that the cooling demand of the building can be adjusted in the energy management process without disturbing the temperature comfort level due to the thermal mass of the building. Therefore, the cooling energy generated by the electric chiller is stored in the building when the electricity price is low (i.e., the electric chiller is started in advance or the power consumption of the electric chiller is increased). In that case, the VESS is charged seen from the BC (i.e., $\dot{Q}'_{cl,building} < \dot{Q}_{cl,building}$). In the same way, the cooling energy generated by the electric chiller is discharged in the building when the electricity price is high (i.e., the electric chiller is shut down in advance or the power consumption of the electric chiller is decreased). In that case, the VESS is discharged seen from the BC (i.e., $\dot{Q}'_{cl,building} < \dot{Q}_{cl,building}$). The charging/discharging power of the VESS, as shown in Eq. (5.8), is obtained following Eq. (5.1). The indoor temperature comfort zone and temperature set point are considered in the model of the VESS to maintain the customer's comfort level. We have the following equation:

$$\dot{Q}_{VESS,t} = \dot{Q}'_{cl,building,t} - \dot{Q}_{cl,building,t} \tag{5.8}$$

where $\dot{Q}_{cl,building}$ is the cooling load of the building with the VESS being scheduled (kW) and $\dot{Q}'_{cl,building}$ is the cooling load of the building without the VESS being scheduled (kW).

5.2.3 Mathematical models of DER

Diesel engine. For the diesel engine, the fuel cost depends on the power generation and fuel cost coefficients, shown in Eq. (5.9):

$$f_{fuel}(P_{DE,t}) = aP_{DE,t}^2 + bP_{DE,t} + c \tag{5.9}$$

where a, b, c are the fuel cost coefficients of the diesel engine and $P_{DE,t}$ is the power generation of the diesel engine.

Fuel cell. For the fuel cell, the fuel cost depends on the power generation and efficiency η_{FC}, shown in Eq. (5.10):

$$f_{fuel}(P_{FC,t}) = C_{gas} \times P_{gas} = C_{gas} \times \left(P_{FC,t}\Delta t / \eta_{FC}\right) \tag{5.10}$$

where C_{gas} is the natural gas purchasing price ($/kWh), P_{gas} is the natural gas consumption (kW), $P_{FC,t}$ is the power generation of the fuel cell, and η_{FC} is the efficiency of the fuel cell.

Electric chiller. The electricity consumption of the electric chiller is determined by the cooling demand and the coefficient of performance, shown in Eq. (5.11):

$$Q_{EC,t} = P_{EC,t} \times EER_{EC} \tag{5.11}$$

where $P_{EC,t}$ is the electric power consumption by the electric chiller (kW) and EER_{EC} is the energy efficiency ratio of the electric chiller.

Battery. The state of charge (SOC) of the battery refers to the ratio of the residual energy to the rated energy. The SOC at dispatch time interval

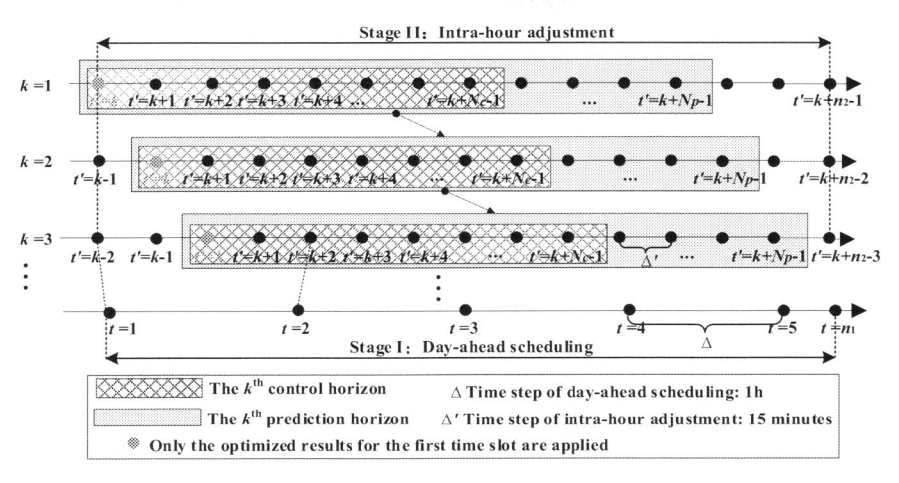

Figure 5.2 Schematic illustration of the scheduling method.

t is described in Eq. (5.12):

$$SOC_t = \begin{cases} SOC_{t-1}(1-\delta) - P_{bt,t}\Delta t \eta_{ch}/CAP_{bt} & P_{bt,t} \leq 0 \\ SOC_{t-1}(1-\delta) - P_{bt,t}\Delta t/(\eta_{dis}CAP_{bt}) & P_{bt,t} > 0 \end{cases} \qquad (5.12)$$

where SOC and δ are the SOC and self-discharge ratio of the battery, $P_{bt,t}$ is the charging/discharging power of the battery (kW), η_{ch} and η_{dis} are the charging/discharging efficiency of the battery, and CAP_{bt} is the rated capacity of the battery (kWh).

5.3 The multi-time scale and MPC-based scheduling method

5.3.1 Framework of the scheduling method

The proposed scheduling method is illustrated in Fig. 5.2, which includes two stages.

With the hourly electric load demand, outdoor temperature, solar radiation, and renewable generation forecasting values, the hourly schedules at the day-ahead stage over the n_1 time slots are obtained by the BCEMS and the BEMSs. This stage works at a slow time scale, namely Δt. The hourly schedules include optimal schedules of the smart buildings (i.e., the optimized power consumption profiles of the electric chiller and the indoor temperature schedules), optimal schedules of the DREs (i.e., unit commitment of the DGs and the charging/discharging schedules of the battery), and day-ahead set points of electric tie-line power.

Based on the day-ahead schedules, computed at the beginning of each scheduling day, a reference profile of the power exchange with the utility grid for the full day ahead is agreed with the utility grid and should be followed to avoid penalties and additional cost. Due to the forecasting errors of the day-ahead forecasting data, there are errors between the schedules of the BC at the day-ahead stage and real-time stage. This induces fluctuations of the electric tie-line power of the BC. Therefore, an MPC-based two-layer real-time dispatch stage is conducted to smooth the fluctuations of the electric tie-line power. A faster time scale is set in this stage with a shorter timestep $\Delta t'$. As shown in Fig. 5.2, Δt is further divided into four steps with an interval of $\Delta t'$. The whole scheduling timeline at the real-time stage is divided into n_2 time slots, and each time slot is allocated with 15 minutes.

The MPC-based real-time rolling adjustment approach works as follows. At the current time slot $t' = k \ \& \ k = 1$, the BEMSs and BCEMS get the current forecasting data over the time horizon from k to $k + N_p - 1$ (i.e., the kth prediction horizon). The BCEMS then solves a forward-looking optimization problem to minimize the errors between the tie-line power at the day-ahead stage and real-time stage over the N_c-slot time horizon (i.e., the kth control horizon). The optimized results are a sequence of adjustments of power output of the DGs, charging/discharging power of the battery, and the power consumption of the electric chillers. Only the optimized results for the first time slot at t', as highlighted in blue in Fig. 5.2, are applied on the BCEMS and issued to the corresponding DG controllers and the BEMSs. The unit commitment of the DGs and the charging/discharging status of the battery at the real-time stage are kept the same as that at the day-ahead stage. Then at the next time slot $t' = k \ \& \ k = 2$, the BEMSs and BCEMS get updated forecasting data for the next N_p time slots and the forward-looking optimization problem over the next control horizon is solved again. In addition, the optimized results for the first time slot are applied as the optimal adjustments for the current time slot. The time horizon moves forward by one time slot for the new forward-looking optimization until all control schedules are determined during the whole timeline.

The current state of the BC at each time slot is used as the initial set points of the MPC at each time slot in this chapter, which is a commonly used method to determine the initial set points of the MPC [35]. The current state of the BC is determined as follows. At each time slot (i.e., $t' = k \ \& \ k \neq 1$), the current state of the BC is updated according to the prediction model of the BC and the information of state space of the BC at the previous time slot (i.e., $t' = k-1$). At the first time slot (i.e., $t' = k \ \& \ k = 1$), the current state of the BC cannot be obtained according to the prediction model of the BC due to lack of information of state space of the BC at the previous time slot. Therefore, day-ahead scheduling results at the first time slot of the day-ahead stage are used as the initial set points for MPC at the first time slot of the real-time stage.

5.3.2 Formulations of the scheduling method

(1) *Stage I.* With the hourly electric load demand, outdoor temperature, solar radiation, and renewable generation forecasting values, the hourly schedules of the day-ahead stage over the horizon $T = \{t=1, t=2, \ldots, t=n_1\}$ are obtained by using an optimal dynamic scheduling program. The optimization problem of stage I is formulated as follows:

- *Objective.* The objective function depicted in Eq. (5.13) is to minimize the total daily operating cost for the BC:

$$
\min \sum_{t \in T} \left\{ \left(\frac{C_{ph,t} + C_{se,t}}{2} P_{grid,t} + \frac{C_{ph,t} - C_{se,t}}{2} \left| P_{grid,t} \right| \right) \right.
$$
$$
+ \left(\sum_{i \in DG} \left[f_{fuel}\left(P_{DG,i,t} \right) + \rho_{DG,i} \times P_{DG,i,t} + \rho_{su,i} \times U'_{DG,i,t} \right] \right)
$$
$$
\left. + \left(\rho_{WT} P_{WT,t} + \rho_{PV} P_{PV,t} + \rho_{bt} \left| P_{bt,t} \right| + \sum_{n \in EC} \rho_{EC} P_{EC,n,t} \right) \right\}
$$

$$(5.13)$$

where C_{ph} and C_{se} are the electricity purchasing and selling prices (\$/kWh); P_{grid} is the electric power exchange with the utility grid (kW); P_{DG} is the power generation of DG (kW); U'_{DG} is the startup status of DG, which is "1" for startup and "0" otherwise; P_{PV} and P_{WT} are electric power generated by photovoltaic system and wind system (kW); and ρ is the maintenance cost of the energy devices (\$/kWh). The first term in Eq. (5.13) represents the cost for electricity purchase from the utility grid; the second term represents the fuel costs depicted by fuel cost function f_{fuel} ( ), maintenance costs, and the startup costs of all controllable DGs; and the third term is the maintenance costs of other devices of the BC.

(2) *Constraints.* Electrical power balance:

$$
P_{grid,t} + \sum_{i \in DG} P_{DG,i,t} + P_{WT,t} + P_{PV,t} + P_{bt,t} - \sum_{m \in BD} P_{el,m,t}
$$
$$
- P_{load,t} - \sum_{n \in EC} P_{EC,n,t} - P_{loss,t} = 0 \qquad \forall t \in T
$$

$$(5.14)$$

where P_{loss} is the power loss of the electric network (kW).

Constraint of electric power exchange.

$$\underline{P_{grid}} < P_{grid,t} < \overline{P_{grid}}, \quad \forall t \in T \tag{5.15}$$

where $\overline{P_{grid}}$ and P_{grid} are the upper and lower limits of electric power exchange with the utility grid of the BC (kW).

Constraint of DGs. In the optimal scheduling model, the constraints from the controllable DGs are introduced to consider the inherent link among the scheduling time intervals. For each controllable DG, the power generation is constrained by the upper and lower power output limits. The power generations between two successive dispatch time intervals are constrained by ramp-up (ramp-down) rates as well as startup (shutdown) rates, as shown in Eq. (5.16):

$$\begin{cases} \underline{P}_{DG,i,t} U_{DG,i,t} \leq P_{DG,i,t} \leq \overline{P}_{DG,i,t} U_{DG,i,t} \\ P_{DG,i,t} - P_{DG,i,t-1} \leq R_{u,i} \Delta t U_{DG,i,t-1} + S_{u,i} \Delta t U'_{DG,i,t} \\ P_{DG,i,t-1} - P_{DG,i,t} \leq R_{d,i} \Delta t U_{DG,i,t} + S_{d,i} \Delta t U''_{DG,i,t} \\ U'_{DG,i,t} = \max\left(0, U_{DG,i,t} - U_{DG,i,t-1}\right) \\ U''_{DG,i,t} = \max\left(0, U_{DG,i,t-1} - U_{DG,i,t}\right) \\ \forall i \in DG, \forall t \in T \end{cases} \tag{5.16}$$

where $\overline{P_{DG}}$ and $\underline{P_{DG}}$ are the upper and lower limits of power generation of DG (kW); U_{DG} is the operation status of DG, where "1" represents the ON state and "0" represents the OFF state; U'_{DG} is the startup status of DG, which is "1" for startup and "0" otherwise; and U''_{DG} is the shutdown status of DG, which is "1" for shutdown and "0" otherwise.

The controllable DGs are also constrained by the minimum up- and downtime limits, as shown in Eq. (5.17):

$$\begin{cases} T^{on}_{i,t} \geq UT_i\left(U_{DG,i,t} - U_{DG,i,t-1}\right) \\ T^{off}_{i,t} \geq DT_i\left(U_{DG,i,t-1} - U_{DG,i,t}\right) \\ \forall i \in DG, \forall t \in T \end{cases} \tag{5.17}$$

where $T^{on}_{i,t}$ and $T^{off}_{i,t}$ are the number of successive ON/OFF time periods of DG i at time t (h).

Constraint of the battery. For the battery, the charging/discharging power and the SOC are constrained by the upper and lower limits shown in Eqs. (5.18 through 5.20), and for energy balance, the stored energy inside the battery is set the same as the initial stored energy, as shown in (5.21):

$$\underline{P_{bt}} \leq P_{bt,t} \leq \overline{P_{bt}}, \qquad \forall t \in T \tag{5.18}$$

$$\underline{SOC} \leq SOC_t \leq \overline{SOC}, \qquad \forall t \in T \tag{5.19}$$

$$SOC_t = \frac{E_{bt,t}}{CAP_{bt}} \tag{5.20}$$

$$\sum_{t \in T} P_{bt,t} = 0 \tag{5.21}$$

where $\overline{P_{bt}}$ and $\underline{P_{bt}}$ are the upper and lower limits of charging/discharging power of the battery (kW), $\overline{SOC}$ and $\underline{SOC}$ are the upper and lower limits of the SOC value of the battery and $E_{bt,t}$ is the capacity of the battery at time t (kWh).

Constraint of the electric chiller.

$$0 \leq Q_{EC,t} \leq \overline{Q}_{EC}, \qquad \forall t \in T \tag{5.22}$$

where $\bar{Q}_{EC}$ is the upper limit of the cooling power output of the electric chiller (kW).

Constraints of the buildings. The constraints of the buildings include the cooling demand constraint, as shown in Eq. (5.23); the building thermal equilibrium equation, as shown in Eq. (5.24); and the indoor temperature set–point constraint, as shown in Eq. (5.25):

$$\dot{Q}_{EC,t} = EER_{EC} \times P_{EC,t} = \dot{Q}_{cl,\,building,t}, \qquad \forall t \in T \tag{5.23}$$

$$\Delta t \left[\dot{Q}_{wall,\,t} + \dot{Q}_{win,\,t} + \dot{Q}_{sw,\,t} + \dot{Q}_{sg,\,t} + \dot{Q}_{in,\,t} - \dot{Q}_{EC,\,t} \right]$$
$$- \rho CV \left(T_{in,\,t+1} - T_{in,\,t} \right) = 0, \qquad \forall t \in T \tag{5.24}$$

$$\underline{T_{in}} < T_{in,t} < \overline{T_{in}}, \qquad \forall t \in T \tag{5.25}$$

where $\overline{T_{in}}$ and $\underline{T_{in}}$ are the upper and lower limits of the indoor temperature set points of the building (°C).

(3) *Stage II.* The MPC strategy with operational constraints is proposed to reschedule the smart buildings and DERs at real-time stage. The BEMSs and the local controllers of the DERs not only have to consider local information, but also exchange the state information with the BCEMS. Therefore, the prediction model of the BC is developed to inform the BCEMS, BEMSs and DER controllers for rescheduling of the smart buildings and DERs. Then, the rolling optimization problem is formulated and the implementation of the MPC based rescheduling method is introduced.

(4) *Prediction models*

Prediction model of the BC. According to the preceding system dynamics shown in Eqs. (5.1), (5.12), and (5.16), constraints of the BC, DERs, and smart buildings, the prediction model of the BC is formulated using state space, as shown in Eq. (5.26). The state of the BC can be predicted by the iteration for the state-space model repeatedly with the updated forecasting data:

$$\begin{cases} \boldsymbol{x}(t'+1) = A \cdot \boldsymbol{x}(t') + B \cdot \boldsymbol{u}(t') + C \cdot \boldsymbol{r}(t') \\ \boldsymbol{y}(t') = D \cdot \boldsymbol{x}(t') \end{cases} \tag{5.26}$$

where

$$\boldsymbol{x}(t') = \left[P_{grid}(t'), \boldsymbol{P_{DG}}(t'), P_{bt}(t'), SOC(t'), \boldsymbol{P_{EC}}(t') \right]^T \tag{5.27}$$

$$\boldsymbol{u}(t') = \left[\boldsymbol{\Delta P_{DG}}(t'), \Delta P_{bt}(t'), \boldsymbol{\Delta P_{EC}}(t') \right]^T \tag{5.28}$$

$$\boldsymbol{r}(t') = \left[\Delta P_{PV}(t'), \Delta P_{WT}(t'), \boldsymbol{\Delta P_{el}}(t') \right]^T \tag{5.29}$$

$$\boldsymbol{y}(t') = P_{grid}(t') \tag{5.30}$$

Here, $x(t')$ is the state vector of the BC at current time slot t', which consists of power exchange with the utility grid $(P_{grid}(t'))$, vector of power output of controllable DGs $(P_{DG}(t'))$, charging/discharging power of the battery $(P_{bt}(t'))$ and its SOC value $(SOC(t'))$, and vector of power consumption of electric chillers $(P_{EC}(t'))$, as shown in (5.27). $u(t')$ is the control vector of the BC at current time slot t', which manages the increments of $P_{DG}(t')$, $P_{bt}(t')$, and $P_{EC}(t')$, as shown in Eq. (5.28). The unit commitment of the controllable DGs and the charging/discharging modes of the battery at the real-time stage are kept the same as that at the day-ahead stage. Therefore, they are not managed and controlled at the real-time stage. $r(t')$ is the input vector that influences the BC at current time slot t', which consists of day-ahead forecasting errors of PV generation $(\Delta P_{PV}(t'))$, wind generation $(\Delta P_{el}(t'))$, and electric loads of the buildings $(\Delta P_{el}(t'))$. $y(t')$ is the output of the prediction model of the BC at current time slot t', which is the power exchange with the utility grid $(P_{grid}(t'))$. The matrices $A, B, C,$ and D are the relevant state-space matrices, as shown in Eq.s (5.31) through (5.34):

$$A = \begin{bmatrix} 1 & 0 & 0 & 0 & 0 \\ 0 & E_n & 0 & 0 & 0 \\ 0 & 0 & 1 & 0 & 0 \\ 0 & 0 & \varphi & 1-\delta & 0 \\ 0 & 0 & 0 & 0 & E_m \end{bmatrix} \tag{5.31}$$

$$B = \begin{bmatrix} -1 & -1 & 1 \\ E_n & 0 & 0 \\ 0 & 1 & 0 \\ 0 & \varphi & 0 \\ 0 & 0 & E_m \end{bmatrix} \tag{5.32}$$

$$C = \begin{bmatrix} -1 & -1 & E_n \\ 0 & 0 & 0 \\ 0 & 0 & 0 \\ 0 & 0 & 0 \\ 0 & 0 & 0 \end{bmatrix} \tag{5.33}$$

$$D = \begin{bmatrix} 1 & \mathbf{0} & 0 & 0 & \mathbf{0} \end{bmatrix} \tag{5.34}$$

where E_n and E_m are identity matrices, and φ is the recurrence coefficient for SOC value calculation of the battery, as shown in Eq. (5.35).

$$\varphi = \begin{cases} -\dfrac{\Delta t \eta_{ch}}{CAP_{bt}} & P_{bt,t} \le 0 \\ -\dfrac{\Delta t}{\eta_{dis} CAP_{bt}} & P_{bt,t} > 0 \end{cases} \tag{5.35}$$

According to the model in Eq. (5.26), the predicted output of the BC at time slot $t' + p$ can be calculated by the iteration based on the state at time slot t', as shown in Eq. (5.36).

$$\begin{aligned} y(t' + p|t') = D \cdot x(t' + p|t') = D \cdot [A^p \cdot x(t') \\ + A^{p-1} B \cdot u(t') + A^{p-1} C \cdot r(t') + , ..., \\ + B \cdot u(t' + p - 1|t') + C \cdot r(t' + p - 1|t')] \end{aligned} \tag{5.36}$$

Then, the predicted output of the BC over the kth prediction horizon can be formulated by the augmented vector, as shown in Eq. (5.37).

$$\begin{aligned} Y = \begin{bmatrix} y(k|k), & y(k+1|k), & ..., & y(k+N_p - 1|k) \end{bmatrix}^T, \\ k = 1, 2, ..., n_2 \end{aligned} \tag{5.37}$$

Prediction model of indoor temperatures of a building. The recurrence equation to predict the indoor temperature of a building can be obtained according to the building model, as shown in Eq. (5.38).

$$T_{in}(t'+1) - T_{in}(t') = \frac{\Delta t\left(\widehat{Q}_{wall} + \widehat{Q}_{win} + \widehat{Q}_{sw} + \widehat{Q}_{sg} + \widehat{Q}_{in} - \widehat{Q}_{EC}\right)}{\rho CV}$$

(5.38)

$\widehat{Q}$ is the updated value with the updated forecasting data of a building $(\lambda(t'))$. $\lambda(t')$ consists of forecasting errors of internal heat gain $(\Delta \dot{Q}_{in}(t'))$, outdoor temperature $(\Delta T_{out}(t'))$, and solar radiation $(\Delta I_T(t'))$:

$$\lambda(t') = \left[\Delta \dot{Q}_{in}(t'), \Delta T_{out}(t'), \Delta I_T(t')\right]$$

(5.39)

According to Eqs. (5.38) and (5.39), the indoor temperature of a building can be predicted by the iteration for the recurrence equation repeatedly with the updated forecasting data and the prior knowledge of state information of the BC, as shown in Eqs. (5.36) and (5.37).

(5) *Rolling optimization.* The objective is to keep the predicted output of the BC (i.e., power exchange with the utility grid $P_{grid}(t')$) close to the day-ahead schedules at each control horizon. The predicted output of the BC over the kth control horizon is shown in Eq. (5.40).

$$\begin{aligned} Y &= [y(k|k), y(k+1|k), ..., y(k+N_c-1|k)]^T, \\ k &= 1, 2, ..., n_2 \end{aligned}$$

(5.40)

The day-ahead schedules of the power exchange of the BC with the utility grid over the kth control horizon are generated in stage I and described as augmented vector, as shown in Eq. (5.41).

$$\begin{aligned} Y_{dh} &= \left[P_{grid}^{dh}(k|k), P_{grid}^{dh}(k+1|k), ..., P_{grid}^{dh}(k+N_c-1|k)\right]^T, \\ k &= 1, 2, ..., n_2 \end{aligned}$$

(5.41)

Then the optimization problem for each control horizon can be formulated as follows:

$$\begin{aligned} \min f &= (Y_{dh} - Y)^T (Y_{dh} - Y) \\ &+ \sum_{t'=k}^{t'=k+N_c-1} \theta_{bt}\left[(\Delta P_{bt}(t'))^2\right] \cdot \Delta t' \end{aligned}$$

(5.42)

$$s.t. \ (14) - (29)$$

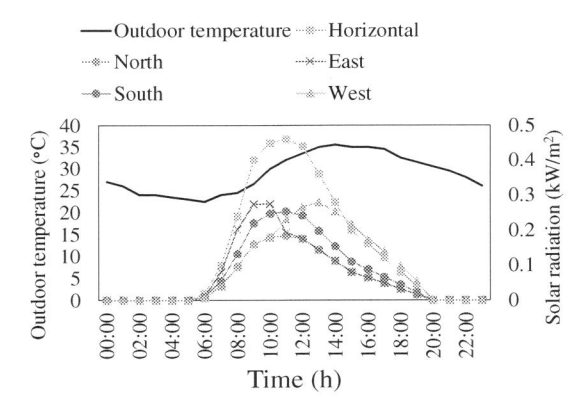

Figure 5.3 Day-ahead forecasting values of the outdoor temperature and solar radiation.

where θ_{bt} is the penalty coefficient that limits the frequent charge and discharge operation of the battery.

5.4 Results and discussions

5.4.1 5.4.1. Case setting

A BC shown in Fig. 5.1 containing smart buildings and DERs is used to verify the proposed scheduling method. Diesel engine 1, diesel engine 2, fuel cell, PV and battery, and wind generator are connected to the BC. Four smart buildings are connected to the BC. Day-ahead forecasting values of the outdoor temperature, solar radiations on the horizontal surface, electric loads and internal heat gains of the four buildings, and renewable generations are shown in Fig. 5.3 through Fig. 5.5. The capacity of the installed PV and wind–based DGs are set to be 300 kW and 500 kW, respectively. All day-ahead forecasting data are collected from Jin et al. [7]. The corresponding incident solar radiation on the walls/windows surface at south, west, north, and east orientations are calculated based on the method presented by Duffie and Beckman [33], as shown in Fig. 5.3. The electricity purchasing prices are shown in Fig. 5.6, and the price for selling electricity back to the utility grid is set to be 0.8 times the price for purchasing electricity. The short-term forecasting data over the prediction horizon at the real–time stage need forecasting techniques. Instead, we assume that the forecasting errors of all data at the real-time stage follow the uniform distribution [36, 37], as shown in Eq. (5.43):

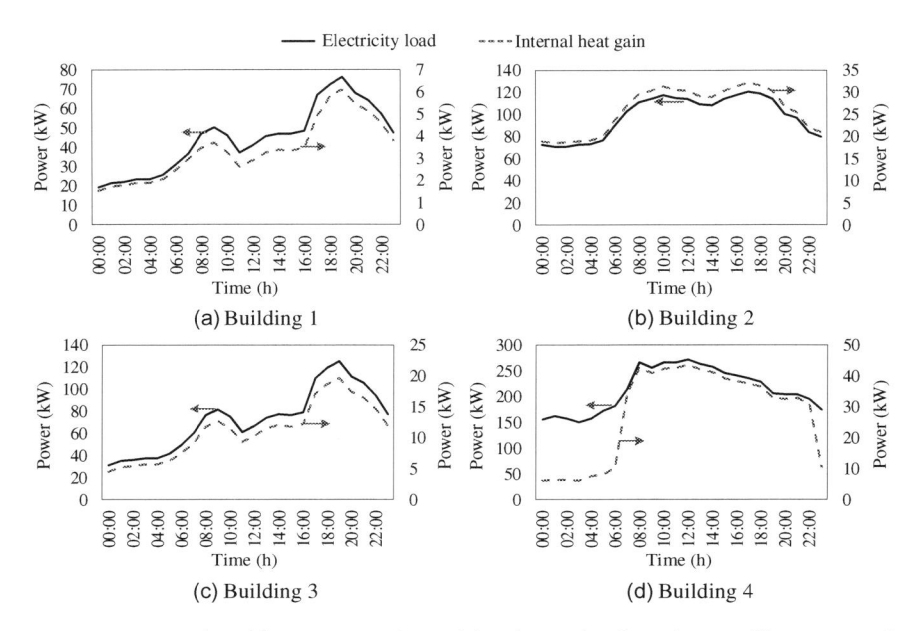

Figure 5.4 Day-ahead forecasting values of the electric loads and internal heat gains of the buildings.

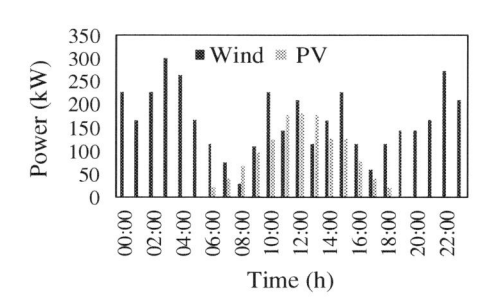

Figure 5.5 Day-ahead forecasting values of the renewable generations.

$$\begin{cases} P_{WT}(t') = P_{WT}(t) \cdot \left[1 + E_{WT}^{\max} \cdot R(t)\right] \\ P_{PV}(t') = P_{PV}(t) \cdot \left[1 + E_{PV}^{\max} \cdot R(t)\right] \\ P_{el}(t') = P_{el}(t) \cdot \left[1 + E_{el}^{\max} \cdot R(t)\right] \\ T_{out}(t') = T_{out}(t) \cdot \left[1 + E_{t}^{\max} \cdot R(t)\right] \end{cases} \quad (5.43)$$

where $E_{WT}^{\max} E_{PV}^{\max} E_{el}^{\max}$, and $E_{t}^{\max}$ are the threshold values of the forecasting errors under different uncertainty levels, and their values are listed in

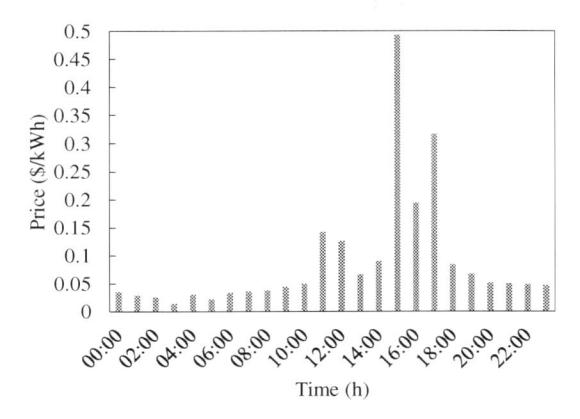

Figure 5.6 Electricity purchasing prices.

Table 5.1 Threshold values of the forecasting errors.

Forecasting data	Value
Wind power	0.20
PV power	0.12
Load	0.04
Outdoor temperature	0.02

Table 5.1; $R(t)$ is a random value that follows the uniform distribution: $R(t) \sim U(-1, 1)$.

The building studied in this chapter is represented by a parallelepiped with a squared floor. The thermal parameters and the occupied hours of the buildings are given in Table 5.2 [7]. The values of the parameters of the air mass density ρ and air specific heat ratio C are set to be 1.2 kg/m^3 and 1000 J/(kg·°C), respectively. The acceptable indoor temperature set–point range for human occupancy in a building varies under different operational scenarios. In this study, the indoor temperature set-point during occupied hours is set to be 22.5°C without the VESS being dispatched. The indoor temperature set-point range during occupied hours is set to be from 20 to 25°C with the VESS being dispatched. It is worth noting that the electric chillers of the buildings are switched off during unoccupied hours for cost savings. In this context, the indoor temperatures are not optimized and no specific indoor temperature set-point range is assigned by the BEMSs during the unoccupied hours. The technical and economic parameters of the DERs are shown in Table 5.3 and Table 5.4. The other operational parameters of

Table 5.2 Thermal parameters of the four buildings.

No.	U_{wall} [W/(m²·K)]	F_{wall} /m²	U_{win} [W/(m²·K)]	Window-to-Wall Ratio (%)	Long Side (m)	Short Side (m)	Height (m)	Occupied Hours
A	1.092	1000	2.800	45	30	20	9	00:00–09:00 & 18:00–23:00
B	0.908	2400	2.750	75	40	20	30	08:00–20:00
C	1.146	1500	2.800	60	30	20	20	Whole day
D	0.820	2700	2.500	65	50	30	20	10:00–22:00

Table 5.3 Technical parameters of the controllable DGs.

DG Type	$\underline{P}_{DG}$ (kW)	$\overline{P}_{DG}$ (kW)	R_u(kW/min)	R_d(kW/min)	S_u(kW/min)	S_d(kW/min)	UT (h)	DT (h)	Number of DG
Diesel engine	40	200	8	8	8	8	2	1	2
Fuel cell	15	150	8	8	8	8	2	1	1

Table 5.4 Economic parameters of the DERs.

	Diesel Engine	Fuel Cell	PV	Wind	Battery	Electric Chiller
Startup Cost ($)	0.24	0.32	—	—	—	—
Maintenance Cost ($)	0.0033	0.0046	0.001	0.001	0.001	0.001

Table 5.5 Parameters of the battery and electric chiller.

Parameters	Value
COP_{EC}	4
$\overline{P}_{EC,1}/\underline{P}_{EC,1}$	120 kW/0 kW
$\overline{P}_{EC,2}/\underline{P}_{EC,2}$	200 kW/0 kW
$\overline{P}_{EC,3}/\underline{P}_{EC,3}$	120 kW/0 kW
$\overline{P}_{EC,4}/\underline{P}_{EC,4}$	600 kW/0 kW
$\overline{P}_{ex}/\underline{P}_{ex}$	450 kW/−450 kW
CAP_{bt}	600 kWh
$\overline{SOC}/\underline{SOC}$	0.8/0.2
$\overline{P}_{bt}/\underline{P}_{bt}$	40 kW/−40 kW
η_{ch}/η_{dis}	0.95/0.95
Δ	0.04
η_{FC}	0.55

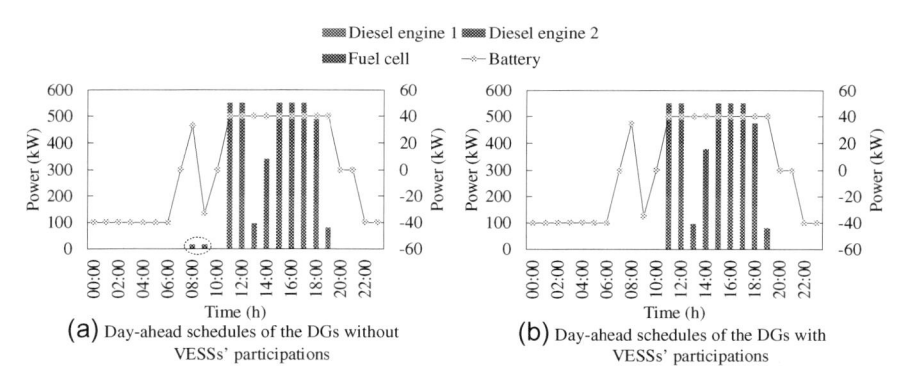

Figure 5.7 Day-ahead schedules of the DERs.

the DERs are shown in Table 5.5. The fuel cost coefficients of the diesel engine are set as $a = 44$ ($/h/MW2), $b = 65.34$ ($/h/MW) and $c = 1.1825$ ($/h). The natural gas price is 42.5 $/MWh. All parameters regarding the DERs are from other works [38–42].

5.4.2 Day-ahead scheduling results

The day–ahead schedules of the DERs are shown in Fig. 5.7. The results show that all controllable DGs are committed and dispatched at their maximum capacities during high electricity purchasing price periods (i.e., 11:00–12:00 and 14:00–18:00). They are switched off or the power generations are reduced for cost savings during low electricity purchasing price periods (i.e., 01:00–10:00 and 19:00–23:00). However, due to the constraint of electric power exchange, as shown in Eq. (5.15), required electric power cannot be imported from the utility grid at 08:00–09:00. Therefore, the fuel cell is committed and scheduled at 08:00–09:00 to cover the power shortage without VESSs' participation in day-ahead scheduling, as shown in Fig. 5.7(a). It is worth noting that only the fuel cell is scheduled at 08:00–09:00 because of its low-cost coefficient compared with the diesel engine. However, the cooling demands of the buildings are adjusted to reduce the power consumptions of the electric chillers considering VESSs' participation, which results in no power shortage at 08:00–09:00. Therefore, unit commitment of the DGs is not needed at 08:00–09:00 and more cost savings are achieved.

The day-ahead cooling schedules of the buildings are shown in Fig. 5.8. We can observe that the indoor temperatures are adjusted within the indoor temperature comfort range (20 to 25°C) during the occupied hours by introducing VESS to the day-ahead scheduling stage. The indoor temperatures are kept at the set points (22.5°C) during the occupied hours without considering VESSs' participation. In this case, the daily operating cost of the BC is reduced from \$327.93 to \$298.88, which is reduced by 8.86%. It can be concluded that the VESSs' participation in day-ahead scheduling can reduce the daily operating cost of the BC.

The day-ahead dispatch results of the VESS and the cooling demand of buildings considering VESSs' participation ($\dot{Q}_{cl,building,t}$) and those not considering VESSs' participation ($\dot{Q}'_{cl,building,t}$) are shown in Fig. 5.9. When $\dot{Q}_{cl,building,t}$ is larger than $\dot{Q}'_{cl,building,t}$, the VESS operates in charging mode; when $\dot{Q}_{cl,building,t}$ is smaller than $\dot{Q}'_{cl,building,t}$, the VESS operates in discharging mode. The dispatch results of the VESS show that the VESS can be charged or discharged according to the optimal dispatch of the BC. The dispatch results also show that the VESS tends to operate in discharging mode to reduce the daily operating cost. However, due to the constraint of the indoor temperature set point and energy dissipation characteristics of the building, the VESS is charged after discharging for several time intervals, as shown in Fig. 5.9.

5.4.3 Real-time dispatch results

The dispatch results of the DERs with MPC method at the real-time stage are shown in Fig. 5.10. The results show that the power outputs of the DGs

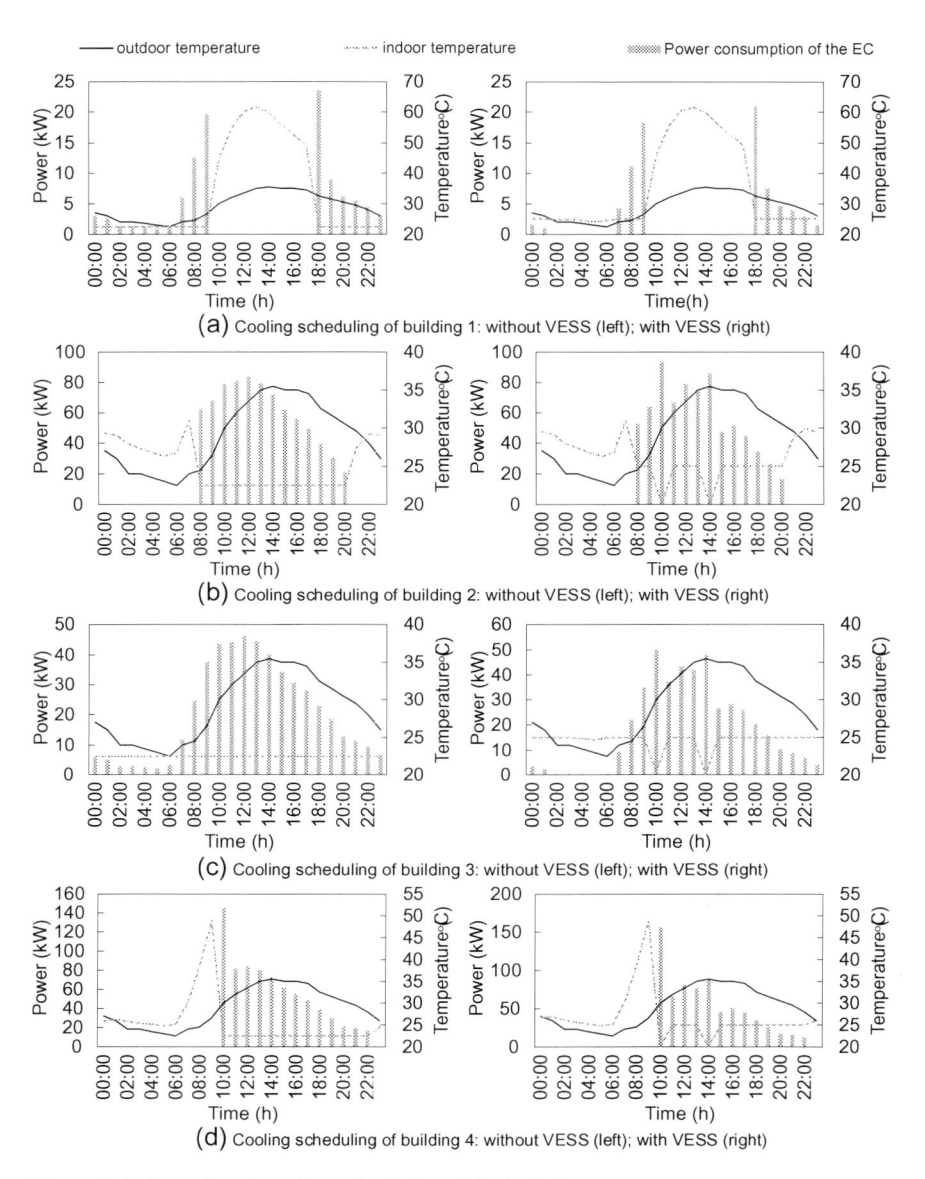

Figure 5.8 Day-ahead cooling schedules of the buildings.

and battery are adjusted to smooth the fluctuations of tie-line power, whereas the unit commitment of the DGs and the charging/discharging status of the battery are kept the same as that at the day-ahead stage.

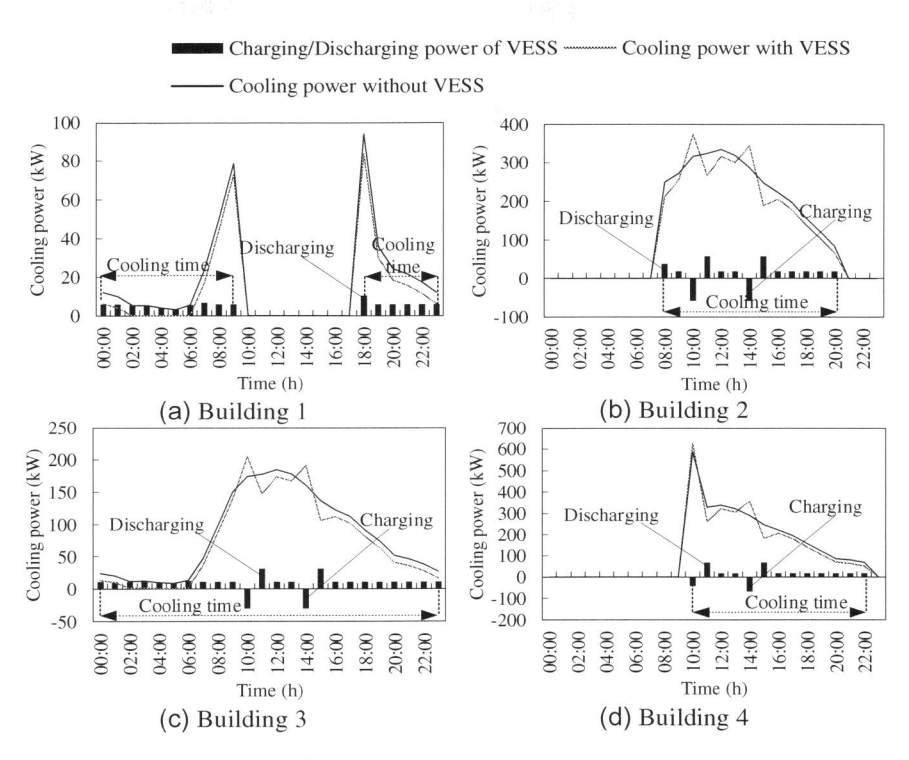

Figure 5.9 Day-ahead schedules of VESSs.

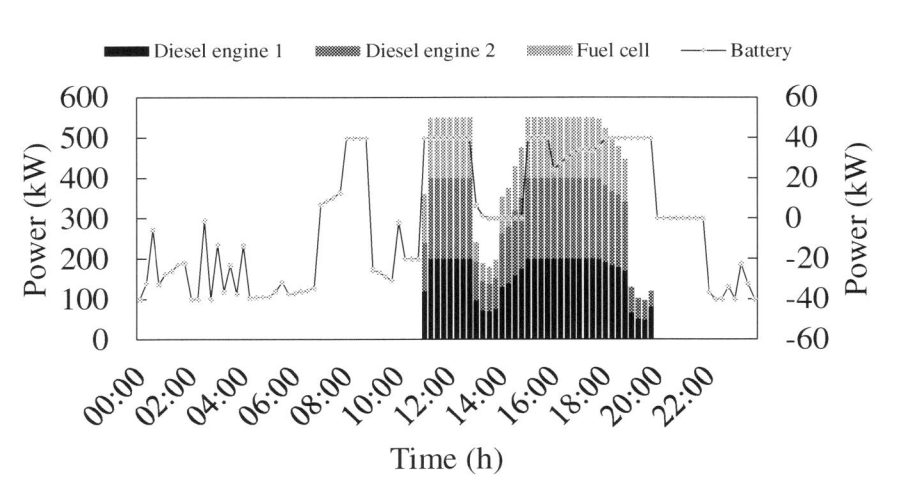

Figure 5.10 Real-time dispatch results of the DERs.

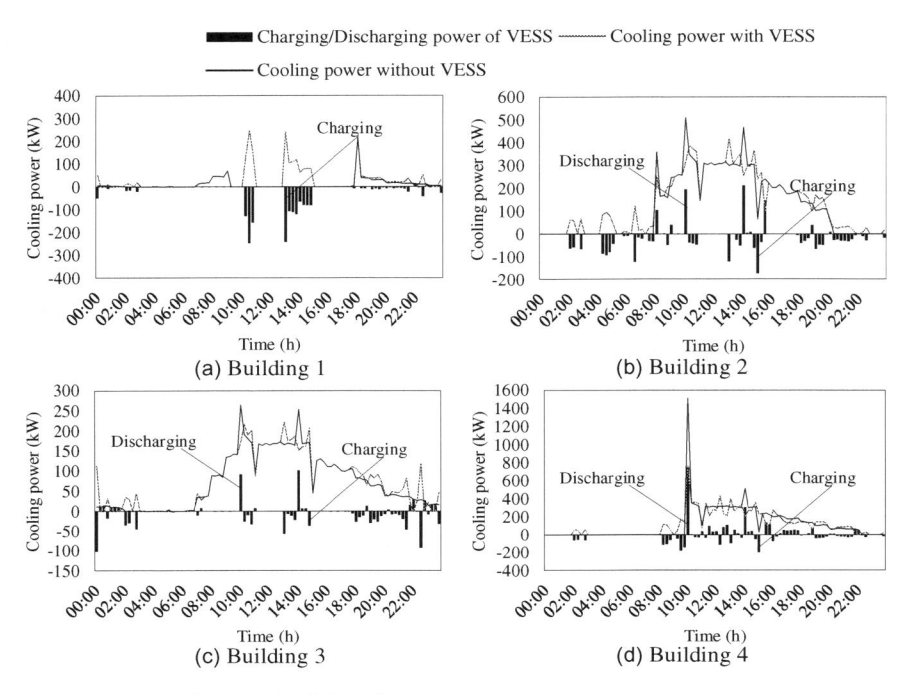

Figure 5.11 Real-time schedules of VESSs.

The schedules of VESSs at the real-time dispatch stage are shown in Fig. 5.11. Compared to the day–ahead scheduling results of VESSs in Fig. 5.9, it can be observed that the schedules of VESSs (the charging/discharging power of the VESS) are adjusted at the real-time stage to smooth the tie-line power fluctuations. Then the cooling schedules of the buildings at the real-time stage are adjusted accordingly, as shown in Fig. 5.12. Compared to the day-ahead cooling schedules in Fig. 5.8, it can be concluded that the indoor temperatures of the buildings and power consumptions of the electric chillers are adjusted considering VESSs' participation at the real-time stage (as shown in the right-hand side of Fig. 5.12).

The day-ahead programming (DA-P) strategy used by Jin et al. [7] and single optimization used in another work by Jin et al. [22] are employed to compare with the MPC-based strategy developed in this chapter. Then the effectiveness of the proposed MPC-based energy management method in the real-time stage is further verified.

DA-P strategy [7]. The DERs and VESSs are not rescheduled under the DA-P strategy in the real-time stage. Therefore, the errors between the

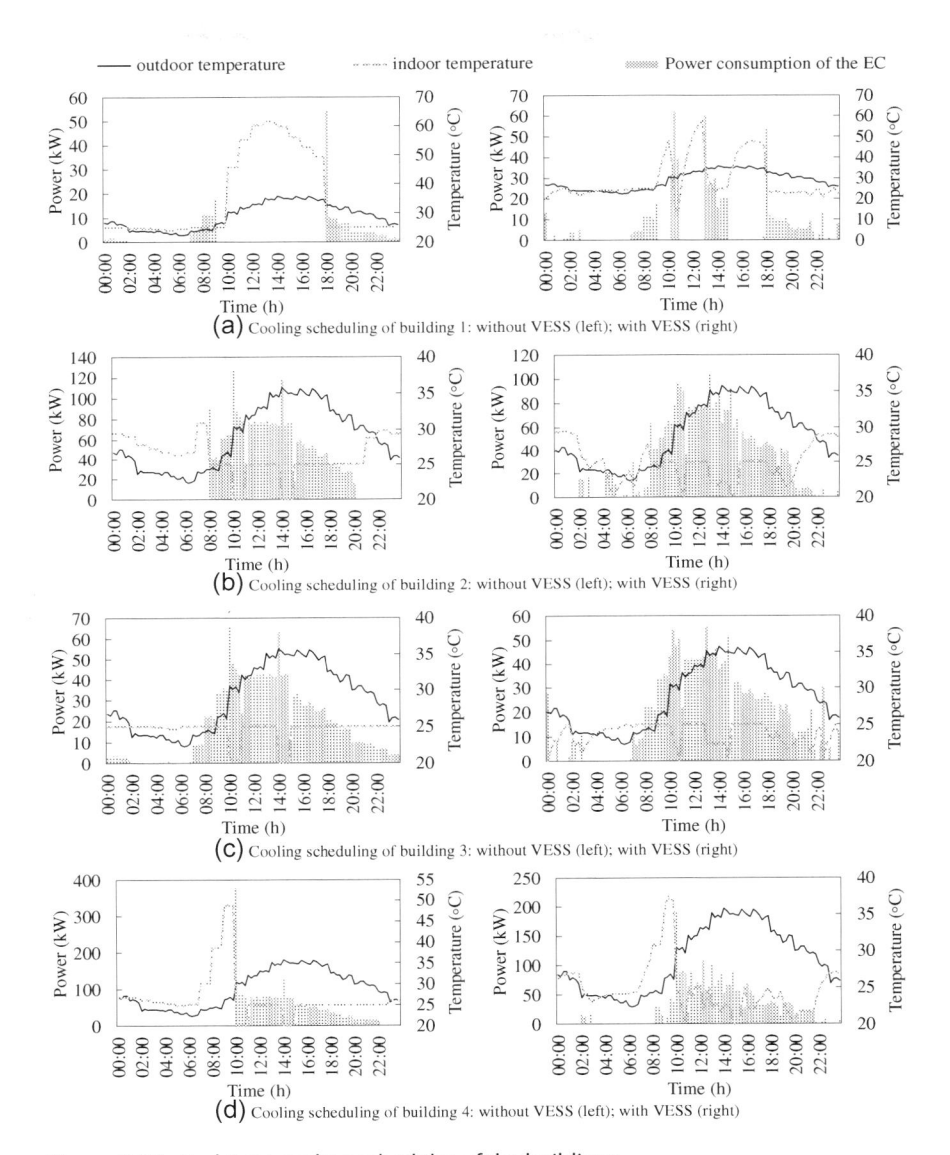

Figure 5.12 Real-time cooling schedules of the buildings.

schedules of the day–ahead stage and real–time stage caused by day–ahead forecasting errors would be balanced by the utility grid.

Single optimization [22]. The DERs and VESSs are rescheduled to cope with the day–ahead forecasting errors in the real–time stage. However, the rescheduled results of the DERs and VESSs are optimized based on the

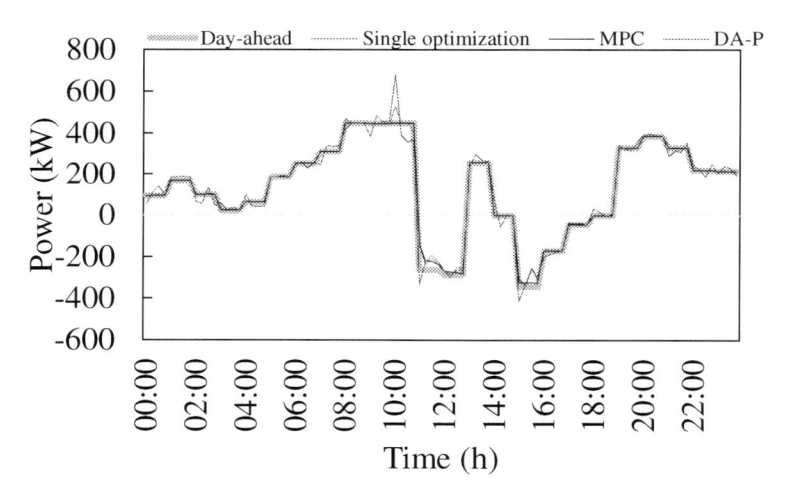

Figure 5.13 Electric tie-line powers of the BC with different methods in the real-time stage.

operating status of the BC and forecasting data at the current operational period rather than the predictive operating status and forecasting data over a future time horizon.

Different from the DA-P strategy and the single-period-based strategy, the MPC-based strategy reschedules the DERs and VESSs based on the predictive operating status and forecasting data over a future time horizon that runs the embedded optimization model repeatedly with updated forecasts. The prediction horizon and control horizon are set to be 4 hours [25] (i.e., $N_p = N_c = 16$) in the real-time dispatch stage in this chapter.

The electric tie-line powers of the BC with different methods are shown in Fig. 5.13. Since all mismatches between the energy demand and supply are balanced by electric power from the utility grid under the DA-P method, all forecasting errors are mainly reflected in the electric tie-line power.

It can be observed from Fig. 5.13 that the performance of electric tie-line power tracking with the MPC method is close to that of single-optimization methods during periods of smooth tie-line power (e.g., 06:00–09:00). However, better performance of tie-line power tracking can be obtained with the MPC method during periods with large tie-line power fluctuations (e.g., 10:00–12:00). This observation was attributed to the fact that the MPC-based optimization method can handle the future behavior of the BC over a future time horizon. In this case, the DERs and VESSs can be adjusted in advance to smooth the upcoming large tie-line power

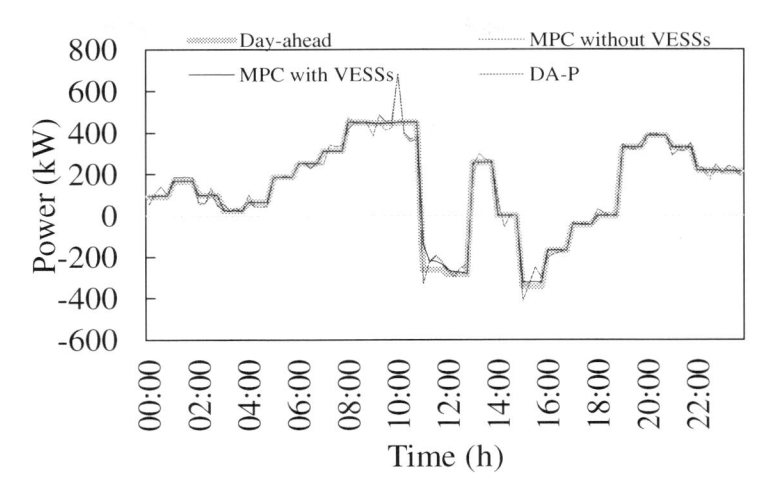

Figure 5.14 Electric tie-line powers of the BC with and without VESSs in the real-time stage.

fluctuations. In other words, timely adjustments of the DERs and VESSs can be obtained with the MPC-based optimization method. However, traditional single optimization cannot handle the future behavior of the BC, the DERs and VESSs are not adjusted in advance accordingly. In this case, the problems of the insufficient adjustments and untimely adjustments of the DERs and VESSs occur naturally due to their technical limits. Therefore, it can be concluded that the MPC-based optimization method can improve the control performance for tie-line power tracking.

Another comparative scenario is set up further to highlight the effect of VESSs' participation at the real-time dispatch stage as MPC without VESSs. In this scenario, only the DERs are rescheduled using the MPC-based optimization strategy, whereas the VESSs' participation is not considered. The electric tie-line powers are shown Fig. 5.14. The following conclusions were derived based on the results:

1) The performance of electric tie-line power tracking with MPC while considering VESSs' participation is better than that with MPC while not considering VESSs' participation due to VESSs' participation in the real-time stage.

2) The performance of electric tie-line power tracking with MPC while not considering VESSs' participation is close to that in tracking with MPC while considering VESSs' participation when the DGs are in the

ON state (e.g., 11:00–18:00) due to DGs' participation in the real–time stage.

3) The performance of electric tie–line power tracking with MPC while not considering VESSs' participation is as poor as that with the DA-P strategy when the DGs are in the OFF state (e.g., 02:00–03:00, 09:00–10:00, and 22:00–23:00). This is because tthe schedulable capacity of the BC is very limited without DGs' and VESSs' participation.

5.5 Discussions

It is shown by the case study that the proposed optimal scheduling method is able to reduce the daily operating cost at the day-ahead stage and smooth the fluctuations of the electric tie–line power of the BC caused by the day-ahead forecasting errors at the real–time stage. The DERs and VESSs are scheduled in an optimal way with the proposed method at both stages.

At the day-ahead stage, the optimal schedules of the DERs are mainly determined by the electricity prices and their own technical constraints for operating cost reduction of the BC. The optimal results of VESSs are closely related to the occupied hours of the buildings, the thermal parameters of the buildings, and the electricity prices. Due to the different occupied hours and different parameters of the buildings, the VESSs are dispatched during different time periods and present different charging/discharging characteristics. Furthermore, the VESS tends to be discharged with higher power during high electricity purchase price periods and be charged with higher power before the discharging process for daily operating cost reduction.

At the real-time stage, the rescheduling results of the DERs and VESS are mainly related to the mismatches between the energy demand and supply caused by day-ahead forecasting errors, as well as their own technical constraints. It is shown from the comparative analysis under different scenarios that the DERs and VESSs can be adjusted in advance to smooth the upcoming large tie–line power fluctuations with the MPC-based strategy. In other words, timely adjustments of the DERs and VESSs can be guaranteed with the MPC-based strategy in a time-varying context.

Several benefits can be achieved by considering VESSs in the optimal scheduling method: (1) the daily operating cost of the BC is reduced at the day-ahead stage, (2) the total startup and shutdown times of the controllable DGs are decreased at the day-ahead stage, and (3) the mismatches between the energy demand and supply can be balanced better at the real-time stage.

5.6 Conclusion

This chapter proposed a two-stage energy management framework to schedule the DERs and smart buildings of a BC in an optimal way. Optimal schedules are generated for the BC at the day-ahead scheduling stage to reduce the daily operating cost. In the real-time dispatch stage, an MPC-based adjustment method is proposed to reschedule the DERs and VESSs to smooth the fluctuations of the electric tie-line power, taking the uncertainties from renewable generation, electric load demand, outdoor temperature, and solar radiation into consideration. The following conclusions were derived based on the numerical studies:

(1) The flexibility of the building with its thermal mass can be used effectively by scheduling the VESS in the day-ahead stage and rescheduling the VESS in the real-time stage.

(2) By using the iterative rolling optimization with finite horizon (i.e., MPC-based optimization) instead of traditional single optimization, the problems of the insufficient adjustments and untimely adjustments of the DERs and VESSs at the real-time stage can be solved. The MPC-based optimization method can improve the control performance in a time-varying context with uncertainties associated with forecasting data.

(3) Various practical constraints from the buildings, DERs, and the electric network of the BC can be considered with the proposed scheduling model.

References

[1] United Nations Department of Economic and Social Affairs Population Division, World Urbanization Prospects: The 2014 Revision, United Nations. New York, 2015.
[2] International Energy Agency,Energy Technology Perspectives 2017: excerpt informing energy sector transformations, 2017. https://www.iea.org/reports/energy-technology-perspectives-2017 (Accessed 6 June 2017).
[3] International Energy Agency, World Energy Outlook 2017: a world in transformation, International Energy Agency, Paris, 2017.
[4] J Yao, GT Costanzo, G Zhu, B Wen, Power admission control with predictive thermal management in smart buildings, IEEE Trans Ind Electron 62 (4) (2015) 2642–2650.
[5] H Li, AT Eseye, J Zhang, D Zheng, Optimal energy management for industrial microgrids with high-penetration renewables, Protect. Control Mod Power Syst 2 (1) (2017) 12.

[6] D Zhang, N Shah, .G Papageorgiou, Efficient energy consumption and operation management in a smart building with microgrid, Energy Conv Manage 74 (2013) 209–222.

[7] X Jin, Y Mu, H Jia, J Wu, T Jiang, X Yu, Dynamic economic dispatch of a hybrid energy microgrid considering building based virtual energy storage system, Appl Energy 194 (2017) 386–398.

[8] JV Roy, N Leemput, F Geth, J Buscher, R Salenbien, J. Driesen, Electric vehicle charging in an office building microgrid with distributed energy resources, IEEE Trans Sustain Energy 5 (4) (2014) 1389–1396.

[9] B Hu, H. Wang, S Yao, Optimal economic operation of isolated community microgrid incorporating temperature controlling devices, Protect. Control Mod Power Syst 2 (1) (2017) 6.

[10] Z Wang, L Wang, AI Dounis, R Yang, Integration of plug-in hybrid electric vehicles into energy and comfort management for smart building, Energy Build 47 (47) (2012) 260–266.

[11] PH Shaikh, NBM Nor, P Nallagownden, I Elamvazuthi, T Ibrahim, Intelligent multi-objective control and management for smart energy efficient buildings, Int J Electr Power Energy Syst 74 (2016) 403–409.

[12] X Wu, X Hu, X Yin, SJ Moura, Stochastic optimal energy management of smart home with PEV energy storage, IEEE Trans Smart Grid 9 (3) (2018) 2065–2075.

[13] D Thomas, O Deblecker, CS Ioakimidis, Optimal operation of an energy management system for a grid-connected smart building considering photovoltaics' uncertainty and stochastic electric vehicles' driving schedule, Appl Energy 210 (2018) 1188–1206.

[14] Y Tian, L Fan, Y Tang, K Wang, G Li, H Wang, A coordinated multi-time scale robust scheduling framework for isolated power system with ESU under high RES penetration, IEEE Access (6) (2018) 9774–9784.

[15] X Ai, J Wen, T Wu, W-J Lee, A discrete point estimate method for probabilistic load flow based on the measured data of wind power, IEEE Applications Society Meeting, (2012).

[16] C Wang, Y Zhou, J Wu, J Wang, Y Zhang, D Wang, Robust-index method for household load scheduling considering uncertainties of customer behavior, IEEE Trans Smart Grid 6 (4) (2015) 1806–1818.

[17] C Wang, Y Zhou, B Jiao, Y Wang, W Liu, D Wang, Robust optimization for load scheduling of a smart home with photovoltaic system, Energy Conv Manage 102 (2015) 247–257.

[18] Y Li, T Zhao, P Wang, HB Gooi, Z Ding, K Li, et al., Flexible scheduling of microgrid with uncertainties considering expectation and robustness, 54, IEEE Trans Indust Appl, 2018, pp. 1–7.

[19] Z Bao, Q Zhou, Z Yang, Q Yang, L Xu, T Wu, A multi time-scale and multi energy-type coordinated microgrid scheduling solution—part I: model and methodology, IEEE Trans Power Syst 30 (5) (2015) 2257–2266.

[20] C Wang, B Jiao, L Guo, Z Tian, J Niu, S Li, Robust scheduling of building energy system under uncertainty, Appl Energy 167 (2016) 366–376.

[21] M Rahmani-Andebili, Scheduling deferrable appliances and energy resources of a smart home applying multi-time scale stochastic model predictive control, Sustain Cities Soc 32 (2017) 338–347.

[22] X Jin, J Wu, Y Mu, M Wang, X Xu, H Jia, Hierarchical microgrid energy management in an office building, Appl Energy 208 (2017) 480–494.

[23] Y Liu, FF Wu, Generator bidding in oligopolistic electricity markets using optimal control: fundamentals and application, IEEE Trans Power Syst 21 (3) (2006) 1050–1061.

[24] L Martinez, S Soares, Comparison between closed-loop and partial open-loop feedback control policies in long term hydrothermal scheduling, IEEE Trans Power Syst 17 (2) (2002) 330–336.

[25] C Chen, J Wang, Y Heo, S Kishore, MPC-based appliance scheduling for residential building energy management controller, IEEE Trans Smart Grid 4 (3) (2013) 1401–1410.

[26] Y Guo, J Xiong, S Xu, W Su, Two-stage economic operation of microgrid-like electric vehicle parking deck, IEEE Trans Smart Grid 7 (3) (2017) 1703–1712.

[27] M De Rosa, V Bianco, F Scarpa, LA Tagliafico, Heating and cooling building energy demand evaluation; a simplified model and a modified degree days approach, Appl Energy 128 (2014) 217–229.

[28] C Cristofari, R Norvaisiene, JL Canaletti, G Notton, Innovative alternative solar thermal solutions for housing in conservation-area sites listed as national heritage assets, Energy Build 89 (2015) 123–131.

[29] IH Yang, MS Yeo, KW Kim, Application of artificial neural network to predict the optimal start time for heating system in building, Energy Conv Manage 44 (17) (2003) 2791–2809.

[30] ISO, ISO 13790: Energy Performance of Buildings—Calculation of Energy Use for Space Heating and Cooling, ISO, Geneva, Switzerland, 2008.

[31] ISO, ISO 6946: Building Components and Building Elements—Thermal Resistance and Thermal Transmittance—Calculation Method, ISO, Geneva, Switzerland, 2007.

[32] M Ozel, K Pihtili, Optimum location and distribution of insulation layers on building walls with various orientations, Build Environ 42 (8) (2007) 3051–3059.

[33] JA Duffie, WA Beckman, Solar Engineering of Thermal Process, Wiley, New York, 1991.

[34] OA Zainal, R Yumrutaş, Validation of periodic solution for computing CLTD (cooling load temperature difference) values for building walls and flat roofs, Energy 82 (2015) 758–768.

[35] DQ Mayne, JB Rawlings, CV Rao, POM Scokaert, Constrained model predictive control: stability and optimality, Automatica 36 (6) (2000) 789–814.

[36] JL Melo, JL Teng, S Sun, Online demand response strategies for non-deferrable loads with renewable energy, IEEE Trans Smart Grid 9 (5) (2018) 527–5235.

[37] Y Zhang, T Zhang, Y-J Liu, B Guo, Optimal energy management of a residential local energy network based on model predictive control, Proc CSEE 35 (14) (2015) 3656–3666.

[38] X Wu, X Wang, C Qu, A hierarchical framework for generation scheduling of microgrids, IEEE Trans Power Deliv 29 (6) (2014) 2448–2457.

[39] P Li, D Xu, Z Zhou, W-J Lee, B Zhao, Stochastic optimal operation of microgrid based on chaotic binary particle swarm optimization, IEEE Trans Smart Grid 7 (1) (2016) 66–73.

[40] R Jabbari-Sabet, SM Moghaddas-Tafreshi, SS Mirhoseini, Microgrid operation and management using probabilistic reconfiguration and unit commitment, Int J Electr Power Energy Syst 75 (2016) 328–336.

[41] W Alharbi, K Raahemifar, Probabilistic coordination of microgrid energy resources operation considering uncertainties, Electr Power Syst Res 128 (2015) 1–10.

[42] AK Basu, Microgrids: planning of fuel energy management by strategic deployment of CHP-based DERs—an evolutionary algorithm approach, Int J Electr Power Energy Syst 44 (1) (2013) 326–336.

CHAPTER 6

Adaptive robust energy and reserve co-optimization of an integrated electricity and heating system considering wind uncertainty

6.1 Introduction

The combined heat and power (CHP) unit produces electricity and heat simultaneously and hence connects the electric power system (EPS) with the district heating system (DHS). Due to its high efficiency, it is widely used in countries and regions with cold winters, such as Denmark and northern China [1]. However, since the electric power and heat production of CHP units mainly depends on heat demands, a large amount of wind power may be curtailed. For instance, during nights in the winter, the electricity demands are quite low, whereas the wind power generation is high. To meet the heat demand, CHP units are dispatched and hence the corresponding electricity is generated proportionally, which results in a limited integration of wind power [2].

To improve the integration of wind power, many efforts have been made to enhance the flexibility of CHP units. Thermal energy storages are used to convert surplus wind power to heat energy and store it in the heat tank, then discharge the heat into the DHS when heat demand is high [3]. Chen et al. [4] and Zhang et al. [5] show the benefits of electrical boilers (EBs) and heat pumps in the DHS on relaxing the coupling between heat and power generation of CHP units, which could facilitate the integration of wind power. Nonetheless, the investment costs for energy storage and power-to-heat devices are quite high. Without installing new components, the heat storage capacity of DHNs was utilized to accommodate more wind power by taking into account the time delay of the temperature change of district heating networks (DHNs) [6]. Beside the measures from source and network sides, the demand side also has huge potential to provide flexibility with the

Optimal operation of integrated multi-energy systems under uncertainty. Copyright © 2022 Elsevier Inc.
DOI: https://doi.org/10.1016/B978-0-12-824114-1.00013-5   All rights reserved.

consideration of thermal dynamic characteristics and acceptable temperature tolerance of heat consumers [7, 8].

In addition to improve the flexibility in energy schedules to reduce the wind spillage mentioned earlier, the requirement for reserve capacities is also increasing due to the uncertainty and limited predictability of wind power [9]. The increasing amount of reserve capacities is used to enable power plants to adjust their production in real time and thus offset wind forecast errors [10]. Different from imposing all reserve requirements on non-CHP thermal units [11, 12], the DHS can provide certain reserve capacity for the EPS by introducing the aforementioned measures to improve the flexibility of CHP units [13, 14]. Thus, the scheduling of energy and reserve in the integrated electricity and heat systems (IEHSs) needs to be optimized jointly by using the reserve capacity from CHP units [15, 16].

Generally, CHP units are assumed to provide full reserve capacity as thermal units without accounting for real-time operation constraints of the DHS. However, the real-time regulation might be infeasible after the reserve optimized in the day-ahead stage is deployed [17]. The two-stage optimization framework that includes real-time operation is an effective method. It is constructed in the unit commitment under gas supply uncertainty and the day-ahead energy and reserve scheduling under wind uncertainty, respectively [18,19]. Therefore, this chapter adopts the two-stage optimization framework to ensure the feasibility of the first-stage decisions (i.e., day-ahead operation) with the consideration of operation constraints in the second stage (i.e., real-time operation). Moreover, there is another solution to properly utilize the feasible reserve capacity and regulation of CHP units, which builds a proper regulation model for CHP units. Pan et al. [20] proposed a feasible region model for the DHS taking into account the building thermal inertia, which can be directly used in the central dispatch. A regulating region method to describe the heating-restricted reserve capacity of CHP units was proposed by Zhou et al. [21].

Given the increasing penetration of wind power, it is essential to determine the reserve allocation adaptive to wind uncertainty, which reduces the conservativeness of deterministic determination according to operators' experiences. There is some decision making under uncertainty used for energy and reserve optimization. One is the fuzzy-based method, but it is difficult for system operators to choose a proper fuzzy membership function to represent the bound of uncertainty set [22]. Stochastic optimization is another widely used method, and it was used by Zhang et al. [23] to optimize the energy and reserve scheduling jointly based on a set of scenarios representing the wind uncertainty, but the computational burden is heavy to ensure the performance of solutions. Robust optimization, without requiring distribution information of uncertain parameters, has been extensively studied in the optimal operation of power systems, such as security-constrained unit commitment [24, 25] and security-constrained economic dispatch [26].

This chapter aims to construct a two-stage adaptive robust energy and reserve co-optimization scheme for the IEHS to improve economic efficiency while maintaining operation security under the worst realization of wind power by utilizing feasible reserve capacity provided by the DHS. The mathematical model of the two-stage adaptive robust energy and reserve co-optimization for the IEHS are described in Section 6.2, including the objective function, uncertainty set description, and the day–ahead and real-time operation constraints of the IEHS. The model is solved by the column-and-constraint generation (C&CG) algorithm, which is presented in Section 6.3. In Section 6.4, the case studies and simulation results are given and discussed. The conclusions are drawn in Section 6.5.

6.2 Mathematical formulation of adaptive robust energy and reserve co-optimization for the IEHS

The symbols and notations described in the following mathematical model are explained here for quick reference.

Indices and sets

$\phi^{N/G/WD}$	Set of buses/thermal units/wind farms in the EPS
$\phi^{CHP/EB}$	Set of CHP units/EBs
$\phi^{HN/HS/HL}$	Set of heat nodes/heat sources/heat loads in the DHS
$\phi^{S/R,\,pipe}$	Set of supply pipelines/return pipelines in the DHS
$\Omega_n^{CHP/EB}$	Set of CHP units/electrical boilers connected to bus n
$\Omega_n^{G/WD/EL/N}$	Set of thermal units/wind farms/electric loads/buses connected to bus n
$\Omega_j^{HN/CHP/EB}$	Set of heat nodes/CHP units/EBs connected to heat source j
$\Omega_{nd}^{HS/HL}$	Set of heat sources/ heat loads connected to heat node nd

$\Omega_{nd}^{S/R,pipe}$	Set of supply pipelines/return pipelines connected to heat node nd
T	Set of hours
$DA/RT/HS/HL$	Day-ahead stage/real-time stage/heat source/heat load

Variables

$P_{h,t}^{CHP}$, $P_{g,t}^{G}$	Electricity output of the CHP/non-CHP thermal unit
$P_{i,t}^{EB}$	Electricity consumption of the electric boiler
$R_{h,t}^{CHP,U/D}$ $R_{g,t}^{G,U/D}$	Upward/downward reserve capacity of the CHP/non-CHP thermal unit
$r_{h,t}^{CHP,U/D}$, $r_{g}^{G,U/D}$	Upward/downward regulation of the CHP/non-CHP thermal unit
$L_{l,t}^{shed}$	Load shedding
$\Delta W_{w,t}$	Deviation of wind power
$W_{w,t}^{spill}$	Wind spillage
$\delta_{n,t}$	Phase angle of buses
$H_{h,t}^{CHP}$ $H_{i,t}^{EB}$	Heat output of the CHP unit/electric boiler
$\tau_{nd,t}^{S/R}$	Temperature of node nd in the supply/return network
$\tau_{p,t}^{S/R,in}$, $\tau_{p,t}^{S/R,out}$	Mass flow temperature at the inlet/outlet of pipeline p in the supply/return network
$H_{l,t}^{HL}$	Heat supply at the heat load aggregator
$H_{l,t}^{Build}$	Actual heat consumption of buildings at the heat load aggregator
$T_{l,t}^{in}$	Indoor temperature of buildings at the heat load aggregator

Parameters

C_{h}^{CHP}	Energy offer price of CHP units
C_{g}^{G}	Energy offer price of non-CHP thermal units
$C_{h}^{+/-}$, $C_{g}^{+/-}$	Upward/downward reserve offer price of CHP units/non-CHP thermal units
C_{l}^{shed}	Cost coefficient for a lost load
B_{nm}	Susceptance of transmission line nm
$P_{w,t}^{we}$	Predicted wind power
$P_{l,t}^{load}$	Electric load

$PR_{h,t}^{CHP,up/down}$, $PR_{g,t}^{G,up/down}$	Upward/downward ramping rate limit of the CHP/non-CHP thermal unit
$P_{h,t}^{CHP,max}$, $P_{g,t}^{G,max}$, $P_{i,t}^{EB,max}$, $\Delta W_{w,t}^{max}$	Maximum value of CHP output/thermal unit output/power consumption of electric boiler/wind power deviation
$P_{h,t}^{CHP,min}$, $P_{g,t}^{G,min}$	Minimum value of CHP output/thermal unit output
η_h^{CHP}, η_i^{EB}	Heat-to-power coefficient of CHP/power to heat the coefficient of the electric boiler
c, l_p	Specific heat capacity of water/length of pipelines
$m_{j,t}^{HS}$, $m_{l,t}^{HL}$, $m_{p,t}^{S/R,pipe}$	Mass flow rate of heat sources/heat loads/supply pipes/return pipes
τ_t^a, $T_{l,t}^{in,max/min}$	Ambient temperature of pipelines/maximum/minimum indoor temperature
c_{air}, ρ_{air}	Air specific heat/indoor air density
K, F, V	Average thermal conductivity/external surface area/volume of buildings

6.2.1 Objective function

A two-stage robust energy and reserve co-optimization scheme for the IEHS is formulated, which consists of the day-ahead operation stage and the real-time regulation stage. The first-stage problem is to minimize the day-ahead energy and reserve cost and to determine the optimal schedule with predicted wind power. The second-stage problem corresponds to the real-time regulation to address the wind power forecast error. Taking into account the worst-case scenario of wind uncertainty, an adaptive robust approach is adopted to formulate this type of two-stage optimization problem [27].

In this regard, the objective function of the proposed robust optimization scheme is formulated as the following min-max-min form [28]. The objective is to minimize the total system cost of the IEHS in both day-ahead and real-time stages under the worst-case realization of wind power within an uncertainty set, as shown in Eq. (6.1). The lower level is a real-time multiperiod economic regulation of the IEHS. The middle level is to determine the worst-case realization of wind power that results in the largest cost. The upper level provides optimal anticipated decisions to minimize the total system cost against the largest wind forecast error. In the proposed model, both non-CHP units and CHP units provide reserve capacity for the power system to cope with uncertainties. Moreover, the operation constraints of the EPS and DHS in the real-time operation are also considered to ensure the feasibility of energy and reserve scheduling.

$$
\min \sum_{t \in T} \left(\sum_{h \in \Phi^{\mathrm{CHP}}} C_h^{G,CHP} P_{h,t}^{CHP} + C_h^{H,CHP} H_{h,t}^{CHP} + \sum_{g \in \Phi^G} C_g^G P_{g,t}^G \right.
$$

$$
+ \sum_{h \in \Phi^{\mathrm{CHP}}} \left(C_h^+ R_{h,t}^{CHP,U} + C_h^- R_{h,t}^{CHP,D} \right) + \sum_{g \in \Phi^G} \left(C_g^+ R_{g,t}^{G,U} + C_g^- R_{g,t}^{G,D} \right) \Bigg)
$$

$$
+ \max_{\Delta W_{w,t}} \min \sum_{t \in T} Bigg(\sum_{h \in \Phi^{\mathrm{CHP}}} C_h^{G,CHP} \left(r_{h,t}^{CHP,U} - r_{h,t}^{CHP,D} \right)
$$

$$
+ C_h^{H,CHP} \left(H_{h,t}^{CHP,RT} - H_{h,t}^{CHP} \right) + \sum_{g \in \Phi^G} C_g^G \left(r_{g,t}^{G,U} - r_{g,t}^{G,D} \right) + \sum_{l \in \Phi^{EL}} C_l^{shed} L_{l,t}^{shed} \Bigg)
$$

$$
\tag{6.1}
$$

6.2.2 Mathematical formulation of EPS and DHS operations in two stages

The scheduled day-ahead energy and reserve are optimized in the first stage, which are kept to constrain the reserve deployments in the second stage (i.e., real-time regulation). Since the EPS and DHS are coupled in these two stages, the day-ahead and real-time operation constraints of the IEHS consist of both the EPS and DHS constraints.

(1) *Day-ahead operation constraints of the EPS.* In the day-ahead operation of the EPS, the scheduling of electric power production and the reserve capacity are optimized. The DC power flow without losses is employed in this optimization model.

- *Electric power balance.* At each bus in the EPS, the power balance between the total power production and electric loads must be maintained at each hour.

$$
\sum_{h \in \Omega_n^{CHP}} P_{h,t}^{CHP} + \sum_{g \in \Omega_n^G} P_{g,t}^G + \sum_{w \in \Omega_n^{WD}} P_{w,t}^{we} - \sum_{i \in \Omega_n^{EB}} P_{i,t}^{EB} - \sum_{l \in \Omega_n^{EL}} P_{l,t}^{load}
$$

$$
= \sum_{m \in \Omega_n^N} B_{nm}(\delta_{n,t} - \delta_{m,t}), \quad \forall n \in \Phi^N, \forall t \in T \tag{6.2}
$$

- *Transmission line constraints.* For each transmission line, the power flow should be lower than the corresponding line capacity.

$$
-f_{nm}^{\max} \le B_{nm}(\delta_{n,t} - \delta_{m,t}) \le f_{nm}^{\max}, \quad \forall n \in \Phi^N, \forall t \in T \tag{6.3}
$$

- *Capacity limits of CHP units and non-CHP thermal units.* Constraints (6.4) and (6.5) denote capacity constraints for the coupling of power production and reserve capacity provision of CHP units and non-CHP thermal units, respectively.

$$P_{h,t}^{CHP,\min} \le P_{h,t}^{CHP} - R_{h,t}^{CHP,D}, \quad P_{h,t}^{CHP} + R_{h,t}^{CHP,U} \le P_{h,t}^{CHP,\max},$$
$$\forall h \in \Phi^{CHP}, \ \forall t \in T \tag{6.4}$$

$$P_{g,t}^{G,\min} \le P_{g,t}^{G} - R_{g,t}^{G,D}, \quad P_{g,t}^{G} + R_{g,t}^{G,U} \le P_{g,t}^{G,\max}, \quad \forall g \in \Phi^{G}, \forall t \in T \tag{6.5}$$

- *Reserve capacity limits.* Constraints (6.6) and (6.7) ensure that the upward and downward reserve capacities provided by each unit should not be larger than its ramping rate.

$$0 \le R_{h,t}^{CHP,U} \le PR_{h}^{CHP,up}, \quad 0 \le R_{h,t}^{CHP,D} \le PR_{h}^{CHP,down},$$
$$\forall h \in \Phi^{CHP}, \forall t \in T \tag{6.6}$$

$$0 \le R_{g,t}^{G,U} \le PR_{g}^{G,up}, \quad 0 \le R_{g,t}^{G,D} \le PR_{g}^{G,down}, \quad \forall g \in \Phi^{G}, \forall t \in T \tag{6.7}$$

- *Ramping limits.* The power output changes of CHP units and non-CHP thermal units between two consecutive hours are limited by their ramping rates.

$$-PR_{h}^{CHP,down} \Delta t \le P_{h,t}^{CHP} - P_{h,t-1}^{CHP} \le PR_{h}^{CHP,up} \Delta t,$$
$$\forall h \in \Phi^{CHP}, \forall t \in T \tag{6.8}$$

$$-PR_{g}^{G,down} \Delta t \le P_{g,t}^{G} - P_{g,t-1}^{G} \le PR_{g}^{G,up} \Delta t, \quad \forall g \in \Phi^{G}, \forall t \in T \tag{6.9}$$

- *Operation limits of EBs.* Constraint (6.10) denotes the minimum and maximum limits for the electricity consumption of EBs.

$$0 \le P_{i,t}^{EB} \le P_{i}^{EB,\max}, \forall i \in \Phi^{EB}, \forall t \in T \tag{6.10}$$

(2) *Day-ahead operation constraints of the DHS.* The DHS consists of heat sources, the DHN, and end users. The DHN consists of two types of pipelines to transmit hot water: supply pipelines delivering heat to demand nodes and return pipelines carrying mass flow back to heat source nodes.

(3) *Heat sources.* CHP units and EBs are considered as heat sources linking the EPS and DHS in this chapter. Among various technologies of CHP units, the back-pressure CHP unit with a linear relationship between electric power and heat production is adopted.

$$H_{h,t}^{CHP} = \eta_h^{CHP} P_{h,t}^{CHP}, \quad \forall h \in \Phi^{CHP}, \forall t \in T \tag{6.11}$$

EBs produce heat energy by consuming electric power, and thus they are regarded as electric loads in the EPS. The model of the EB is described as follows.

$$H_{i,t}^{EB} = \eta_i^{EB} P_{i,t}^{EB}, \quad \forall i \in \Phi^{EB}, \forall t \in T \tag{6.12}$$

The heat energy from heat sources is used to warm up water or steam and is then transferred to end users. The relationship between heat production and the temperatures of supply and return pipelines is shown next.

$$\sum_{h \in \Omega_j^{CHP}} H_{h,t}^{CHP} + \sum_{i \in \Omega_j^{EB}} H_{i,t}^{EB} = c \cdot m_{j,t}^{HS} \cdot \left(\tau_{nd,t}^{S,HS} - \tau_{nd,t}^{R,HS} \right),$$

$$\forall j \in \Phi^{HS}, \ nd = \Omega_j^{HS}, \forall t \in T \tag{6.13}$$

(4) *District heating network.* The steady state energy flow model is considered in this chapter. The continuity of mass flow rate should be hold (i.e., for each node in the DHN, the mass flows entering the node are equal to the mass flows leaving the node [29]), which is depicted as follows.

$$\sum_{j \in \Omega_{nd}^{HS}} m_{j,t}^{HS} + \sum_{p \in \Omega_{nd}^{S,pipe}} a_{nd,p} m_{p,t}^{S,pipe} - \sum_{l \in \Omega_{nd}^{HL}} m_{l,t}^{HL} = 0, \ \forall nd \in \Phi^{HN}, \forall t \in T$$

$$\tag{6.14}$$

$$\sum_{l \in \Omega_{nd}^{HL}} m_{l,t}^{HL} - \sum_{p \in \Omega_{nd}^{R,pipe}} a_{nd,p} m_{p,t}^{R,pipe} - \sum_{j \in \Omega_{nd}^{HS}} m_{j,t}^{HS} = 0, \ \forall nd \in \Phi^{HN}, \forall t \in T$$

$$\tag{6.15}$$

In the preceding equations, $a_{nd,p}$ is defined in the following matrix $\boldsymbol{A}$, which is the network matrix that relates the nodes to the pipes in all supply pipelines [30]. The possible values of $a_{nd,p}$ are $+1, -1$, and 0, which represent the mass flow injects into or leaves from the node nd through pipe p, and pipe p does not connect to node nd, respectively.

$$A = \begin{bmatrix} a_{1,1} & a_{1,2} & \cdots & a_{1,p} \\ a_{2,1} & a_{2,2} & \cdots & a_{2,p} \\ \vdots & \vdots & \ddots & \vdots \\ a_{nd,1} & a_{nd,2} & \cdots & a_{nd,p} \end{bmatrix} \tag{6.16}$$

There are heat losses when mass flow passes through pipelines, resulting in temperature drops along the pipeline. The expression of heat losses is given in Eqs. (6.17) and (6.18), which keeps consistent with the flow direction. Since most pipeline networks are located underground and the temperature variation underground is relatively small, the ambient temperature is set constantly at $10°C$ [31].

$$\tau_{p,t}^{S,out} - \tau_t^a = \left(\tau_{p,t}^{S,in} - \tau_t^a\right)e^{-k_p \pi d_p l_p / c \cdot m_{p,t}^{S,pipe}}, \forall p \in \Phi^{S,pipe}, \forall t \in T \tag{6.17}$$

$$\tau_{p,t}^{R,out} - \tau_t^a = \left(\tau_{p,t}^{R,in} - \tau_t^a\right)e^{-k_p \pi d_p l_p / c \cdot m_{p,t}^{R,pipe}}, \forall p \in \Phi^{R,pipe}, \forall t \in T \tag{6.18}$$

The temperature of the mixed fluid is determined according to the energy conservation law when the mass flow of pipelines enters a node.

$$\sum_{p \in \Omega_{nd}^{S,pipe}} \left(m_{p,t}^{S,pipe} \cdot \tau_{p,t}^{S,out}\right) + \sum_{j \in \Omega_{nd}^{HS}} \left(m_{j,t}^{HS} \cdot \tau_{j,t}^{S,HS}\right)$$
$$= \tau_{nd,t}^{S,HN} \left(\sum_{p \in \Omega_{nd}^{S,pipe}} m_{p,t}^{S,pipe} + \sum_{j \in \Omega_{nd}^{HS}} m_{j,t}^{HS}\right), \forall nd \in \Phi^{HN}, \forall t \in T \tag{6.19}$$

$$\sum_{p \in \Omega_{nd}^{R,pipe}} \left(m_{p,t}^{R,pipe} \cdot \tau_{p,t}^{R,out}\right) + \sum_{l \in \Omega_{nd}^{HL}} \left(m_{l,t}^{HL} \cdot \tau_{l,t}^{R,HL}\right)$$
$$= \tau_{nd,t}^{R,HN} \left(\sum_{p \in \Omega_{nd}^{R,pipe}} m_{p,t}^{R,pipe} + \sum_{l \in \Omega_{nd}^{HL}} m_{l,t}^{HL}\right), \forall nd \in \Phi^{HN}, \forall t \in T \tag{6.20}$$

The temperature of pipelines flowing out of a node is equal to the mixed temperature at this node.

$$\tau_{p,t}^{S,in} = \tau_{nd,t}^{S,HN}, \forall nd \in \Phi^{HN}, \forall p \in \Phi^{S,pipe}, \forall t \in T \tag{6.21}$$

$$\tau_{p,t}^{R,in} = \tau_{nd,t}^{R,HN}, \forall nd \in \Phi^{HN}, \forall p \in \Phi^{R,pipe}, \forall t \in T \tag{6.22}$$

(5) *Heat load.* Similar to the relationship between heat production and temperatures in heat sources, the heat consumption at one node is calculated as follows.

$$H_{l,t}^{\mathrm{HL}} = c \cdot m_{l,t}^{HL} \cdot \left(\tau_{nd,t}^{S,HL} - \tau_{nd,t}^{R,HL} \right), \ \forall l \in \Phi^{\mathrm{HL}}, \ nd = \Omega_l^{HL}, \ \forall t \in T \tag{6.23}$$

In this chapter, a heat load aggregator managing all buildings that connect to the same heat substation is assumed, which is denoted as one consumption node. Given the large thermal inertia of buildings, the heat supply and demand can be temporally decoupled in the DHS, namely the indoor temperature could be maintained for a while when thermal supply changes. Thus, the thermal inertia of buildings regarding space heating is used, which reflects the thermal dynamic process of buildings [31]. To simplify the thermal dynamic process, it is modeled based on the model described by Zhang et al. [32], as shown in Eq. (6.24). The indoor temperature of buildings is centrally controlled and assumed to be the same. The left–hand side of Eq. (6.24) represents the variation of the indoor air heat energy. A more detailed description can be found in the work of Xu et al. [33].

$$c_{air}\rho_{air}V\frac{d\,T_{l,t}^{in}}{dt} = H_{l,t}^{\mathrm{HL}} - H_{l,t}^{\mathrm{Build}}, \ \forall l \in \Phi^{\mathrm{HL}}, \ \forall t \in T \tag{6.24}$$

To make the dynamic model computationally tractable, the preceding differential equation is converted into a discrete difference equation with the time interval of 1 hour.

$$c_{air}\rho_{air}V\left(T_{l,t+1}^{in} - T_{l,t}^{in}\right) = H_{l,t}^{\mathrm{HL}} - H_{l,t}^{\mathrm{Build}}, \ \forall l \in \Phi^{\mathrm{HL}}, \ \forall t \in T \tag{6.25}$$

The needed heat demands for space heating in buildings depends on the difference between the indoor and outdoor temperatures of buildings. It is expressed as follows.

$$H_{l,t}^{\mathrm{Build}} = KF\left(T_{l,t}^{in} - T_{l,t}^{env}\right), \ \forall l \in \Phi^{\mathrm{HL}}, \ \forall t \in T \tag{6.26}$$

The indoor temperature of buildings should be maintained within human thermal comforts.

$$T_l^{in,\min} \le T_{l,t}^{in} \le T_l^{in,\max}, \quad \forall l \in \Phi^{\mathrm{HL}}, \forall t \in T \tag{6.27}$$

Since some operation constraints in the second stage are similar to the corresponding constraints in the first stage, inequalities (6.3), (6.8) through (6.10), and (6.27) are transformed into a compact form as in (6.28), and equalities (6.11) through (6.26) are transformed into constraint (6.29) to simplify the discussion in the following sections.

$$h^{DA}\left(P_{h,t}^{CHP},\ P_{g,t}^{G},\ P_{i,t}^{EB},\ \delta_{n,t},\ T_{l,t}^{in}\right) \leq 0 \tag{6.28}$$

$$g^{DA}\left(H_{h,t}^{CHP},\ H_{i,t}^{EB},\ H_{l,t}^{HL},\ \tau_{nd,t}^{S/R,HS},\ \tau_{nd,t}^{S/R,HL},\ \tau_{p,t}^{S/R,in},\ \tau_{p,t}^{S/R,out},\ \tau_{nd,t}^{S/R,HN}\right) = 0 \tag{6.29}$$

(1) *Real-time regulation constraints of the EPS and DHS.* The second-stage problem is formulated to handle the wind power forecast error in the real-time operation. Based on the dispatch schemes from the first stage and uncertainty set, the output of thermal units, CHP units, and EBs in the IEHS are adjusted adaptively.

The construction of a suitable uncertainty set determines the conservativeness of the solution of robust optimization and is thus critical in robust optimization. In the second stage, the uncertainty of each wind power is revealed and is represented by an interval uncertainty set that contains all possible deviations from the day-ahead wind forecast. The whole uncertainty set is described as follows and is a polyhedron.

$$\mathbf{U} = \{\mathbf{\Delta}W| -\Delta W_{w,t}^{\max} \leq \Delta W_{w,t} \leq \Delta W_{w,t}^{\max}, \quad \forall w \in \Phi^{WD}, \forall t \in T\} \tag{6.30}$$

Since the production of generators are regulated in the real-time operation to cope with the wind power forecast error, it is formulated as the second-stage problem. In this chapter, the outputs of thermal units, CHP units, and EBs in the IEHS are adjusted adaptively according to the uncertainty set and day-ahead reserve scheduling from the first stage [34]. Constraints (6.31) through (6.34) show that the regulation of generators in real-time operation should be lower than scheduled reserve capacity in the day-ahead stage.

$$0 \leq r_{h,t}^{CHP,U} \leq R_{h,t}^{CHP,U} \quad \forall h \in \Phi^{CHP}, \forall t \in T \tag{6.31}$$

$$0 \leq r_{h,t}^{CHP,D} \leq R_{h,t}^{CHP,D} \quad \forall h \in \Phi^{CHP}, \forall t \in T \tag{6.32}$$

$$0 \leq r_{g,t}^{G,U} \leq R_{g,t}^{G,U} \quad \forall g \in \Phi^{G}, \forall t \in T \tag{6.33}$$

$$0 \leq r_{g,t}^{G,D} \leq R_{g,t}^{G,D} \quad \forall g \in \Phi^{G}, \forall t \in T \tag{6.34}$$

The power rebalance should be satisfied in the real–time operation.

$$\sum_{h\in\Omega_n^{CHP}} \left(r_{h,t}^{CHP,U} - r_{h,t}^{CHP,D}\right) + \sum_{g\in\Omega_n^G} \left(r_{g,t}^{G,U} - r_{g,t}^{G,D}\right)$$
$$+ \sum_{w\in\Omega_n^{WD}} \left(\Delta W_{w,t} - W_{w,t}^{spill}\right) - \sum_{i\in\Omega_n^{EB}} \left(P_{i,t}^{EB,RT} - P_{i,t}^{EB}\right) + \sum_{l\in\Omega_n^{LD}} L_{l,t}^{shed} \qquad (6.35)$$
$$= \sum_{m\in\Omega_n^N} B_{nm}\left(\delta_{n,t}^{RT} - \delta_{m,t}^{RT} - \delta_{n,t} + \delta_{m,t}\right) \quad \forall n\in\Phi^N, \forall t\in T$$

Constraints (6.36) and (6.37) denote the limits of the wind spillage and load shedding, respectively.

$$0 \le W_{w,t}^{spill} \le P_{w,t}^{wd} + \Delta W_{w,t}, \quad \forall w\in\Phi^{WD}, \forall t\in T \qquad (6.36)$$

$$0 \le L_{l,t}^{shed} \le P_{l,t}^{load} \quad \forall l\in\Phi^{EL}, \forall t\in T \qquad (6.37)$$

Constraints (6.38) and (6.39) are the compact form of the operation constraints of the EPS and DHS under uncertainty in the second stage, which are similar to the corresponding day–ahead operation const(6.28) and (6.29).

$$h^{RT}\left(P_{h,t}^{CHP,RT}, P_{g,t}^{G,RT}, P_{i,t}^{EB,RT}, \delta_{n,t}^{RT}, T_{l,t}^{in,RT}\right) \le 0 \qquad (6.38)$$

$$g^{RT}\left(H_{h,t}^{CHP,RT}, H_{i,t}^{EB,RT}, H_{l,t}^{HL,RT}, \tau_{nd,t}^{S/R,HS,RT}, \tau_{nd,t}^{S/R,HL,RT},\right.$$
$$\left. \tau_{p,t}^{S/R,in,RT}, \tau_{p,t}^{S/R,out,RT}, \tau_{nd,t}^{S/R,HN,RT}\right) = 0 \qquad (6.39)$$

6.2.3 Robust compact formulation

For ease of expression, the compact matrix form of the proposed robust energy and reserve co-optimization scheme is adopted and described as follows:

$$\min_{\mathbf{x}} \mathbf{c}^T\mathbf{x} + \max_{\mathbf{u}\in U} \min_{\mathbf{y}\in F(\mathbf{x},\mathbf{u})} \mathbf{d}^T\mathbf{y} \qquad (6.40)$$

$$s.t. \ \mathbf{Ax} \le \mathbf{b}, \quad \mathbf{Bx} = \mathbf{a}, \quad \mathbf{x}\in\mathbb{R}_+^n \qquad (6.41)$$

where

$$F(\mathbf{x}, \mathbf{u}) = \{\mathbf{y}|\mathbf{Cy} + \mathbf{Dx} \le \mathbf{h} \qquad (6.42)$$

$$\mathbf{Gy} + \mathbf{Mu} = \mathbf{g} \qquad (6.43)$$

$$\mathbf{y}\in\mathbb{R}_+^n\} \qquad (6.44)$$

where $\mathbf{x}$ is the decision vector in the first stage representing the continuous variables reflecting the operation conditions of the EPS and DHS, $\mathbf{y}$ is the decision vector in the second stage representing the real-time regulation variables, and $\mathbf{u}$ is the uncertain variables of wind power. Constraint (6.41) represents all constraints in the first stage (i.e., (6.2)-(6.27)). Constraint (6.42) denotes the inequalities in the second stage (i.e., (6.30)-(6.34) and (6.36)-(6.38)). Constraint (6.43) refers to (6.35) and (6.39). Constraint (6.44) states that the $\mathbf{y}$ are positive variables.

6.3 Solution methodology

The proposed two-stage adaptive robust optimization is a tri-level problem, which cannot be solved by a commercial solver directly. Thus, the C&CG algorithm, which is a cutting plane–based method, is adopted to solve it [35]. First, the proposed model is decomposed into a master problem (MP) and a subproblem (SP). Second, the max-min SP is transformed into a mixed-integer linear programming with its Karush-Kuhn-Tucker (KKT) conditions. Third, the MP and SP are solved iteratively until an acceptable error is reached. In each iteration, the optimal solution of the SP is considered as a significant scenario and new variables and corresponding constraints are added to the MP.

The MP minimizes the total system operation cost under the worst-case wind realization $\mathbf{u}_l^*$ obtained from the SP in the previous iteration. It is described as follows:

$$\text{MP} : \min_{\mathbf{x},\eta} \mathbf{c}^T\mathbf{x} + \eta \tag{6.45}$$

$$\text{s.t. } \mathbf{Ax} \le \mathbf{b}, \quad \mathbf{Bx} = \mathbf{a} \tag{6.46}$$

$$\eta \ge \mathbf{d}^T\mathbf{y}^l, \ \forall l \in O \tag{6.47}$$

$$\mathbf{Cy}^l + \mathbf{Dx} \le \mathbf{h}, \ \forall l \in O \tag{6.48}$$

$$\mathbf{Gy}^l + \mathbf{Mu}_l^* = \mathbf{g}, \ \forall l \in O \tag{6.49}$$

$$\mathbf{x} \in \mathbb{R}_+^n, \mathbf{y}^l \in \mathbb{R}_+^n, \ \forall l \in O \tag{6.50}$$

where η is the auxiliary variable, $\mathbf{y}^l$ are the new variables generated from the SP and added to the MP, O is the index set for wind uncertainty scenarios l, and $\mathbf{u}_l^*$ is the optimal value obtained from the SP in the last iteration, which is considered as the current worst-case realization.

The SP is to identify the worst-case scenario with the optimal result $\mathbf{x}^*$ obtained from the MP.

$$\text{SP} : \max_{\mathbf{u} \in U} \min_{\mathbf{y} \in F(\mathbf{x}^*, \mathbf{u})} \mathbf{d}^T \mathbf{y} \tag{6.51}$$

$$\text{s.t. } \mathbf{Cy} + \mathbf{Dx}^* \leq \mathbf{h} \tag{6.52}$$

$$\mathbf{Gy} + \mathbf{Mu} = \mathbf{g} \tag{6.53}$$

$$\mathbf{y} \in \mathbb{R}_+^n \tag{6.54}$$

To convert the original max-min problem into an single-level problem that can be solved by commercial solvers, the strong duality and KKT conditions are used [36] given that the inner min problem in the SP is a linear problem. Reformulated with the KKT conditions, the max-min SP is transferred into an equivalent single-level problem as follows. It includes the primary feasibility constraints (6.52) through (6.54), stationarity constraint (6.56), and complementary slackness conditions (6.57) through (6.59):

$$\text{KKT} - \text{SP} : \max_{\mathbf{u} \in U, \mathbf{y} \in F(\mathbf{x}^*, \mathbf{u}), \lambda, \mu, \nu} \mathbf{d}^T \mathbf{y} \tag{6.55}$$

s.t. (6.52)-(6.54)

$$\mathbf{d} + \mathbf{C}\lambda + \mathbf{G}\mu - \nu = 0 \tag{6.56}$$

$$(\mathbf{Cy} + \mathbf{Dx}^* - \mathbf{h})_i \lambda_i = 0, \quad \forall i \tag{6.57}$$

$$y_j \nu_j = 0, \quad \forall j \tag{6.58}$$

$$\lambda_i \geq 0, \ \nu_j \geq 0, \ \mu \text{ is free} \tag{6.59}$$

where λ, μ, ν are dual variables corresponding to constraints (6.52) through (6.54), and i and j are the indices of the corresponding constraints. In the KKT-SP problem, most of constraints are linear except the bi-linear complementary constraints (6.57) through (6.58), which are linearized by the big-M method [35]. Thus, the KKT-SP is converted into a mixed-integer linear programming problem.

The reformulated MP and SP are solved by the C&CG algorithm iteratively, which is described in Fig. 6.1.

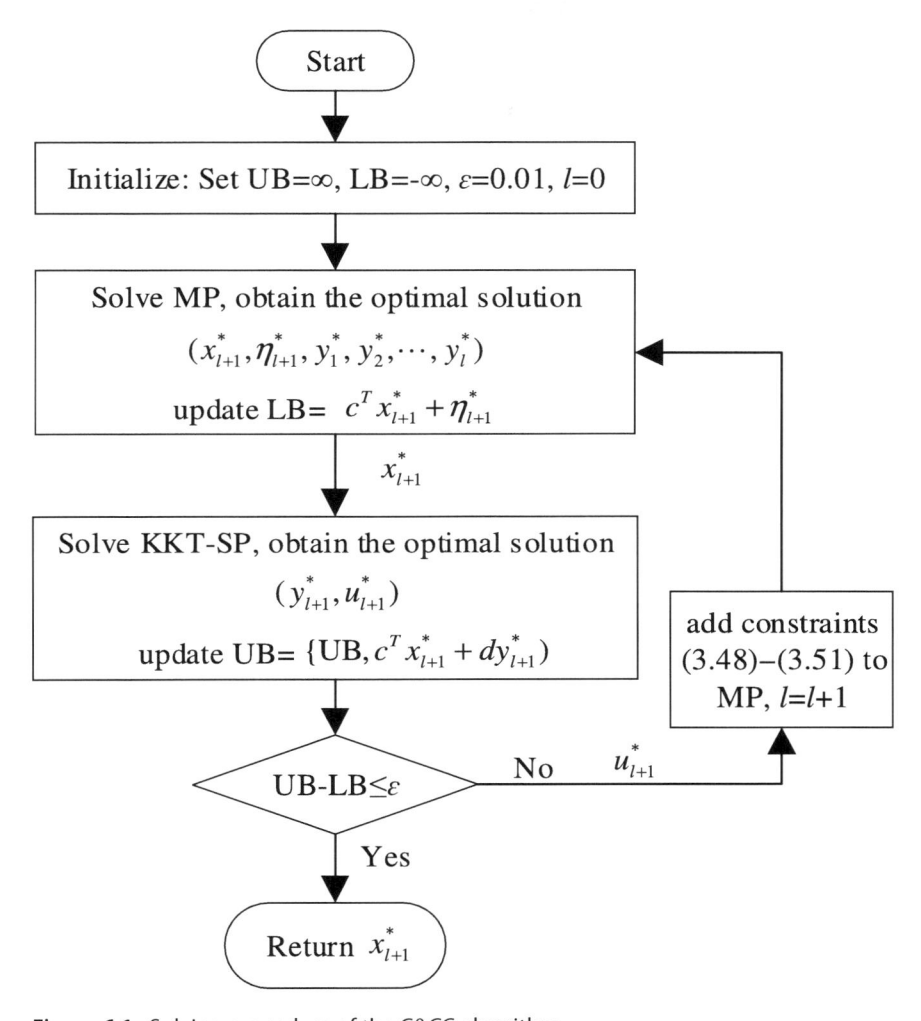

Figure 6.1 Solving procedure of the C&CG algorithm.

(1) Initialization. Set the upper bound as ∞ and the lower bound as $-\infty$, convergence error $[= 0.01$, iteration index $l = 0$, and $O = \varnothing$.

(2) Solve the MP (i.e., (6.45)–(6.50)) and derive the optimal solution $x_{l+1}^*, \eta x_{l+1}^*, y_1^*, y_2^*, ..., y_l^*$ and update the lower bound $LB = \{c^T x_{l+1}^* + y_{l+1}^*\}$.

(3) Solve the KKT-SP (i.e., (6.55) and (6.57)–(6.59)) with the optimal solution x_{l+1}^* obtained in Step (2), and get the optimal solution (y_{l+1}^*, u_{l+1}^*), then update the upper bound $UB = \{UB, c^T x_{l+1}^* + d^T y_{l+1}^*\}$.

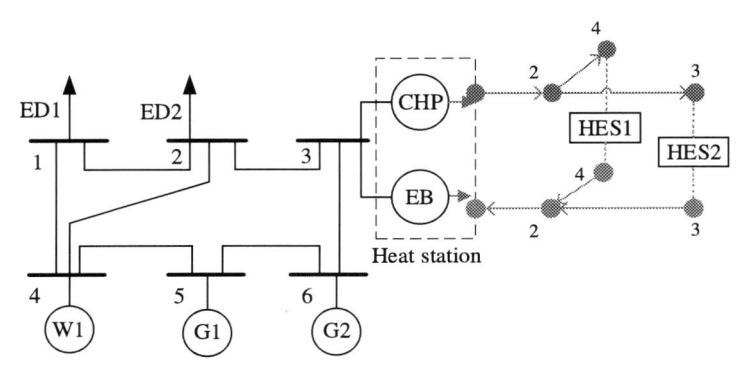

Figure 6.2 Test system with 6-bus EPS and 4-node DHS.

(4) Convergence check. If UB–LB $\leq \varepsilon$, return y^*_{l+1} and terminate. Otherwise, generate new variables y_{l+1} and add corresponding new constraints (6.47) through (6.50) to the MP, and update $l = l + 1$, $O = O \cup \{l + 1\}$, then go to Step (2).

6.4 Simulation results

The effectiveness of the proposed robust energy and reserve co-optimization scheme is verified based on the numerical simulation on an IEHS test system. The optimization model is coded in the general algebraic modeling system and is solved using the CPLEX.

6.4.1 System configuration

Fig. 6.2 shows the configuration of the IEHS test system. The system consists of a modified 6-bus EPS [12] and a 4-node DHS [37]. The EPS consists of two thermal units (G1 and G2) located at Bus 5 and Bus 6, one CHP unit located at Bus 3, one wind farm (W1) located at Bus 4, and two electric demands (ED1 and ED2) located at Bus 1 and Bus 2, respectively. An EB is located at Bus 3 with 12–MW capacity, which has a fixed electricity-to-heat ratio of 1. The physical and economic parameters of generators are listed in Table 6.1.

The DHS operates in the mode with constant mass flow and variable temperature. To guarantee the quality of heat supply, the temperature of the mass flow from heat sources is set at 80°C. The temperature in supply pipes is controlled within 60 to 80°C, whereas the temperature in return pipes depends on the absorption of heat demands. The heat demands of each

Table 6.1 Study cases.

Unit	Pmax (MW)	Hmax (MW)	Ramp Rate (MW/h)	Electricity Cost ($/MWh)	UR Price ($/MWh)
CHP	200	300	50	16	10
G1	100	/	20	20	8
G2	100	/	30	12	15

aggregator are assumed to be the same and are used for space heating in buildings. The standard indoor temperature is set at $20°C$, and the thermal comfort temperature of end users is $\pm 2°C$ deviated from the standard temperature [38]. Fig. 6.3(a) depicts the hourly electric load and heat load during the whole day. Fig. 6.3(b) shows the predicted and actual deviation of the outputs of two wind farms.

6.4.2 Benchmark cases

First, a conventional single-stage energy and reserve optimization model M1 is considered as a benchmark to demonstrate the effectiveness of the proposed two-stage adaptive robust optimization scheme M2. The reserves are optimized with a predefined system reserve requirement without considering the regulation constraints in the real-time stage, where CHP units provide full reserve capacity without considering the feasibility of real-time regulation. The deterministic optimized model is formulated as follows [39]. Instead, in the proposed two-stage robust energy and reserve optimization model M2, the CHP unit provides available reserve capacity that is constrained by EPS and DHS operation constraints in both day–ahead and real-time stages.

$$
\min \sum_{t \in T} \left(\sum_{h \in \Phi^{\mathrm{CHP}}} C_h^{G,CHP} P_{h,t}^{CHP,DA} + C_h^{H,CHP} H_{h,t}^{DA} + \sum_{g \in \Phi^G} C_g^G P_{g,t}^{G,DA} \right.
$$

$$
\left. + \sum_{h \in \Phi^{\mathrm{CHP}}} (C_h^+ R_{h,t}^U + C_h^- R_{h,t}^D) + \sum_{g \in \Phi^G} \left(C_g^+ R_{g,t}^{G,U} + C_g^- R_{g,t}^{G,D} \right) \right) \quad (6.60)
$$

$$
\sum_{h \in \Phi^{\mathrm{CHP}}} R_{h,t}^{CHP,U} + \sum_{g \in \Phi^G} R_{g,t}^{G,U} \geq RU_t,
$$

$$
\sum_{h \in \Phi^{\mathrm{CHP}}} R_{h,t}^{CHP,D} + \sum_{g \in \Phi^G} R_{g,t}^{G,D} \geq RD_t \quad \forall t \in T \quad (6.61)
$$

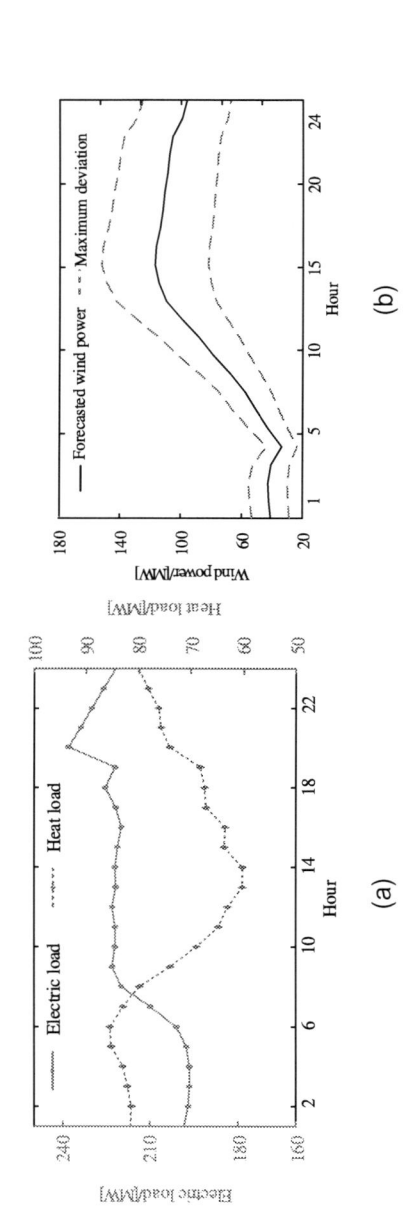

Figure 6.3 Profiles: (a) hourly electric demand and total heat demand and (b) the total predicted generation and deviation of two wind farms.

Table 6.2 Study cases.

Case	CHP as Reserve	EB	Thermal Inertia	Optimized Method
1	×	×	×	M2
2	√	√	×	M1
3	√	√	×	M2
4	√	√	√	M1
5	√	√	√	M2

In this chapter, five cases are designed to verify the advantages of the proposed optimization scheme, of which the details are listed in Table 6.2. Case 1 is a base schedule without exploiting any flexibility from the DHS. EBs and CHP units are considered to provide reserve capacities in Cases 2 and 3. Based on Cases 2 and 3, the thermal inertia of buildings is additionally introduced in Cases 4 and 5. Cases 2 and 4 are optimized by M1, whereas Cases 1, 3, and 5 are optimized by M2.

6.4.3 Reserve sufficiency analysis in real-time regulation

In this section, the advantage of the proposed two-stage robust model (M2) in ensuring the feasibility of real-time regulation is demonstrated. It is noted that there is no downward reserve allocated, because no cost is set for wind power spillage.

The optimal results of real-time power regulation in Cases 2 through 5 are presented in Fig. 6.4. In all four cases, all generators and EBs are scheduled to provide upward reserve in the day-ahead operation, but the real-time regulation is not always sufficient in these four cases. Fig. 6.4(a) shows that the real-time regulation in Case 2 is not sufficient. Even though CHP units are scheduled to provide upward reserves in the day-ahead operation, CHP units cannot increase their outputs in the real-time operation in order to keep a heat energy balance without considering either EBs or thermal inertia. Hence, the actual available upward regulation that CHP units can provide is 0. Although the thermal inertia of buildings is introduced in Case 4 allowing the changes of CHP units' generation to some extent, CHP units cannot provide sufficient real-time regulation either, as shown in Fig. 6.4(c). Thus, a large amount of load shedding in Cases 2 and 4 occurs due to the lack of upward regulation, which leads to the high load shedding cost in the real-time operation. However, the reserve capacities in Cases 3 and 5 allocated in the day-ahead stage are both sufficient for corresponding real-time regulation, as shown in Fig. 6.4(b) and (d). The comparison results demonstrate that in M1 without considering the regulation ability

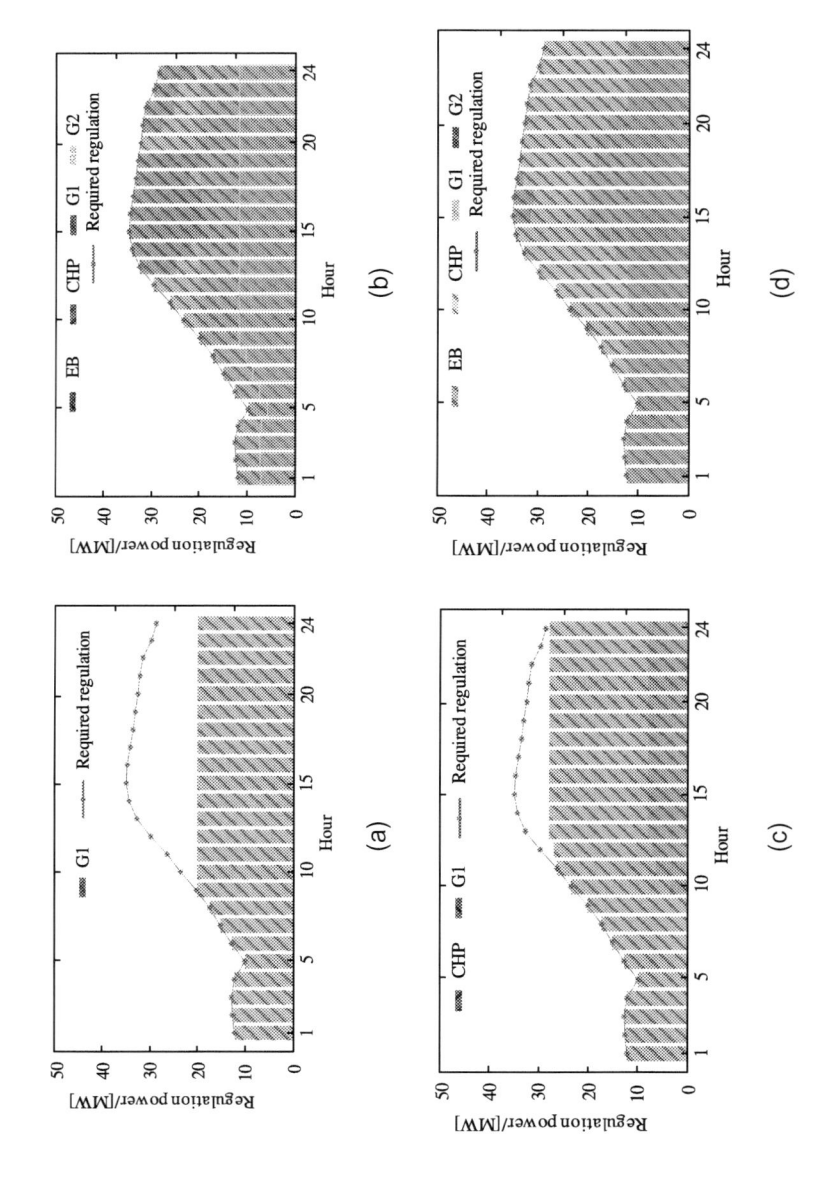

Figure 6.4 Required regulation and actual available regulation of generating units: (a) Case 2, (b) Case 3, (c) Case 4, (d) Case 5.

Table 6.3 Total costs of three cases.

Cost ($)	Case 1	Case 3	Case 5
Energy	50,712.88	53,308.02	52,491.81
Reserve capacity	7315.78	3861.84	2663.79
Regulation (worst case)	9072.37	6319.55	6276.51
Total	67,101.03	63,489.40	61,432.11

of CHP units, the assumption that CHP units can provide full reserve capacity as thermal units is overoptimistic. By taking into account the real-time operation constraints of the DHS in the proposed two-stage robust optimization model (M2), the feasibility of day-ahead reserve allocation is guaranteed in M2.

6.4.4 Economic benefits of introducing DHS flexibility

To demonstrate the economic benefits of introducing additional reserve capacity provided by the DHS, Cases 1, 3, and 5 are selected. Table 6.3 lists the total system operation costs of the three cases, including the day-ahead energy generation cost, day-ahead reserve allocation cost, and real-time regulation cost. Fig. 6.5 illustrates the optimal scheduling of day-ahead heat energy and reserve capacity in Cases 1, 3, and 5.

Case 1 is the benchmark that does not consider the reserve provision from CHP units and EBs, as well as the utilization of thermal inertia of buildings. Since there is no flexibility in the DHS, CHP units cannot adjust their outputs to fulfill heat demand, and hence the upward reserves are provided by non-CHP thermal units, as shown in Fig. 6.5(b). Compared with Case 1, Table 6.3 shows that the reserve capacity cost in Case 3 decreases from $7315.78 to $3861.84. There are two reasons. First, with the introduction of EBs in the IEHS in Case 3, CHP units can accordingly adjust their outputs and then provide reserve capacity, as shown in Fig. 6.5(d). Some reserves provided by G2 are replaced by the reserves from CHP units, of which the reserve price is lower than that of thermal unit G2. Thus, the total reserve capacity cost in Case 3 decreases compared with Case 2. Second, in the two-stage robust optimization, EBs are also utilized to provide upward reserves to offset the wind power forecast error. Different from generators, the EB provides upward reserves by lowering its electricity consumption. Compared with the reserve capacity provided by CHP units and thermal units, the price of the reserve capacity provided by EBs is relatively cheaper. Therefore, EBs are scheduled in a priority as shown in Fig. 6.5(c). This leads to an increased energy cost in Case 3 due to the replacement of heat supply from CHP units

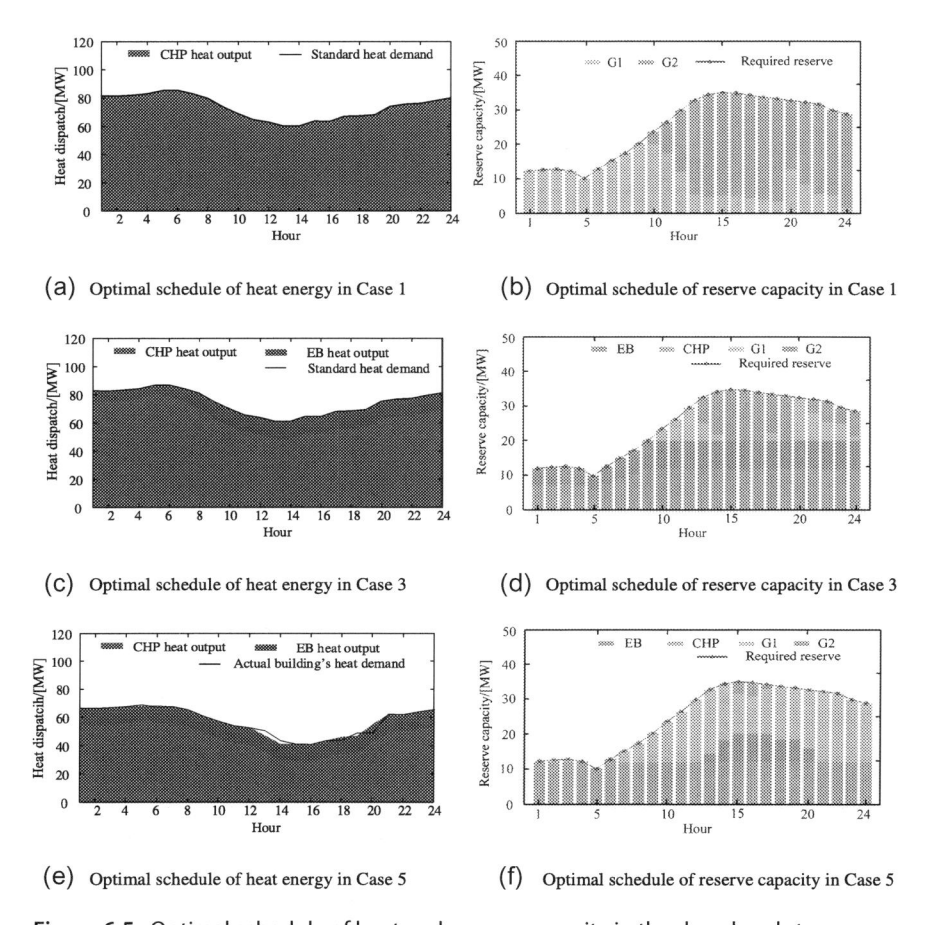

(a) Optimal schedule of heat energy in Case 1

(b) Optimal schedule of reserve capacity in Case 1

(c) Optimal schedule of heat energy in Case 3

(d) Optimal schedule of reserve capacity in Case 3

(e) Optimal schedule of heat energy in Case 5

(f) Optimal schedule of reserve capacity in Case 5

Figure 6.5 Optimal schedule of heat and reserve capacity in the day-ahead stage.

by more costly heat supply from EBs. Even so, the total system operation cost in Case 3 decreases by $3611.63 overall. All of these observations verify the effectiveness of introducing EBs in improvement of available reserve capacity and the reduction of the operation cost of the IEHS.

In addition to Case 3, Case 5 considers the building thermal inertia and thermal comfort in the optimization, where the variation of indoor temperature is acceptable. Fig. 6.5(e) shows that in Case 5, the actual heat consumption in buildings can be temporarily different from the heat supply with utilizing the thermal inertia of buildings. This indicates that the building can be considered as a heat storage and storing excess heat when the heat supply is more than heat demand with a bit of increase of the indoor temperature. Taking into account the thermal inertia and thermal comfort, the flexibility of the DHS is further improved and EBs provide more reserve capacity in Case 5, which is shown in Fig. 6.5(f). This also leads to a lower reserve capacity cost in Case 5, which decreases by $1198.05. It can be concluded that EBs can provide more reserve capacity and the total system operation cost can be further reduced by introducing the building thermal inertia.

6.5 Summary and conclusion

Utilizing the flexibility of DHSs can improve the secure and reliable operation of integrated electricity and heat systems. This chapter proposed a two-stage adaptive robust energy and reserve co-optimization scheme for the integrated electricity and heat system to cope with wind power uncertainty, which utilizes the additional reserve capacity provided by improved flexibility from the DHS. We utilized the flexibility introduced by electrical boilers and the building thermal inertia in the DHS to provide additional reserve capacity to the EPS. In addition, we incorporated the operation constraints of the EPS and DHS in the real-time stage into the optimization, which guarantees the feasibility of the optimal schedule of energy and reserve capacity of the integrated electricity and heat system. The adaptive robust model is reformulated with the KKT conditions and is solved by the C&CG algorithm. Compared with the conventional single-stage optimization, the proposed scheme can reduce both reserve capacity cost and total system operation cost while maintaining system secure operation.

References

[1] H Lund, Large-scale integration of wind power into different energy systems, Energy 30 (13) (2005) 2402–2412.
[2] X Chen, C Kang, M O'Malley, Q Xia, J Bai, C Liu, et al., Increasing the flexibility of combined heat and power for wind power integration in China: modeling and implications, IEEE Trans. Power Syst 30 (4) (2015) 1848–1857.
[3] M Leśko, W Bujalski, K Futyma, Operational optimization in district heating systems with the use of thermal energy storage, Energy 165 (2018) 902–915.
[4] X Chen, C Kang, M O'Malley, Q Xia, J Bai, C Liu, et al., Increasing the flexibility of combined heat and power for wind power integration in China: Modeling and implications, IEEE Trans. Power Syst 30 (4) (2015) 1848–1857.
[5] N Zhang, X Lu, MB McElroy, CP Nielsen, X Chen, Y Deng, et al., Reducing curtailment of wind electricity in China by employing electric boilers for heat and pumped hydro for energy storage, Appl Energy 184 (2016) 987–994.

[6] Z Li, W Wu, M Shahidehpour, J Wang, B Zhang, Combined heat and power dispatch considering pipeline energy storage of district heating network, IEEE Trans Sustain Energy 7 (1) (2016) 12–22.

[7] H Cai, C Ziras, S You, R Li, K Honoré, HW Bindner, Demand side management in urban district heating networks, Appl Energy 230 (2018) 506–518.

[8] Y Yang, K Wu, H Long, J Gao, X Yan, T Kato, et al., Integrated electricity and heating demand-side management for wind power integration in China, Energy 78 (2014) 235–246.

[9] T Sadamoto, A Chakrabortty, T Ishizaki, JI Imura, Retrofit control of wind-integrated power systems, IEEE Trans Power Syst 33 (2018) 2804–2815.

[10] D Wang, S Parkinson, W Miao, H Jia, C Crawford, N Djilali, Hierarchical market integration of responsive loads as spinning reserve, Appl Energy 104 (2013) 229–238.

[11] C Lin, W Wu, B Zhang, Y Sun, Decentralized solution for combined heat and power dispatch through benders decomposition, IEEE Trans Sustain Energy 8 (4) (2017) 1361–1372.

[12] Z Li, W Wu, J Wang, B Zhang, T Zheng, Transmission-constrained unit commitment considering combined electricity and district heating networks, IEEE Trans Sustain Energy 7 (2) (2016) 480–492.

[13] S Rinne, S Syri, The possibilities of combined heat and power production balancing large amounts of wind power in Finland, Energy 82 (2015) 1034–1046.

[14] P Sorknæs, H Lund, AN Andersen, P Ritter, Small-scale CHP as a balancing reserve for wind—the case of participation in the German secondary control reserve, Int J. Sustain Energy Plan Manage 4 (2014) 31–42.

[15] M Caramanis, E Ntakou, WW Hogan, A Chakrabortty, J Schoene, Co-optimization of power and reserves in dynamic T&D power markets with nondispatchable renewable generation and distributed energy resources, Proc IEEE 104 (2016) 807–836.

[16] G Li, R Zhang, T Jiang, H Chen, L Bai, H Cui, et al., Optimal dispatch strategy for integrated energy systems with CCHP and wind power, Appl Energy 192 (2017) 408–419.

[17] W Wei, F Liu, S Mei, Y Hou, Robust energy and reserve dispatch under variable renewable generation, IEEE Trans. Smart Grid 6 (1) (2015) 369–380.

[18] B Zhao, AJ Conejo, R Sioshansi, Unit commitment under gas-supply uncertainty and gas-price variability, IEEE Trans Power Syst 32 (2017) 2394–2405.

[19] M Zugno, AJ Conejo, A robust optimization approach to energy and reserve dispatch in electricity markets, Eur J Oper. Res 247 (2015) 659–671.

[20] Z Pan, Q Guo, H Sun, Feasible region method based integrated heat and electricity dispatch considering building thermal inertia, Appl Energy 192 (2017) 395–407.

[21] Y Zhou, W Hu, Y Min, Y Dai, Integrated power and heat dispatch considering available reserve of combined heat and power units, IEEE Trans Sustain Energy 10 (3) (2018) 1300–1310.

[22] C Liu, A Botterud, Z Zhou, P Du, Fuzzy energy and reserve co-optimization with high penetration of renewable energy, IEEE Trans Sustain Energy 8 (2017) 782–791.

[23] M Zhang, X Ai, J Fang, W Yao, W Zuo, Z Chen, A systematic approach for the joint dispatch of energy and reserve incorporating demand response, Appl Energy 230 (2018) 1279–1291.

[24] D Bertsimas, E Litvinov, XA Sun, J Zhao, T Zheng, Adaptive robust optimization for the security constrained unit commitment problem, IEEE Trans Power Syst 28 (1) (2012) 52–63.

[25] N Amjady, S Dehghan, A Attarha, AJ Conejo, Adaptive robust network-constrained AC unit commitment, IEEE Trans Power Syst 32 (2017) 672–683.

[26] H Ye, Z Li, Robust security-constrained unit commitment and dispatch with re-course cost requirement, IEEE Trans Power Syst 31 (5) (2015) 3527–3536.

[27] S Chen, Z Wei, G Sun, KW Cheung, D Wang, H Zang, Adaptive robust day-ahead dispatch for urban energy systems, IEEE Trans Ind Electron 66 (2) (2019) 1379–1390.

[28] NG Cobos, JM Arroyo, N Alguacil, J Wang, Robust energy and reserve scheduling considering bulk energy storage units and wind uncertainty, IEEE Trans Power Syst 33 (2018) 5206–5216.

[29] Y Cao, W Wei, L Wu, S Mei, M Shahidehpour, Z Li, Decentralized operation of interdependent power distribution network and district heating network: a market-driven approach, IEEE Trans Smart Grid 10 (5) (2018) 5374–5385.

[30] X Liu, J Wu, N Jenkins, A Bagdanavicius, Combined analysis of electricity and heat networks, Appl Energy 162 (2016) 1238–1250.

[31] F Brahman, M Honarmand, S Jadid, Optimal electrical and thermal energy management of a residential energy hub, integrating demand response and energy storage system, Energy Build 90 (2015) 65–75.

[32] Y Zhang, K Lin, Q Zhang, H Di, Ideal thermophysical properties for free-cooling (or heating) buildings with constant thermal physical property material, Energy Build 38 (10) (2006) 1164–1170.

[33] X Xu, Y Zhang, K Lin, H Di, R Yang, Modeling and simulation on the thermal performance of shape-stabilized phase change material floor used in passive solar buildings, Energy Build 37 (10) (2005) 1084–1091.

[34] AJ Conejo, M Carrion, JM Morales, Decision Making Under Uncertainty in Electricity Markets, International Series in Operations Research and Management Science, Springer, New York, 2010.

[35] B Zeng, L Zhao, Solving two-stage robust optimization problems using a column-and-constraint generation method, Oper Res Lett 41 (5) (2013) 457–461.

[36] S Boyd, L Vandenberghe, Convex Optimization, Cambridge University Press, Cambridge, UK, 2004.

[37] A Turk, Q Zeng, Q Wu, A Hejde, Optimal operation of integrated electrical, district heating and natural gas system in wind dominated power system, Int J Smart Grid Clean Energy 9 (2) (2018) 237–246.

[38] H Cai, C Ziras, S You, R Li, K Honoré, H.W Bindner, Demand side management in urban district heating networks, Appl Energy 230 (2018) 506–518.

[39] J Wang, H Zhong, Q Xia, C Kang, E Du, Optimal joint-dispatch of energy and reserve for CCHP-based microgrids, IET Gener. Transm. Distrib 11 (3) (2017) 785–794.

CHAPTER 7

Decentralized robust energy and reserve co-optimization for multiple integrated electricity and heating systems

7.1 Introduction

The increasing integration of uncertain and fluctuating wind power requires more reserve capacity to cope with wind power uncertainty [1]. Given the rapid development of modern energy systems, there are several approaches to increase the available reserves in electric power systems (EPSs) instead of installing new generators to provide flexibility. For instance, thanks to the intensified coupling between EPSs and district heating systems (DHSs), the DHS is able to provide additional reserve capacity to the EPS considering the joint operation of an integrated electricity and heating system (IEHS) [2]. Moreover, the energy and reserve resources can be shared in multiple interconnected IEHSs owing to the increasing interconnection among them in the modern energy system. In this regard, there are more energy and reserves available to address wind power uncertainty with the efficient coordination among multiple IEHSs.

As introduced in Chapter 6, installing electric boilers (EBs), heat accumulators, and utilizing the thermal inertia of buildings can decouple the power and heat production of combined heat and power (CHP) units. Thus, it can enhance the flexibility of CHP units, and accordingly, the DHS can provide a certain amount of reserves for the EPS to handle system uncertainties [3]. To obtain an optimal reserve allocation with feasible real-time regulation implementation, two-stage adaptive robust optimization is adopted, taking into account the real-time regulation constraints of the DHS, where the real-time adjustments of generators are carried out according to the reserve scheduled in the day–ahead operation [4].

In addition, since the interconnection of multiple IEHSs enables the interregional resources sharing [5], the overall system economic efficiency

Optimal operation of integrated multi-energy systems under uncertainty. Copyright © 2022 Elsevier Inc.
DOI: https://doi.org/10.1016/B978-0-12-824114-1.00001-9   All rights reserved.

can be improved and each IEHS can better accommodate the uncertain wind power taking into account the energy and reserve coordination among multiple IEHSs. Accordingly, it is of significant importance to investigate a cost-efficient energy and reserve scheduling scheme for multiple interconnected IEHSs that utilizes additional reserve capacity in the whole system via tie lines to minimize the overall system operation cost.

The ideal operation for multiple IEHSs is to coordinate all IEHSs with a central operator, which achieves the minimum operation cost for the whole system [6, 7]. However, there are actually different entities managing different IEHSs. It is impractical that one central operator would schedule all IEHSs together and take charge of all data of generating units, transmission networks, and demands from different IEHSs. In this regard, a decentralized optimization framework is necessary to achieve the synergistic but independent operation for all IEHSs with information privacy protected. The decentralized algorithms based on the augmented Lagrangian relaxation and the first-order optimal conditions are two typical methods in EPS decompositions [8]. The alternating direction method of multipliers (ADMM) based on the augmented Lagrangian relaxation is widely used on the decentralized operation of multiarea EPSs due to the low amount of data exchange and application simplicity. In the work of Li et al. [9], the decentralized generation unit and tie-line scheduling in a multiarea system is realized with the ADMM algorithm. He et al. [10] applied an iterative ADMM scheme to the optimal energy flow operation of multiarea electricity–natural gas systems.

To exploit the available reserves in multiple IEHSs in a more cost-effective way, this chapter investigates an efficient decentralized energy and reserve co-optimization scheme for multiple IEHSs. It not only improves the overall economic efficiency for the energy and reserve scheduling but also preserves the information privacy for each IEHS. In each IEHS, two-stage adaptive robust programming is used to cope with uncertainty and ensure the feasibility of reserve deployments. Section 7.2 introduces the structure and decentralized operation framework of multiple IEHSs first. Then, the mathematical model of decentralized robust energy and reserve co-optimization for multiple IEHS are described described in Section 7.3, including the objective function and the day-ahead and real-time operation constraints of the IEHS. The improved decentralized solution algorithm based on ADMM is presented in Section 7.4. In Section 7.5, the case studies and simulation results are given and discussed. The conclusions are drawn in Section 7.5.

7.2 Structure and decentralized operation framework of multiple IEHSs

The symbols and notations described in the following mathematical model are explained here for quick reference.

Indices and sets

$\Phi_s^{N/G/WF/ED}$	Set of internal buses/thermal units/wind farms/electric demand in IEHS s
Φ_s^{BB}	Set of boundary buses in IEHS s
Φ_s^{FB}	Set of fictitious boundary buses in IEHS s
$\Phi_s^{CHP/EB}$	Set of CHP units/electric boilers in IEHS s
$\Phi_s^{HN/HS/HD}$	Set of heat nodes/heat sources/heat demands in the DHS in IEHS s
$\Phi_s^{S/R,pipe}$	Set of supply/return pipelines in the DHS in IEHS s
Φ_n	Set of adjacent IEHSs connected to bus n
$\Omega_n^{CHP/EB}$	Set of CHP units/electric boilers connected to bus n
$\Omega_n^{G/WF/ED/N}$	Set of thermal units/wind farms/electric demands/internal buses connected to bus n
$\Omega_j^{HN/CHP/EB}$	Set of heat nodes/CHP units/electric boilers corresponding to heat source j
$\Omega_{nd}^{B/E,S/R,pipe}$	Set of supply/return pipelines beginning/ending at node nd
T	Set of hours
DA/RT	Day-ahead stage/Real-time stage

Parameters

$C_s^{DA,EN/RC}$	Cost of day-ahead energy/reserve in IEHS s
$C_s^{PE/RT}$	Penalty term/Cost of real-time regulation
$c_{h,t}^{E/H,CHP}$	Power/heat production cost rates offered by CHP units
$c_{g,t}^{G}$	Production cost rate offered by non–CHP thermal units
$c_{h,t}^{R}$, $c_{g,t}^{R}$	Cost rates offered by CHP units/non–CHP thermal units for reserve capacity
c_l^{shed}	Cost rate for load shedding
B_{nm} f_{nm}^{max}	Susceptance/maximum power of transmission line nm
$P_{w,t}^{wd}$ $P_{l,t}^{ED}$	Forecasted wind power/Electric demand
$RU_{h,t}^{CHP}$, $RD_{h,t}^{CHP}$	Ramping up/down limit of CHP units
$RU_{g,t}^{G}$, $RD_{g,t}^{G}$	Ramping up/down limit of non–CHP thermal units

$P_{h,t}^{CHP,max/min}$	Maximum/minimum generation of CHP units
$P_{g,t}^{G,max/min}$	Maximum/minimum generation of non–CHP thermal units
$P_{i,t}^{EB,max}$	Maximum power consumption of the electric boiler
$\Delta W_{w,t}^{max}$	Maximum deviation of wind power
η_h^{CHP}	Heat-to-power coefficient of CHP units
η_i^{EB}	Power-to-heat coefficient of the electric boiler
c, l_p	Specific heat capacity of water/Length of pipelines
$m^{HS/HD}$	Mass flow rate of heat sources/heat loads
$m^{S/R,pipe}$	Mass flow rate of supply/return pipelines
τ_t^a	Ambient temperature of pipelines
ρ	Penalty parameter in ADMM

Variables

$P_{h,t}^{CHP}, P_{g,t}^{G}$	Day-ahead generation of CHP/non-CHP thermal units
$P_{i,t}^{EB}$	Day-ahead electricity consumption of electric boilers
$R_{h,t}^{CHP} R_{g,t}^{G}$	Day-ahead reserve capacity scheduled for CHP/non-CHP thermal units
$r_{h,t}^{CHP,U/D}$	Upward/downward regulation of CHP units
$r_{g,t}^{G,U/D}$	Upward/downward regulation of non–CHP thermal units
$L_{l,t}^{shed}$	Load shedding
$\Delta W_{w,t}, W_{w,t}^{spill}$	Deviation of wind power/Wind spillage
$\delta_{s,n,t}$	Phase angle of buses in IEHS s
$\delta_{n,t}$	Average angle of fictitious boundary buses in adjacent IEHSs
$H_{h,t}^{CHP} H_{i,t}^{EB}$	Heat output of CHP units/electric boilers
$H_{h,t}^{CHP,RT}$	Heat output of CHP units in the real-time stage
$\tau_{nd,t}^{S/R,ND}$	Temperature of node nd in supply/return networks
$\tau_{p,t}^{S/R,in}, \tau_{p,t}^{S/R,out}$	Mass flow temperature at the inlet/outlet of a pipeline in supply/return networks

Fig. 7.1 shows the schematic structure of interconnected multiple IEHSs. Each IEHS consists of an EPS and a DHS. In each IEHS, the electric demand is supplied by conventional thermal units, CHP units, and wind power, whereas heat demand is fulfilled by CHP units and electric

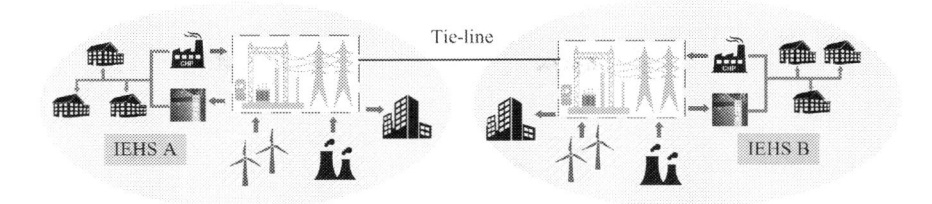

Figure 7.1 Schematic structure of two IEHSs.

boilers. In the EPS, electric power and heat are generated by different energy sources, transmitted by power and district heating networks, and finally delivered to consumers. Different from electric power that can be transmitted over long distances, heat energy is generated, transferred, and consumed locally in the DHS. Thus, in this chapter, neighboring IEHSs are only connected via electric power tie-lines, as illustrated in Fig. 7.1. Generally, the operation of all of these IEHSs is optimized by minimizing the total system operation cost, which is managed by a central system operator.

However, different IEHSs are actually cleared in different markets and there is no central operator managing the whole system. Thus, a decentralized framework is required to optimize the independent but coordinated energy and reserve scheduling for multiple interconnected IEHSs. Here, interconnected IEHSs are decoupled by relaxing power tie-lines (i.e., introducing the fictitious boundary buses of tie-lines). Fig. 7.2 represents the decomposed IEHSs with the introduction of fictitious boundary buses, which are the copies of the boundary buses in adjacent IEHSs and represented by the dotted lines in Fig. 7.2. In the decentralized operation, each IEHS operates independently and the adjacent IEHSs exchange the phase angles of their boundary buses with each other. The decentralized process is finished until the phase angles of these fictitious boundary buses are the same as those of the real boundary buses (i.e., $\delta_{A,i'}$ $\delta_{A,1}$ in IEHS A and $\delta_{B,i'}$ in IEHS B $\delta_{C,k}$ are equal to $\delta_{B,i}$ in IEHS B and $\delta_{C,1}$ $\delta_{A,i}$ $\delta_{A,j}$ in IEHS A), which are described by Eqs. (7.1) and (7.2). The whole system is decomposed into several IEHS subsystems by relaxing Eqs. (7.1) and (7.2), which is shown in Fig. 7.2.

$$\delta_{A,j} = \delta_{B,j'} \tag{7.1}$$

$$\delta_{A,i'} = \delta_{B,i} \tag{7.2}$$

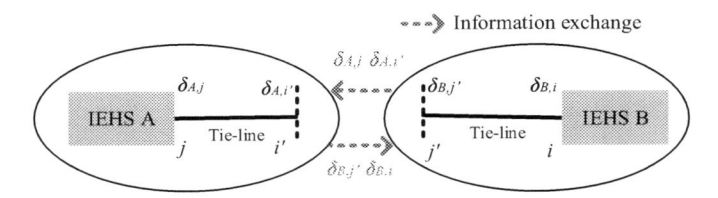

Figure 7.2 Decomposed IEHSs with relaxed boundary buses.

7.3 Mathematical formulation of decentralized robust energy and reserve co-optimization for multiple IEHSs

The energy and reserve capacity are optimized together in the joint market like the PJM market in North America [11]. This chapter aims to realize cost-effective operation along with utilizing the reserve capacity from the DHS and the reserve sharing from interconnected IEHSs. The energy and reserve scheduling for each IEHS is formulated as a two-stage adaptive robust optimization. In the first stage, the day-ahead energy and reserve capacity and exchanged tie-line power flow are determined based on wind power forecasts. In the second stage, the real-time regulation and reserve deployment are obtained with the revealed wind uncertainty. The mathematical formulation of decentralized robust energy and reserve optimization scheme is described as follows.

7.3.1 Objective function

This chapter aims to minimize the total system operation cost composed of day-ahead energy and reserve costs and real-time regulation cost for the IEHS s under the worst case. The operation problem for each IEHS is formulated as an adaptive robust optimization. Eq. (7.3) represents the objective after the decomposition with Lagrangian relaxation in this chapter, which is formulated as an augmented Lagrangian including weighted quadratic penalty terms in Eq. (7.6) after relaxing coupling constraints (7.1) and (7.2). $\lambda_{s,n,t}$ is the dual variable with regard to coupling constraints (7.1) and (7.2). ρ is a parameter. The four terms in Eq. (7.3) represent the day-ahead energy production cost, reserve cost, penalty term, and real-time regulation cost of IEHS s, which are formulated in detail in Eqs. (7.4) through (7.7). Eq. (7.8) denotes the averaged phase angle of all boundary buses connected to bus n in IEHS s, and $|\Phi_n|$ is the number of boundary buses connected to bus n.

$$\min(C_s^{\mathrm{DA,EN}} + C_s^{\mathrm{DA,RC}} + C_s^{\mathrm{PE}} + \max_{\Delta W_{w,t}} \min C_s^{\mathrm{RT}}) \tag{7.3}$$

$$C_s^{\mathrm{DA,EN}} = \sum_{t\in T}\left\{ \sum_{h\in\Phi_s^{\mathrm{CHP}}} (c_{h,t}^{\mathrm{E,CHP}} P_{h,t}^{\mathrm{CHP}} + c_{h,t}^{\mathrm{H,CHP}} H_{h,t}^{\mathrm{CHP}}) + \sum_{g\in\Phi_s^G} c_{g,t}^G P_{g,t}^G \right\} \tag{7.4}$$

$$C_s^{\mathrm{DA,RC}} = \sum_{t\in T}\left(\sum_{h\in\Phi_s^{\mathrm{CHP}}} c_{h,t}^R R_{h,t}^{\mathrm{CHP}} + \sum_{g\in\Phi_s^G} c_{g,t}^R R_{g,t}^G \right) \tag{7.5}$$

$$C_s^{\mathrm{PE}} = \sum_{t\in T}\sum_{n\in\Phi_s^{\mathrm{BB}}\cup\Phi_s^{\mathrm{FB}}}\left\{ \lambda_{s,n,t}(\delta_{s,n,t} - \overline{\delta}_{n,t}) + 0.5\rho \left\| \delta_{s,n,t} - \overline{\delta}_{n,t} \right\|_2^2 \right\} \tag{7.6}$$

$$C_s^{\mathrm{RT}} = \sum_{t\in T}\left\{ \sum_{h\in\Phi_s^{\mathrm{CHP}}} c_h^{\mathrm{E,CHP}} (r_{h,t}^{\mathrm{CHP,U}} - r_{h,t}^{\mathrm{CHP,D}}) + \sum_{g\in\Phi_s^G} c_g^G (r_{g,t}^{\mathrm{G,U}} - r_{g,t}^{\mathrm{G,D}}) \right.$$

$$\left. + \sum_{h\in\Phi_s^{\mathrm{CHP}}} c_h^{\mathrm{H,CHP}} (H_{h,t}^{\mathrm{CHP,RT}} - H_{h,t}^{\mathrm{CHP}}) + \sum_{l\in\Phi_s^{EL}} c_l^{\mathrm{shed}} L_{l,t}^{\mathrm{shed}} \right\} \tag{7.7}$$

$$\overline{\delta}_{n,t} = \left(\sum_{s\in\Phi_n} \delta_{s,n,t} \right)\Big/ |\Phi_n| \tag{7.8}$$

7.3.2 Day-ahead operation constraints of IEHSs

The energy and reserves in IEHSs are optimized with wind power forecasts in the day-ahead operation of the IEHS. In addition to the day-ahead operation constraints of the EPS and DHS, the transmission capacity limits of tie-lines are taken into account as well. The operation models of the EPS and DHS are built based on the DC power flow and steady-state model with constant mass flow and variable temperature.

(1) *Day-ahead operation constraints of the EPS.* In the EPS, the electric loads are supplied by CHP units, conventional thermal units, and wind power. The operation constraints regarding generators and power transmissions are formulated as follows.

$$\sum_{h\in\Omega_n^{\mathrm{CHP}}} P_{h,t}^{\mathrm{CHP}} + \sum_{g\in\Omega_n^G} P_{g,t}^G + \sum_{w\in\Omega_n^{\mathrm{WF}}} P_{w,t}^{\mathrm{wd}} - \sum_{i\in\Omega_n^{\mathrm{EB}}} P_{i,t}^{\mathrm{EB}} - \sum_{l\in\Omega_n^{\mathrm{ED}}} P_{l,t}^{\mathrm{load}}$$

$$= \sum_{m\in\Omega_n^N} B_{nm}(\delta_{s,n,t} - \delta_{s,m,t}), \quad \forall n \in \Phi_s^N, \, \forall t \in T \tag{7.9}$$

$$-f_{nm}^{\max} \leq B_{nm}(\delta_{s,n,t} - \delta_{s,m,t}) \leq f_{nm}^{\max}, \ \forall n, m \in \Phi_s^{N} \cup \Phi_s^{FB}, \ \forall t \in T \tag{7.10}$$

$$P_{h,t}^{CHP} + R_{h,t}^{CHP} \leq P_{h,t}^{CHP,\max}, \ \forall h \in \Phi_s^{CHP}, \forall t \in T \tag{7.11}$$

$$P_{h,t}^{CHP,\min} \leq P_{h,t}^{CHP} - R_{h,t}^{CHP}, \ \forall h \in \Phi_s^{CHP}, \forall t \in T \tag{7.12}$$

$$P_{g,t}^{G} + R_{g,t}^{G} \leq P_{g,t}^{G,\max}, \ \forall g \in \Phi_s^{G}, \forall t \in T \tag{7.13}$$

$$P_{g,t}^{G,\min} \leq P_{g,t}^{G} - R_{g,t}^{G}, \ \forall g \in \Phi_s^{G}, \forall t \in T \tag{7.14}$$

$$0 \leq R_{h,t}^{CHP} \leq \min(RU_h^{CHP}, RD_h^{CHP}), \ \forall h \in \Phi_s^{CHP}, \forall t \in T \tag{7.15}$$

$$0 \leq R_{g,t}^{G} \leq \min(RU_g^{G}, RD_g^{G}), \ \forall g \in \Phi_s^{G}, \forall t \in T \tag{7.16}$$

$$-RD_h^{CHP}\Delta t \leq P_{h,t}^{CHP} - P_{h,t-1}^{CHP} \leq RU_h^{CHP}\Delta t, \ \forall h \in \Phi_s^{CHP}, \forall t \in T \tag{7.17}$$

$$-RD_g^{G}\Delta t \leq P_{g,t}^{G} - P_{g,t-1}^{G} \leq RU_g^{G}\Delta t, \ \forall g \in \Phi_s^{G}, \forall t \in T \tag{7.18}$$

Eq. (7.9) is the power balance constraint denoting that electric demands should be kept the same as the total power production at each hour at each bus. Eq. (7.10) denotes that the power flow in each transmission line should be lower than the corresponding line capacity. Eqs. (7.11) through (7.14) show the capacity constraints for the coupling of power production and reserve capacity provision of CHP units and non–CHP thermal units, respectively. Eqs. (7.15) and (7.16) limit the available reserves from CHP units and non–CHP thermal units. Eqs. (7.17) and (7.18) enforce that the power output changes of CHP units and non–CHP thermal units between two consecutive hours cannot be larger than their ramping rates.

(2) *Day-ahead operation constraints of the DHS.* As can be seen from Fig. 7.1, the DHS consists of heat sources, the district heating network, and end users. The district heating network consists of two types of pipelines to transmit hot water: supply pipelines delivering heat to demand nodes and return pipelines carrying mass flow back to heat source nodes. CHP units and EBs are considered as heat sources linking the EPS and DHS in this chapter. Among various technologies of CHP units, the back–pressure CHP unit with a linear relationship between electric power and heat production is adopted. Since the DHS operates at the constant mass flow mode, the hydraulic model regarding mass flows is not formulated here.

$$H_{h,t}^{\text{CHP}} = \eta_h^{\text{CHP}} P_{h,t}^{\text{CHP}}, \quad \forall h \in \Phi_s^{\text{CHP}}, \forall t \in T \tag{7.19}$$

$$H_{i,t}^{\text{EB}} = \eta_i^{\text{EB}} P_{i,t}^{\text{EB}}, \quad \forall i \in \Phi_s^{\text{EB}}, \forall t \in T \tag{7.20}$$

$$0 \le H_{i,t}^{\text{EB}} \le H_i^{\text{EB,max}}, \forall i \in \Phi_s^{\text{EB}}, \forall t \in T \tag{7.21}$$

$$\sum_{h \in \Omega_j^{\text{CHP}}} H_{h,t}^{\text{CHP}} + \sum_{i \in \Omega_j^{\text{EB}}} H_{i,t}^{\text{EB}} = c \cdot m_{j,t}^{\text{HS}} \cdot \left(\tau_{nd,t}^{\text{S,ND}} - \tau_{nd,t}^{\text{R,ND}} \right), \quad \forall j \in \Phi_s^{\text{HS}},$$
$$nd = \Omega_j^{\text{HN}}, \forall t \in T \tag{7.22}$$

$$\tau_{p,t}^{\text{S,out}} - \tau_t^{\text{a}} = \left(\tau_{p,t}^{\text{S,in}} - \tau_t^{\text{a}} \right) e^{-k_p \pi d_p l_p / c \cdot m_{p,t}^{\text{S,pipe}}}, \quad \forall p \in \Phi_s^{\text{S,pipe}}, \forall t \in T \tag{7.23}$$

$$\tau_{p,t}^{\text{R,out}} - \tau_t^{\text{a}} = \left(\tau_{p,t}^{\text{R,in}} - \tau_t^{\text{a}} \right) e^{-k_p \pi d_p l_p / c \cdot m_{p,t}^{\text{R,pipe}}}, \quad \forall p \in \Phi_s^{\text{R,pipe}}, \forall t \in T \tag{7.24}$$

$$\sum_{p \in \Omega_{nd}^{\text{E,S,pipe}}} \left(m_{p,t}^{\text{S,pipe}} \cdot \tau_{p,t}^{\text{S,out}} \right) = \tau_{nd,t}^{\text{S,ND}} \sum_{p \in \Omega_{nd}^{\text{E,S,pipe}}} m_{p,t}^{\text{S,pipe}}, \quad \forall nd \in \Phi_s^{\text{HN}}, \forall t \in T \tag{7.25}$$

$$\sum_{p \in \Omega_{nd}^{\text{E,R,pipe}}} \left(m_{p,t}^{\text{R,pipe}} \cdot \tau_{p,t}^{\text{R,out}} \right) = \tau_{nd,t}^{\text{R,ND}} \sum_{p \in \Omega_{nd}^{\text{E,R,pipe}}} m_{p,t}^{\text{R,pipe}}, \quad \forall nd \in \Phi_s^{\text{HN}}, \forall t \in T \tag{7.26}$$

$$\tau_{p,t}^{\text{S,in}} = \tau_{nd,t}^{\text{S,ND}}, \forall nd \in \Phi_s^{\text{HN}}, \forall p \in \Phi_s^{\text{B,S,pipe}}, \forall t \in T \tag{7.27}$$

$$\tau_{p,t}^{\text{R,in}} = \tau_{nd,t}^{\text{R,ND}}, \forall nd \in \Phi_s^{\text{HN}}, \forall p \in \Phi_s^{\text{B,R,pipe}}, \forall t \in T \tag{7.28}$$

$$H_{l,t}^{\text{HD}} = c \cdot m_{l,t}^{\text{HD}} \cdot \left(\tau_{nd,t}^{\text{S,ND}} - \tau_{nd,t}^{\text{R,ND}} \right), \quad \forall l \in \Phi_s^{\text{HD}}, nd = \Omega_l^{\text{HN}}, \forall t \in T \tag{7.29}$$

Eq. (7.19) denotes the coupling relationship of back-pressure CHP units. EBs are another heat source in the DHS. Eq. (7.20) calculates the heat energy generation from EBs by consuming electricity. Eq. (7.21) enforces the operation limits of EBs. Eqs. (7.22) represents that the generated heat is used to heat water or steams and thus is related to the temperature difference between supply and return pipelines. Given that there are heat losses along pipelines during the heat delivery, Eqs. (7.23) and (7.24) describe the temperature drop in the supply and return pipelines. Eqs. (7.25) and (7.26) are the temperature mixture at each node according to the energy conservation law, which indicate that the total heat energy in outlet pipelines is equal to that in inlet pipelines. Eqs. (7.27) and (7.28) mean that the temperature of pipelines flowing out of one node is equal to the mixed temperature at this node. Eq. (7.29) calculates the heat consumption of heat demands, which determine the temperature of return pipelines.

To simplify the expressions in the real-time operation, part of the inequalities and equalities in the day-ahead operation are rewritten as compact forms, which are similar to the corresponding constraints in the second stage. Eqs. (7.10), (7.17) and (7.18), and (7.21) are formulated as inequality (7.30). Equalities (7.9), (7.19) and (7.20), and (7.22) through (7.29) are transformed into constraint (7.31).

$$h_s^{\mathrm{DA}}(P_{h,t}^{\mathrm{CHP}}, P_{g,t}^{\mathrm{G}}, P_{i,t}^{\mathrm{EB}}, \delta_{s,n,t}) \leq 0 \tag{7.30}$$

$$g_s^{\mathrm{DA}}(H_{h,t}^{\mathrm{CHP}}, H_{i,t}^{\mathrm{EB}}, \tau_{nd,t}^{\mathrm{S/R,ND}}, \tau_{p,t}^{\mathrm{S/R,in}}, \tau_{p,t}^{\mathrm{S/R,out}}) = 0 \tag{7.31}$$

7.3.3 Real-time regulation constraints of the EPS and DHS

The real-time regulation is uncertain due to the uncertainty of wind power generation. Thus, it is considered as the second stage in the two-stage robust optimization to accommodate the wind power forecast error. In the robust optimization, wind power uncertainty is modeled as an uncertainty set (7.32). More specifically, it is defined as a polyhedron consisting of deviation intervals of different wind farms. Based on the uncertainty set, the middle level in the robust optimization is to determine the worst case among all possible realizations of wind power generation.

In this second stage, the outputs of thermal units, CHP units, and EBs in the IEHS are adjusted adaptively according to the uncertainty set day-ahead reserve scheduling from the first stage. The corresponding operation constraints are formulated as follows.

$$\mathbf{U}_s = \{\Delta W | -\Delta W_{w,t}^{\max} \leq \Delta W_{w,t} \leq \Delta W_{w,t}^{\max}, \forall w \in \Phi_s^{\mathrm{WF}}, \forall t\} \tag{7.32}$$

$$0 \leq r_{h,t}^{\mathrm{CHP,U}} \leq R_{h,t}^{\mathrm{CHP}}, \quad \forall h \in \Phi_s^{\mathrm{CHP}}, \forall t \in T \tag{7.33}$$

$$0 \leq r_{h,t}^{\mathrm{CHP,D}} \leq R_{h,t}^{\mathrm{CHP}}, \quad \forall h \in \Phi_s^{\mathrm{CHP}}, \forall t \in T \tag{7.34}$$

$$0 \leq r_{g,t}^{\mathrm{G,U}} \leq R_{g,t}^{\mathrm{G}}, \quad \forall g \in \Phi_s^{\mathrm{G}}, \forall t \in T \tag{7.35}$$

$$0 \leq r_{g,t}^{G,D} \leq R_{g,t}^{G}, \quad \forall g \in \Phi_s^G, \forall t \in T \tag{7.36}$$

$$\sum_{h\in\Omega_n^{CHP}} (r_{h,t}^{CHP,U} - r_{h,t}^{CHP,D}) + \sum_{g\in\Omega_n^G} (r_{g,t}^{G,U} - r_{g,t}^{G,D}) + \sum_{w\in\Omega_n^{WF}} (\Delta W_{w,t} - W_{w,t}^{spill})$$

$$- \sum_{i\in\Omega_n^{EB}} (P_{i,t}^{EB,RT} - P_{i,t}^{EB}) + \sum_{l\in\Omega_n^{ED}} L_{l,t}^{shed}$$

$$= \sum_{m\in\Omega_n^{N}} B_{nm}(\delta_{s,n,t}^{RT} - \delta_{s,m,t}^{RT} - \delta_{s,n,t} + \delta_{s,m,t}), \quad \forall n \in \Phi_s^N, \forall t \in T$$

$$\tag{7.37}$$

$$0 \leq W_{w,t}^{spill} \leq P_{w,t}^{wd} + \Delta W_{w,t}, \quad \forall w \in \Phi_s^{WF}, \forall t \in T \tag{7.38}$$

$$0 \leq L_{l,t}^{shed} \leq P_{l,t}^{ED}, \quad \forall l \in \Phi_s^{ED}, \forall t \in T \tag{7.39}$$

$$h_s^{RT}(P_{h,t}^{CHP,RT}, P_{g,t}^{G,RT}, P_{i,t}^{EB,RT}, \delta_{s,n,t}) \leq 0 \tag{7.40}$$

$$g_s^{RT}(H_{h,t}^{CHP,RT}, H_{i,t}^{EB,RT}, \tau_{nd,t}^{S/R,ND,RT}, \tau_{p,t}^{S/R,in,RT}, \tau_{p,t}^{S/R,out,RT}) = 0 \tag{7.41}$$

Eqs. (7.33) through (7.36) show that the regulation of generators in real–time operation should be lower than scheduled reserve capacity in the day–ahead stage. Eq. (7.37) ensures the power rebalance in the real–time operation. Eq. (7.38) enforces that the wind spillage in real–time operation should be lower than wind power generation. Eq. (7.39) denotes the limit of load shedding in the real–time regulation. Eqs. (7.40) and (7.41) are the compact form of the operation constraints of the EPS and DHS under uncertainty in the second stage, which are similar to corresponding day-ahead operation constraints (7.30) and (7.31).

7.4 Solution methodology

To make the solution process concise and clear, the mathematical formulation of the robust energy and reserve optimization for each IEHS described in Section 7.3 is rewritten in a compact form. Then, the solution procedures of a traditional ADMM and an improved ADMM algorithm with locally adaptive penalty parameters are introduced in Sections 7.4.2 and 7.4.3, which realizes the independent yet coordinated operation of multiple IEHSs with limited exchanged information. Finally, the tri-level robust optimization for each IEHS addressing the wind uncertainty is solved by an iterative column–and–constraint generation (C&CG) algorithm.

7.4.1 Compact formulation of decentralized two-stage robust optimization

Incorporating the system operating variables (i.e., variables reflecting the day-ahead operation, phase angles of boundary buses, system uncertainty, and real-time regulation) into different vectors, the compact matrix model of the two-stage robust optimization for IEHS s is described as follows:

$$\min_{\mathbf{x}_s, \boldsymbol{\delta}_s}(\mathbf{c}_s^{\mathrm{T}}\mathbf{x}_s + \boldsymbol{\lambda}_s^{\mathrm{T}}(\boldsymbol{\delta}_s - \overline{\boldsymbol{\delta}}) + 0.5\rho\left\|\boldsymbol{\delta}_s - \overline{\boldsymbol{\delta}}\right\|_2^2 + \max_{\mathbf{u}_s \in \mathbf{U}_s} \min_{\mathbf{y}_s \in F(\mathbf{x}_s, \mathbf{u}_s)} \mathbf{d}_s^{\mathrm{T}}\mathbf{y}_s) \quad (7.42)$$

$$s.t. \quad \mathbf{A}_s(\mathbf{x}_s + \boldsymbol{\delta}_s) \leq \mathbf{b}_s, \ \mathbf{B}_s(\mathbf{x}_s + \boldsymbol{\delta}_s) = \mathbf{a}_s, \ \mathbf{x}_s \in \mathbb{R}_+^n \quad (7.43)$$

$$F(\mathbf{x}_s, \mathbf{u}_s) = \left\{\mathbf{y}_s | \mathbf{C}_s\mathbf{y}_s + \mathbf{D}_s\mathbf{x}_s \leq \mathbf{g}_s \right. \quad (7.44)$$

$$\mathbf{G}_s\mathbf{y}_s + \mathbf{M}_s\mathbf{u}_s = \mathbf{h}_s \quad (7.45)$$

$$\mathbf{y}_s \in \mathbb{R}_+^n \left.\right\} \quad (7.46)$$

where $\mathbf{x}_s$ is the variable vector of all first-stage decision variables $\left\{P_{h,t}^{\mathrm{CHP}}, P_{g,t}^{\mathrm{G}}, P_{i,t}^{\mathrm{EB}}, R_{h,t}^{\mathrm{CHP}}, R_{g,t}^{\mathrm{G}}, H_{h,t}^{\mathrm{CHP}}, H_{i,t}^{\mathrm{EB}}, \tau_{nd,t}^{\mathrm{S/R,ND}}, \tau_{p,t}^{\mathrm{S/R,in}}, \tau_{p,t}^{\mathrm{S/R,out}}\right\}$ reflecting the day-ahead energy and reserve operation states, $\boldsymbol{\delta}_s$ is the vector of boundary buses' phase angles $\{\delta_{s,n,t}\}$, $\mathbf{y}_s$ is the decision vector of all second-stage decision variables $\left\{r_{h,t}^{\mathrm{CHP,U}}, r_{h,t}^{\mathrm{CHP,D}}, r_{g,t}^{\mathrm{G,U}}, r_{g,t}^{\mathrm{G,D}}, P_{i,t}^{\mathrm{EB,RT}}, W_{w,t}^{\mathrm{spill}}, L_{l,t}^{\mathrm{shed}}, H_{h,t}^{\mathrm{CHP,RT}}, H_{i,t}^{\mathrm{EB,RT}}, \tau_{nd,t}^{\mathrm{S/R,ND,RT}}, \tau_{p,t}^{\mathrm{S/R,in,RT}}, \tau_{p,t}^{\mathrm{S/R,out,RT}}\right\}$ representing the real-time regulated states, and $\mathbf{u}_s$ is the uncertain variables $\left\{\Delta W_{w,t}\right\}$. $F(\mathbf{x}_s, \mathbf{u}_s)$ is the feasible regulation set in the real-time operation. Eq. (7.43) indicates all day-ahead operation constraints (7.9) through (7.29). Eqs. (7.44) and (7.45) include real-time operation constraints (7.33) through (7.39). Eq. (7.46) denotes that $\mathbf{y}_s$ is positive.

7.4.2 Standard ADMM decentralized algorithm

The relaxed IEHS operation model described in Section 7.3 is still indecomposable due to the dual variable and penalty terms in Eq. (7.6). To achieve the independent but synergetic operation among multiple IEHSs, the ADMM algorithm is used in this chapter to implement the decentralized energy and reserve optimization for multiple IEHSs. First, the dual variables and the exchanged phase angles of boundary buses belonging to other IEHSs are converted to parameters with a given value. The decomposed operation subproblems (SPs) (7.47) and (7.48) of IEHSs are optimized separately in iteration i and are shown next.

$$\min_{\mathbf{x}_s^{(i)}, \boldsymbol{\delta}_s^{(i)}} \left\{ \mathbf{c}_s^{\mathrm{T}} \mathbf{x}_s^{(i)} + \left(\boldsymbol{\lambda}_s^{(i)}\right)^{\mathrm{T}} \left(\boldsymbol{\delta}_s^{(i)} - \overline{\boldsymbol{\delta}}_s^{(i-1)}\right) + 0.5\rho \left\| \boldsymbol{\delta}_s^{(i)} - \overline{\boldsymbol{\delta}}_s^{(i-1)} \right\|_2^2 + \max_{\mathbf{u}_s \in \mathbf{U}_s} \min_{\mathbf{y}_s^{(i)} \in F_s(\mathbf{x}_s^{(i)}, \mathbf{u}_s)} \mathbf{d}_s^{\mathrm{T}} \mathbf{y}_s^{(i)} \right\}$$

(7.47)

$s.t.$ Eqs. (7.43) through (7.46) $\hspace{6cm}$ (7.48)

Then, the dual variables in each SP are updated according to Eq. (7.49) after obtaining the newest optimal results of exchanged phase angles of boundary buses $\boldsymbol{\delta}_s^{(i)}$ in this iteration. These updated values are adopted in the next iteration.

$$\boldsymbol{\lambda}_s^{(i)} = \boldsymbol{\lambda}_s^{(i-1)} + \rho\left(\boldsymbol{\delta}_s^{(i)} - \overline{\boldsymbol{\delta}}_s^{(i)}\right)$$

(7.49)

Finally, the whole iteration process is finished and the decentralized operation is realized until an agreement among interconnected IEHSs is achieved (i.e., the maximum value of the primal residual $\Delta\boldsymbol{\delta}_s^{\mathrm{prim},(i)}$ and the dual residual $\Delta\boldsymbol{\delta}_s^{\mathrm{dual},(i)}$ is less than the convergence tolerance [), which is expressed as follows.

$$\Delta\boldsymbol{\delta}_s^{\mathrm{prim},(i)} = \left\| \boldsymbol{\delta}_s^{(i)} - \overline{\boldsymbol{\delta}}_s^{(i)} \right\|_2, \quad \Delta\boldsymbol{\delta}_s^{\mathrm{dual},(i)} = \left\| \overline{\boldsymbol{\delta}}_s^{(i)} - \overline{\boldsymbol{\delta}}_s^{(i-1)} \right\|_2$$

(7.50)

$$\max\left\{ \Delta\boldsymbol{\delta}_s^{\mathrm{prim},(i)}, \Delta\boldsymbol{\delta}_s^{\mathrm{dual},(i)} \right\} \leq \varepsilon$$

(7.51)

The solution procedure of the standard ADMM is described as follows. The detailed introduction of ADMM can be found in the work of Boyd et al. [12].

Algorithm: Standard ADMM with fixed penalty parameter

1: **Input:** the convergence tolerance [and the maximum iteration number I of standard ADMM.

2: Initialize the exchange variables $\boldsymbol{\delta}_s^{()}$, the Lagrangian multipliers $\boldsymbol{\lambda}_s^{()}$, and the penalty parameter $\rho^{()}$.

3: Initialize the iteration count $i = 0$.

4: **Repeat:** $i = i + 1$
   With $\boldsymbol{\lambda}_s^{(i-1))}$ and $\overline{\boldsymbol{\delta}}_s^{(i-1)}$ obtained from the previous iteration, each IEHS solves its own problem (i.e., Eqs. (7.47) and (7.48)) and obtains the optimal day-ahead schedule solutions $\mathbf{x}_s^{(i)}$ and $\boldsymbol{\delta}_s^{(i)}$ by using the C&CG algorithm presented in Section 7.4.4.
   Update the Lagrangian multipliers $\boldsymbol{\lambda}_s^{(i)}$ using Eq. (7.49).

5: **Until** $i > I$ or the maximum residue is smaller than [(i.e., Eq. (7.51)).

6: **Output** $\mathbf{x}_s = \mathbf{x}_s^{(i)}, \boldsymbol{\delta}_s = \boldsymbol{\delta}_s^{(i)}$.

7.4.3 Improved locally adaptive ADMM decentralized algorithm

Since the value of penalty parameter ρ has a significant impact on the convergence performance of the ADMM-based decentralized optimization, the convergence of the standard ADMM with a fixed penalty parameter is slow [12]. To accelerate the convergence rate, some variants of ADMM are investigated focusing on adaptive penalty parameters. For instance, He et al. [13] updated the penalty parameters by taking into account the balance between the magnitudes of the primal and dual residuals; however, its convergence rate is not guaranteed. Xu et al. [14] demonstrated that an improved ADMM (LA-ADMM) adaptive to the local sharpness property of the objective function has lower iteration complexity, which is adopted in this chapter to reduce the iteration and computation time.

In LA-ADMM, the penalty parameters are updated. There are multiple stages in the whole iterative process in LA-ADMM. The penalty parameters are updated adaptively in different stages. However, at each stage j, the standard ADMM is called and performed with a fixed penalty parameter ρ^j and is finished within a constant number of iterations K. In each iteration, the optimization model is shown as follows.

$$
\min_{\mathbf{x}_s^{(k)}, \boldsymbol{\delta}_s^{(k)}} \left\{ \mathbf{c}_s^{\mathrm{T}} \mathbf{x}_s^{(k)} + \left(\boldsymbol{\lambda}_s^{(k)}\right)^{\mathrm{T}} \left(\boldsymbol{\delta}_s^{(k)} - \overline{\boldsymbol{\delta}}^{(k-1)}\right) + 0.5\rho^j \left\| \boldsymbol{\delta}_s^{(k)} - \overline{\boldsymbol{\delta}}^{(k-1)} \right\|_2^2 + \max_{\mathbf{u}_s \in \mathbf{U}_s} \min_{\mathbf{y}_s^{(k)} \in F_s(\mathbf{x}_s^{(k)}, \mathbf{u}_s)} \mathbf{d}_s^T \mathbf{y}_s^{(k)} \right\}
$$

$$(7.52)$$

$s.t.$ Eqs. (7.43) through (7.46) $\hspace{4cm}$ (7.53)

Similar to Eq. (7.49), the Lagrangian multipliers are updated to the current stage as follows.

$$
\boldsymbol{\lambda}_s^{(k)} = \boldsymbol{\lambda}_s^{(k-1)} + \rho^j \left(\boldsymbol{\delta}_s^{(k)} - \overline{\boldsymbol{\delta}}^{(k)}\right) \tag{7.54}
$$

Once each stage with the standard ADMM is finished within a constant number of iterations K, the penalty parameter ρ is updated by multiplying a constant factor larger than 1 and is used in the next stage as a parameter. In LA-ADMM, the penalty parameter is allowed to increase to infinity. This process is terminated when Eq. (7.51) is satisfied. The solution procedure of LA-ADMM is described as follows.

Algorithm: LA-ADMM with adaptive penalty parameter

1: Input: the convergence tolerance ε, the maximum stage number J in LA-ADMM, and the maximum iteration number K of standard ADMM in each stage.

2: Initialize the exchange variables $\boldsymbol{\delta}_s^{(0)}$, the Lagrangian multipliers $\boldsymbol{\lambda}_s^{(0)}$, and the penalty parameter ρ^0.

3: **Repeat:** $j = j + 1$

4: Call **Steps 3 to 4** in the standard ADMM algorithm.
Until $k > K$ or the maximum residue is smaller than ε (7.51).
Output $\mathbf{x}_s^j = \mathbf{x}_s^{(k)}$, $\boldsymbol{\delta}_s^j = \boldsymbol{\delta}_s^{(k)}$.

5: Update the penalty parameter with a constant factor α: $\rho^{j+1} = \alpha\rho^j$.

6: **Until** the maximum residue is smaller than ε or $j > J$.

7: **Output** the solution $\mathbf{x}_s^j, \boldsymbol{\delta}_s^j$.

7.4.4 C&CG algorithm for each IEHS

The two-stage problem (Eqs. (7.52) through (7.53)) for IEHS s in each iteration in the decentralized structure is an adaptive robust optimization, and it is solved by the C&CG algorithm [15]. First, the original tri-level min–max–min problem is decomposed into a master problem (MP) with the first level and an SP with the second and third levels, which are solved iteratively in the C&CG algorithm. In each iteration, the MP minimizes the total system operation cost with the added cutting planes from the SP and the added constraints corresponding to the scenario identified in the SP. Based on the obtained values of the variables from the MP, the max–min SP optimizes the real-time regulation cost under the worst-case realization of wind power uncertainty. To make the SP computationally tractable, the max–min SP is reformulated as a single-level problem using Karush–Kuhn–Tucker conditions given that the inner min problem is linear. The iteration process is terminated when the difference between the objective function value of the MP and SP is smaller than the convergence tolerance.

7.5 Simulation results

The topology of the test system is shown in Fig. 7.3, which consists of two IEHSs connected by power tie-lines. In each IEHS, CHP units and EBs are the coupling components connecting a modified six-bus EPS and a four-node DHS with supply and return pipelines. IEHS 1 consists of one CHP unit, one EB, two thermal power units, one wind farm, and two

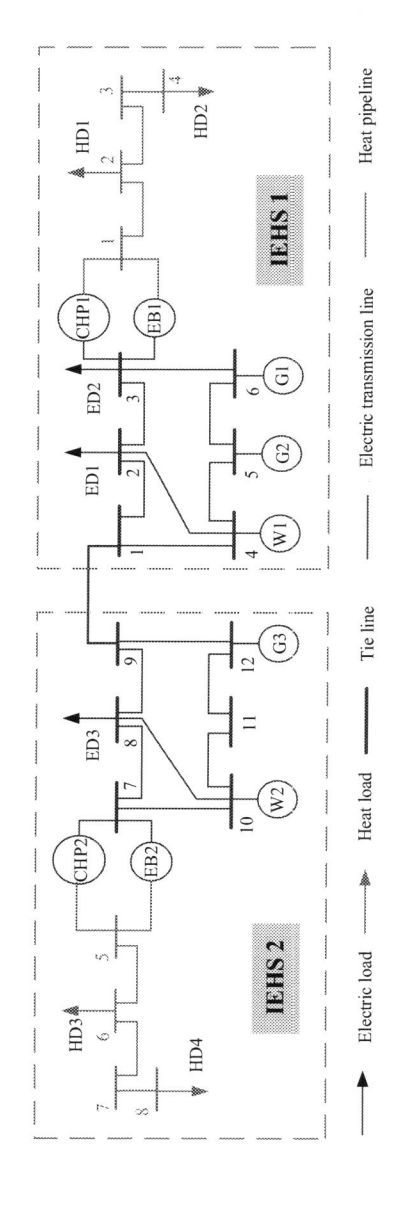

Figure 7.3 Test system with two IEHSs.

Table 7.1 Parameters of generators and EBs.

Energy Generator	CHP 1/2	G1	G2	G3	EB 1/2
Capacity (MW)	200	100	100	200	12
Ramp rate (MW/h)	50	30	30	50	/
Energy price ($/MWh)	16	15	12	20	/
Reserve price ($/MW)	10	8	15	12	/
Efficiency	1.5	/	/	/	1

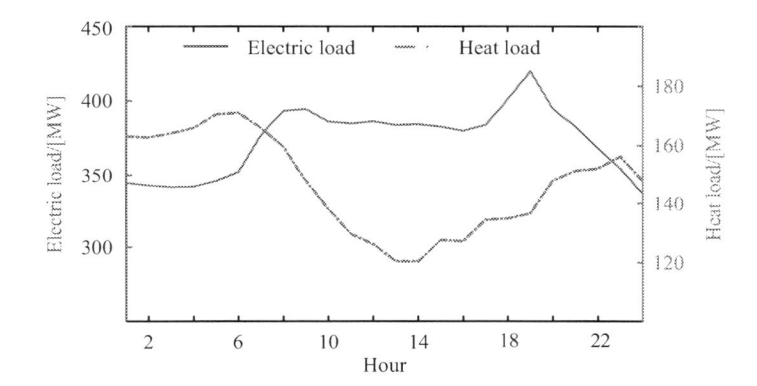

Figure 7.4 Hourly electric load and heat load in the whole system.

electric demands, whereas IEHS 2 consists of one CHP unit, one EB, one thermal power unit, one wind farm, and one electric demand. Table 7.1 lists the parameters of generators and EBs. The cost of load shedding is set as 200 $/MWh. The DHS operates in constant mass flow and variable temperature mode. The output temperature in the supply pipelines of heat sources is set at 80°C. The temperature of supply pipelines is between 60 and 80°C, and the temperature in the return pipelines depends on the heat demands. Fig. 7.4 shows the hourly electric and heat load profiles in the whole system. In the improved LA-ADMM decentralized method, the initial value of penalty parameter ρ^0 and the constant factor to update ρ are chosen as 0.5 and 3, respectively. The convergence tolerance of LA-ADMM is set as 0.001. The case studies were implemented in the general algebraic modeling system using CPLEX as the QCP solver.

7.5.1 Validation of the proposed two-stage robust optimization

To demonstrate the effectiveness of the proposed two-stage adaptive robust optimization scheme, a conventional single-stage optimization scheme

Table 7.2 Study cases.

Case	CHP as Reserve	EB	Optimized Method
1	√	√	M1
2	√	√	M2

Table 7.3 Total system costs of different study cases.

Cost ($)	Case 1	Case 2
Energy	86,183.40	91,153.89
Reserve capacity	11,007.54	6399.77
Regulation (worst case)	152,372.90	12,895.41
Total	249,563.80	110,449.10

without taking into account the regulated operation of DHSs in the second stage is used for comparison. These two comparative cases are described in Table 7.2. In the two cases, CHP units are considered to provide reserve capacity to the EPS with the introduction of EBs, which enables CHP units to adjust their outputs and decouple the heat output of CHP units and heat demand temporally. The operation is conducted in a centralized manner under both of the two cases:

(1) *M1*: Conventional single-stage optimization with the predefined system reserve requirement [16]. The real-time regulation is optimized by taking the optimized day-ahead energy and reserve scheduling as inputs.

(2) *M2*: The proposed two-stage robust optimization model. The day-ahead energy and reserve are optimized by taking into account the operation constraints of EPSs and DHSs in the real-time stage.

Table 7.3 lists the total system operation costs under the two cases. Fig. 7.5 illustrates the optimized day-ahead heat production and reserve allocation, and the reserve deployment in the real-time regulation in Cases 1 and 2. Since there is no cost of curtailed wind power, there are no downward reserves allocated.

Case 1. In this case, M1 is adopted to optimize the day-ahead energy and reserve scheduling. Since the heat demand is not considered to limit the regulation of CHP units when they provide reserves, CHP units are assumed to be able to provide full reserve capacity as thermal units in this case. The optimal heat production and reserve allocation, and the reserve deployment, are presented in Fig. 7.5(a). The "Heat scheduling" subfigure in Fig. 7.5(a) shows that since the cost-efficiency of CHP units is higher than that of EBs, CHP units are scheduled first to fulfill heat demand in this case and thus EBs are not scheduled in the day-ahead stage. Without introducing any

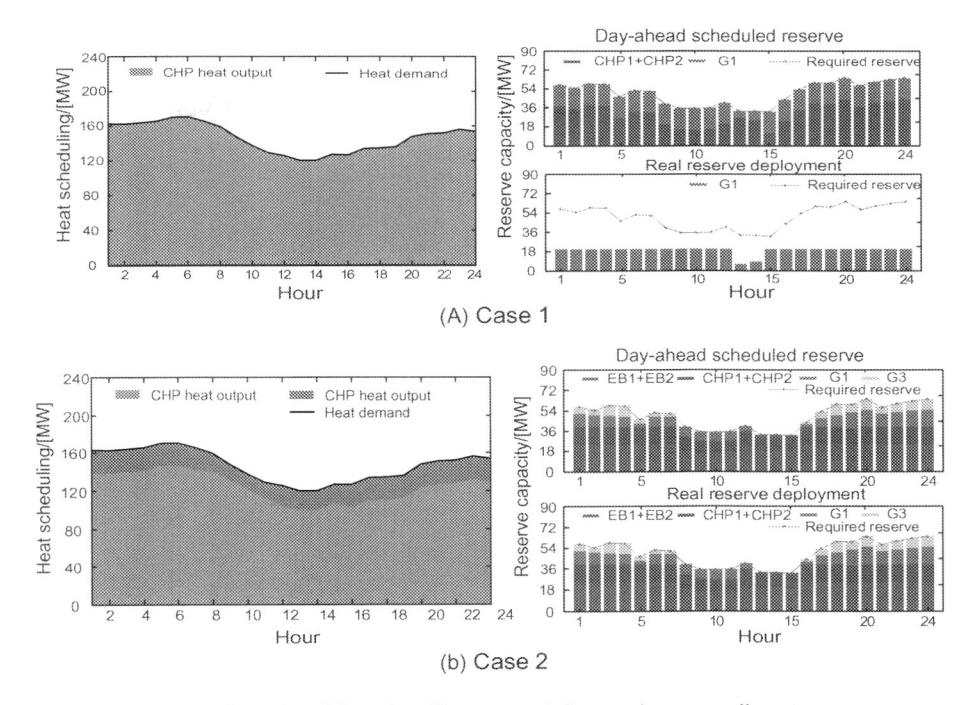

Figure 7.5 Optimal results of day-ahead heat scheduling and reserve allocation.

flexibility from EBs, CHP units cannot change their outputs to keep heat balance, and consequently, the actual available upward regulation that CHP units can provide is 0. Even though the required upward reserves in the day-ahead stage are provided by both CHP units and thermal units, the actual available balancing resources in the real-time operation are insufficient, as shown in the "Real reserve deployment" subfigure in Fig. 7.5(a). Therefore, parts of electric loads have to be shed, which results in the high regulation cost of $152,372.90, as shown in Table 7.3.

Case 2. In this case, the proposed two-stage robust optimization method M2 is adopted by taking into account the real-time operation constraints of the EPS and DHS. The optimal heat production and reserve allocation, and the reserve deployment, are presented in Fig. 7.5(b). Different from the optimal "Heat scheduling" in Case 1, EBs are scheduled to produce heat in the day-ahead stage in this case, which enables CHP units to regulate their outputs to some extent. Taking into account the DHS operation constraints while deploying scheduled reserves in real-time stage, the deployed reserves

Table 7.4 Compared results of different operation frameworks.

Cost ($)	Isolated	Centralized	Decentralized
Energy	97,485.49	91,153.89	91,065.41
Reserve capacity	6395.20	6399.77	6395.20
Regulation (worst case)	128,88.32	12,895.41	12,888.32
Total	116,769.00	110,449.10	110,348.93

are always sufficient and hence the reserve capacity scheduled in the day-ahead stage is always feasible. Moreover, it can be seen from the "Day-ahead scheduled reserve" subfigure in Fig. 7.5(b) that most reserve capacities are provided by CHP units and EBs, as their reserve costs are lower than those of thermal units. Thus, the reserve capacity cost in Case 2 decreases from $11,007.54 in Case 1 to $6,399.77, as illustrated in Table 7.3. Even the energy cost in Case 2 increases a bit compared with Case 1, and the overall operation of the IEHS in Case 2 is more cost effective. This demonstrates that by properly exploiting the reserve capacity from the DHS, the economic efficiency of reserve allocation and overall system operation can be improved.

7.5.2 Comparison of different operation frameworks

In this section, Case 2 is used as an example and the two–stage operation of multiple IEHSs is optimized under isolated, centralized, and LA-ADMM-based decentralized operation frameworks to illustrate the advantages of coordinated operation among multiple IEHSs and to verify the effectiveness of the decentralized algorithm. IEHS 1 and IEHS 2 are operated in a cooperative manner under centralized and decentralized frameworks, whereas IEHS 1 and IEHS 2 are operated separately without power exchange under an isolated operation framework. Under all operation frameworks, the two-stage robust optimization in each IEHS is solved by the C&CG algorithm. Table 7.4 presents the total operation costs under different operation frameworks.

It can be seen that the cost of isolated operation is the highest, as two IEHSs are operated separately without power exchange and resources sharing. Under the centralized and decentralized operation frameworks, IEHS 1 and IEHS 2 are connected with tie-lines, which enables the power exchange and resource sharing between the two IEHSs. Fig. 7.6 shows the tie-line power flow optimized by two IEHSs under the decentralized operation framework, which is scheduled 24 hours ahead. It can be seen that the surplus power generation with lower costs from IEHS 1 is output to IEHS 2 through the tie-line during the whole day. Since the cheapest

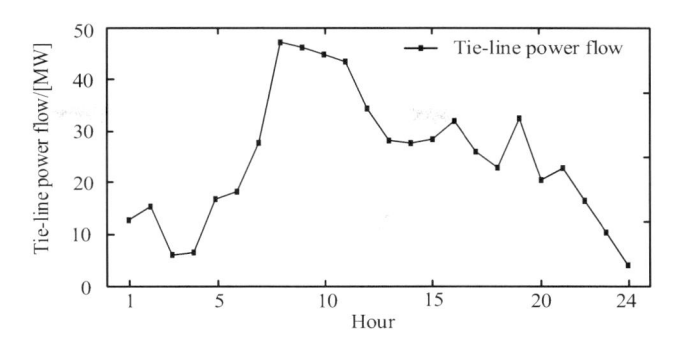

Figure 7.6 Optimized tie-line power flow under decentralized operation.

resource in the whole system is fully used, the energy cost and the total system operation cost in both centralized and decentralized operation are reduced accordingly. Compared with the isolated operation, the operational cost efficiency of the whole system under centralized and decentralized operation is improved, as one IEHS is able to provide backup energy and reserve to the other IEHS through tie-lines.

As can be seen from Table 7.4, the gap of the total costs between centralized and decentralized operation is around 0.09%, showing that the optimal results of the decentralized operation are very close to the centralized one. This means that with limited information exchange, the decentralized operation framework not only preserves the privacy of different systems with limited information exchange but also achieves the synergistic coordination between the two IEHSs without losing economic efficiency.

7.5.3 Performance comparison of different ADMM algorithms

In this section, Case 2 is studied as an example and the standard ADMM is adopted as a benchmark to demonstrate the effectiveness of the improved LA-ADMM. Compared with the standard ADMM, the lower iteration complexity of LA-ADMM with adaptive updated penalty parameters has been proved by Xu et al. [14]. Fig. 7.7 shows the comparison results of the maximum residual versus the iterations during the convergence procedure using two algorithms. The residual using both algorithms decreases along with the increase of iterations until the maximum residual is lower than the convergence tolerance, implying that the tie-line power flow optimized by two IEHSs reaches an agreement in the end. It can be seen from Fig. 7.7 that every time the penalty parameter is updated, the maximum

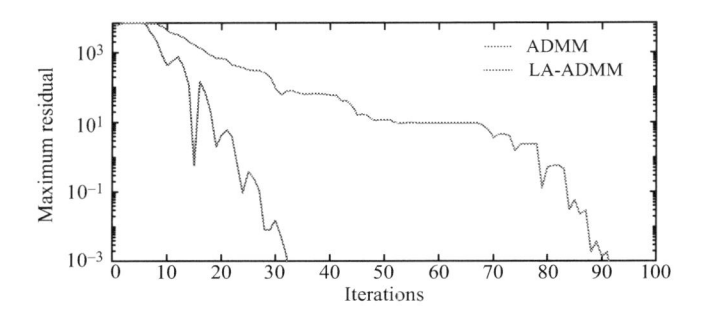

Figure 7.7 Residual in each iteration with different ADMM algorithms.

Table 7.5 Compared costs of different ADMM algorithms.

Cost ($)	Standard ADMM	LA-ADMM
Energy	91,075.27	91,065.41
Reserve capacity	6395.20	6395.20
Regulation (worst case)	12,888.32	12,888.32
Total	110,358.79	110,348.93

residual in LA-ADMM decreases a lot more than that in the standard ADMM. The decentralized optimization based on the standard ADMM converged after 92 iterations, whereas the interactive iteration of the LA-ADMM-based decentralized optimization is 33. This shows that the convergence rate of LA-ADMM has been improved greatly compared with the standard ADMM. As such, the total exchanged data between two IEHSs and the corresponding communication demand are reduced by using the improved LA-ADMM.

Furthermore, Table 7.5 lists the optimal operation costs in Case 2 using the standard ADMM and improved LA-ADMM. It can be seen that the difference in the total costs using two different ADMM algorithms is quite small, which demonstrates that LA-ADMM can improve computation efficiency while achieving a satisfied operation result as the standard ADMM.

7.6 Conclusion

This chapter proposed an improved decentralized robust energy and reserve co-optimization scheme for multiple IEHSs. It utilized the flexibility from DHSs and resource sharing among multiple IEHSs. A decentralized operation based on an improved LA-ADMM algorithm was implemented to achieve the independent yet synergistic operation of multiple IEHSs. In each

IEHS, the two-stage adaptive robust optimization is adopted to cope with wind uncertainty and was solved by the C&CG algorithm to obtain optimal energy and reserve scheduling. Case studies show that the feasibility of day-ahead allocated reserves is guaranteed by taking into the real-time regulation constraints of DHSs. The economic efficiency is improved by properly using the available reserve capacity provided by the DHS and reserve sharing among IEHSs. A coordination of multiple IEHSs can be achieved by the proposed decentralized operation while preserving the privacy of different IEHSs. Compared to the standard ADMM, the convergence of the improved LA-ADMM algorithm with adaptive penalty parameters is increased by 67%.

References

[1] H Lund, Large-scale integration of wind power into different energy systems, Energy 30 (13) (2005) 2402–2412.
[2] G Papaefthymiou, B Hasche, C Nabe, Potential of heat pumps for demand side management and wind power integration in the German electricity market, IEEE Trans Sustain Energy 3 (2012) 636–642.
[3] Y Zhou, W Hu, Y Min, Y Dai, Integrated power and heat dispatch considering available reserve of combined heat and power units, IEEE Trans Sustain Energy 10 (2019) 1300–1310.
[4] J Tan, Q Wu, Q Hu, W Wei, F Liu, Adaptive robust energy and reserve co-optimization of integrated electricity and heating system considering wind uncertainty, Appl Energy 260 (2020) 114230.
[5] A Ahmadi-Khatir, AJ Conejo, R Cherkaoui, Multi-area unit scheduling and reserve allocation under wind power uncertainty, IEEE Trans Power Syst 29 (2014) 1701–1710.
[6] W Wei, F Liu, S Mei, Y Hou, Robust energy and reserve dispatch under variable renewable generation, IEEE Trans Smart Grid 6 (2015) 369–380.
[7] M Zugno, AJ Conejo, A robust optimization approach to energy and reserve dispatch in electricity markets, Eur J Oper Res 247 (2015) 659–671.
[8] A Kargarian, J Mohammadi, J Guo, S Chakrabarti, M Barati, G Hug, et al., Toward distributed/decentralized DC optimal power flow implementation in future electric power systems, IEEE Trans Smart Grid 9 (2018) 2574–2594.
[9] Z Li, M Shahidehpour, W Wu, B Zeng, B Zhang, W Zheng, Decentralized multiarea robust generation unit and tie-line scheduling under wind power uncertainty, IEEE Trans Sustain Energy 6 (2015) 1377–1388.
[10] Y He, M Yan, M Shahidehpour, Z Li, C Guo, L Wu, et al., Decentralized optimization of multi-area electricity-natural gas flows based on cone reformulation, IEEE Trans Power Syst 33 (2018) 4531–4542.
[11] PJM, PJM Manual 11: energy & ancillary services market operations. https://www.pjm.com/directory/manuals/m11/index.html#Sections/Section%201%20Overview%20of%20Energy%20%20Ancillary%20Services%20Market%20Operations.html. (Accessed 18 April 2021).
[12] S Boyd, N Parikh, E Chu, B Peleato, J Eckstein, Distributed optimization and statistical learning via the alternating direction method of multipliers, Found Trends Mach Learn 3 (2010) 1–122.

[13] BS He, H Yang, SL Wang, Alternating direction method with self-adaptive penalty parameters for monotone variational inequalities, J Optim Theory Appl 106 (2000) 337–356.

[14] Y Xu, M Liu, Q Lin, T Yang, ADMM without a fixed penalty parameter: faster convergence with new adaptive penalization, Adv Neural Inf Process Syst 2017 (2017) 1268–1278.

[15] B Zeng, L Zhao, Solving two-stage robust optimization problems using a column-and-constraint generation method, Oper Res Lett 41 (2013) 457–461.

[16] G Li, R Zhang, T Jiang, H Chen, L Bai, H Cui, et al., Optimal dispatch strategy for integrated energy systems with CCHP and wind power, Appl Energy 192 (2017) 408–419.

CHAPTER 8

Chance-constrained energy and multi-type reserves scheduling exploiting flexibility from combined power and heat units and heat pumps

8.1 Introduction

With growing environmental concerns like global climate change, the share of renewable energy in energy systems has been increasing rapidly in past years [1]. The increasing power imbalance introduced by fluctuating renewable energy poses great challenges to the secure and reliable operation of power systems, which in turn changes the operating reserve requirements in electric power systems (EPSs) [2]. On the one hand, the intrinsic uncertainty of renewable energy (e.g., wind power) due to instantaneous weather conditions requires increasing following reserves to deal with wind power forecast errors. On the other hand, the increasing replacement of conventional synchronous generators by low-inertia wind power reduces the total inertia of power systems, which consequently deteriorates the system frequency security under contingencies [3]. This remarkably increases the need of primary frequency response reserves (FRRs) to maintain the frequency security of EPSs.

One effective way to ensure the operational security of EPSs is to access additional flexibility from other energy sectors such as district heating systems (DHSs) via energy coupling devices. Combined heat and power (CHP) units and heat pumps (HPs), as co-generation units and power-to-heat devices connecting EPSs and DHSs, are promising candidates to provide multiple reserves to EPSs [4]. Compared with conventional thermal units, CHP units based on simple-cycle or combined-cycle gas turbines have advantages such as fast ramping rate and fast startup–shutdown speed, which are flexible and feasible to provide reserves for EPSs [5]. Moreover, HPs are

Optimal operation of integrated multi-energy systems under uncertainty. Copyright © 2022 Elsevier Inc.
DOI: https://doi.org/10.1016/B978-0-12-824114-1.00012-3 All rights reserved.

also efficient supplements to provide reserves to EPSs. For example, HPs can provide upward reserves for EPSs by reducing the consumption of electric power. Meesenburg et al. [6] investigated and validated the optimal control strategy of large-scale HPs to provide primary frequency regulation in the smart grid.

Efficient utilization of the flexibility from CHP units and HPs requires optimal coordination between EPSs and DHSs. Due to the coupling between electric power and heat, it is essential to develop a feasible reserve model for CHP units not only considering ramping limits and capacity limits in the EPS sector but also taking into account the heat regulation constraints in the DHS sector. In the work of Xu et al. [7], the maximum flexibility from a DHS was quantified by taking into account all DHS operation limits, but it was not incorporated into the optimal operation of integrated electricity and heating systems (IEHSs). Pan et al. [8] proposed a feasible region method to formulate new DHS models with exploiting their available flexibility, which can be directly incorporated into the dispatch model. Based on the heat-dependency and temporal-coupling features of CHPs, Zhou et al. [9] presented a flexibility region method to properly assess the available reserve capacity of CHP units, which was used in an IEHS dispatch.

In addition to the provision of following reserves for offsetting wind power forecast errors, the negative impact of low-inertia wind power on system frequency security under contingencies should also be taken into account. The decreased system inertia caused by wind power replacing synchronous generators reinforces the need for primary FRRs to maintain system frequency security after the loss of generators [10]. Given that HPs are able to provide additional primary FRRs to EPSs while their cost in this provision is relatively low, it is of significance to coordinate the operation of EPSs and DHSs, and utilize the flexibility from DHSs to alleviate the frequency regulation burden on conventional generators.

Compared with the reserve scheduling using deterministic methods with predefined requirements, decision making under uncertainty offers advantages in the cost-effective integration of wind power. Tan et al. [11] proposed an adaptive robust energy and reserve co-optimization for IEHSs taking into account the worst-case realization of wind power uncertainty, but it suffers from conservativeness. Stochastic programming (SP), which can reduce the conservativeness by modeling the uncertainty via multiple scenarios, is widely used in the optimization of IEHSs as well [12,13]. Even though it can reduce the conservativeness of optimal scheduling, some extreme scenarios could result in a large amount of load shedding with small probability. To achieve a good trade-off between cost efficiency and reserve sufficiency, a chance-constrained program [14] is needed to manage the risk level of load shedding caused by wind power uncertainty.

Therefore, this chapter proposes a two-stage chance-constrained energy and multi-type reserves scheduling scheme for IEHSs utilizing the flexibility from CHP units and HPs to accommodate wind forecast errors and respond to the outage of the largest generator. The following reserves and the primary FRRs from CHP units and HPs are optimized using chance constraints and satisfy the system steady-state frequency requirement, respectively.

8.2 Framework of chance-constrained two-stage energy and multi-type reserves scheduling

The proposed scheme optimizes the energy and multi-type reserves scheduling for IEHSs to accommodate the high penetration of uncertain and low-inertia wind power by exploiting the flexibility from CHP units and HPs. It determines the optimal scheduling of unit commitment, power and heat energy, and multiple reserves simultaneously. In general, there are different types of reserves required to guarantee the reliable and secure operation of EPSs [15]. In this chapter, we consider two types of reserves with different response time scales and purposes—that is, the following reserve for offsetting power imbalances caused by wind power uncertainty and the primary FRR for arresting system frequency decline after the outage of the largest generator.

The framework of the proposed chance-constrained two-stage energy and multi-type reserves scheduling is illustrated in Fig. 8.1. The objective is subject to system operation constraints under normal conditions and following the outage of the largest generator, which consists of three parts: (1) the day-ahead unit commitment, power and heat energy, and reserves scheduling with wind power forecasts under normal conditions; (2) the contingency operation of the IEHS after the outage of the largest generator; and (3) the real-time reserve deployment with revealed wind power uncertainty under normal conditions. These three parts of constraints are interrelated. The first and second parts are coupled by reserve capacity limits, which capture the interdependence between the following reserves and primary FRRs. The first and third parts are coupled by the day-ahead reserve scheduling and real-time reserve deployments. In addition, the EPS and DHS are coupled in all three parts via CHP units and HPs.

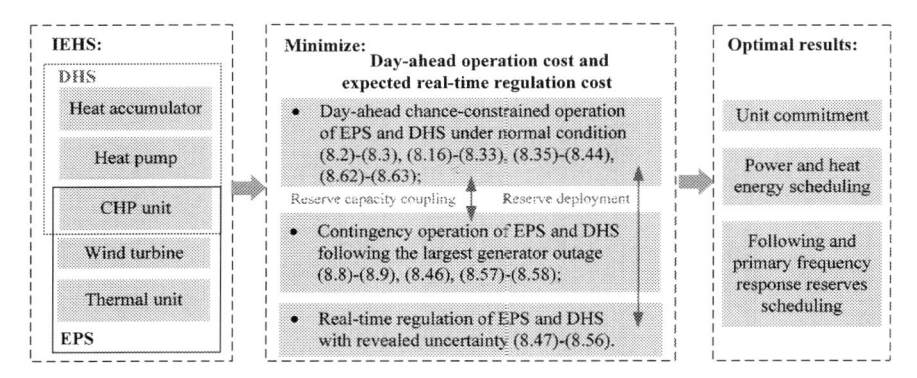

Figure 8.1 Optimization framework of chance-constrained two-stage energy and multi-type reserves scheduling.

8.3 Primary FRR and following reserve provision from CHP units and HPs

The symbols and notations described in the following mathematical model are explained here for quick reference.

Indices and sets

$\Phi^{TU/W/LD}$	Index set of thermal units/wind farms/electric load
Φ^{J}	Index set of segments of linearized fuel cost function of thermal units
$\Phi^{CHP/HP}$	Index set of CHP units/HPs
$\Phi^{L/B}$	Index set of transmission lines/buses in EPSs
$\Phi^{HN/HS/HA/HD}$	Index set of heat nodes/heat sources//heat accumulators/heat demands in DHSs
$\Phi^{S/R,pipe}$	Index set of supply/return pipelines in DHSs
$\Omega_{n}^{CHP/HP}$	Index set of CHP units/HPs connected to bus n
$\Omega_{n}^{TU/W/LD}$	Index set of thermal units/wind farms/electric loads connected to bus n
$\Omega_{j}^{HN/CHP/HP/HA}$	Index set of heat nodes/CHP units/HPs/heat accumulators corresponding to heat source j
$\Omega_{nd}^{S/R,B/E,pipe}$	Index set of supply/return pipelines beginning/ending at node nd
T	Index set of hours
DA/RT	Day-ahead stage/Real-time stage

Parameters

$c_i^{TU/CHP}$	Startup cost of thermal/CHP unit i
$c_i^{min,TU}$	Minimum-load cost of thermal unit i
$L_{i,j}^{TU}$	Slope of segment j of linearized fuel cost function of thermal unit i
$c_i^{TU,FRR/UR/DR}$	Capacity price of primary FRR/upward/ downward following reserve of thermal unit i
$c_i^{TU,+/-}$	Cost of upward/downward following reserve deployment of thermal unit i
$c_{i,k}^{CHP}$	Fuel cost of the kth vertex of feasible operation region of CHP unit i
$c_i^{CHP,FRR/UR/DR}$	Capacity price of primary FRR/upward/ downward following reserve of CHP unit i
$c_i^{TU/CHP,+/-}$	Cost of upward/downward following reserve deployment of thermal/CHP unit i
$c^{W/LS}$	Price for wind power curtailment and load shedding
$P_{i,k}^{CHP} \ H_{i,k}^{CHP}$	Electricity/heat production of the kth vertex of feasible operation region of CHP unit i
$P_i^{TU,max/min}$	Maximum/minimum generation of thermal unit i
$RU_i^{TU/CHP}, RD_i^{TU/CHP}$	Ramping up/down limit of thermal/CHP unit i
$RU_i^{TU/CHP,FRR}$	Primary FRR limit provided by thermal/CHP unit i
$P_i^{HP,max/min}$	Maximum/minimum generation of HP i
$P_i^{W,pre} \ P_i^{LD}$	Predicted wind power/Electric load
$g_{l,i}^{TU/CHP/HP/W/LD}$	Distribution factor from thermal units/CHP units/HPs/wind farms/electric load to line l
$\bar{f}l$	Capacity of transmission line l
p^{OUT}	The outage of the largest generator
ε	Risk level of load shedding
D	Droop parameter of generators
Δf^{max}	Maximum system frequency deviation
η_i	Power to heat coefficient of HP i
$H_{i,t}^{HA,max/min}$	Maximum/minimum heat energy level of HA i at time t
c,λ,l_p	Specific heat capacity of water/Heat transfer coefficient of pipelines/Length of pipelines
$m^{HS/HD}$	Mass flow rate of heat sources/heat loads

$m^{S/R,pipe}$	Mass flow rate of supply/return pipelines
τ_t^a	Ambient temperature of pipelines

Variables

$C^{DA}_{TU/CHP/W}$	Costs of thermal units/CHP units/wind farms in the day-ahead stage
$C^{RT}_{TU/CHP/W/LS}$	Costs of thermal units/CHP units/wind farms/load shedding in the real-time stage
$x^{TU/CHP}_{i,t}$	On-off status of thermal/CHP unit i at time t
$y^{TU/CHP}_{i,t}$	Startup indicator of thermal/CHP unit i at time t
$z^{TU/CHP}_{i,t}$	Shutdown indicator of thermal/CHP unit i at time t
$P^{TU/CHP/HP}_{i,t}$	Day-ahead generation of thermal/CHP units/HPs i at time t
$P^{TU}_{i,j,t}$	Scheduled value of segment k of thermal units i at time t
$P^{DA,WC}_{i,t}$	Day-ahead wind power curtailment of wind farm i at time t
$R^{CHP,flex}_{i,t}$	Flexibility provided by CHP units i at time t
$R^{CHP,UR/DR}_{i,t}$	Day-ahead upward/downward following reserve capacity scheduling of CHP unit i at time t
$R^{TU,UR/DR}_{i,t}$	Day-ahead upward/downward following reserve capacity scheduling of thermal unit i at time t
$R^{HP}_{i,t}$	Day-ahead following reserve capacity scheduling of HP i at time t
$R^{TU/CHP/HP,FRR}_{i,t}$	Day-ahead primary FRR scheduling of thermal/CHP units/HP i at time t
Δf_t	Quasi-steady-state system frequency at time t
$r^{CHP,UR/DR}_{i,t}$	Upward/downward regulation of CHP unit i at time t
$r^{TU,UR/DR}_{i,t}$	Upward/downward regulation of thermal unit i at time t
$r^{HP}_{i,t}$	Regulation of HP i at time t
$P^{RT,WC}_{i,t}$	Real-time wind power curtailment of wind farm i at time t
$P^{RT,LS}_{i,t}$	Real-time load shedding at time t
$H^{CHP/HP}_{i,t}$	Heat output of CHP units/HP i at time t

$\Delta H_{i,t}^{HA}$	Charging or discharging heat energy of HA i at time t
$H_{i,t}^{HA}$	Heat energy level of HA i at time t
$H_{i,t}^{HD}$	Heat energy consumption of heat exchange station i at time t
$H_{h,t}^{CHP,RT}$	Heat output of CHP units in the real-time stage
$\tau_{nd,t}^{S/R,ND}$	Temperature of node nd in supply/return networks
$\tau_{p,t}^{S/R,in}$, $\tau_{p,t}^{S/R,out}$	Mass flow temperature at the inlet/outlet of pipeline p in supply/return networks

8.3.1 Flexibility model of extraction CHP units and HPs

In this chapter, the flexibility is considered as regulation ability (i.e., the electricity magnitude of CHP units and HPs regulating from their operation set points), which is expressed as Eq. (8.1):

$$R_{i,t}^{flex} = P_{i,t}^{act} - P_{i,t}^{sch}, \quad \forall i \in \Phi^{CHP/HP}, \forall t \in T \tag{8.1}$$

where $P_{i,t}^{act}$ is the actual electricity magnitudes of CHP units and HPs after regulation, and $P_{i,t}^{sch}$ is the scheduled electricity magnitudes of CHP units and HPs in the day–ahead operation.

This chapter studies the extraction CHP unit with a convex feasible operation region as an example to illustrate the mathematical model of the flexibility from CHP units. The electricity and heat production are formulated as a convex combination of the extreme points of the feasible operation region [16], which is modeled as follows.

$$P_{i,t}^{CHP} = \sum_{k=1}^{M_i} \alpha_{i,k,t} P_{i,k}^{CHP}, \quad H_{i,t}^{CHP} = \sum_{k=1}^{M_i} \alpha_{i,k,t} H_{i,k}^{CHP}, \quad \forall i \in \Phi^{CHP}, \forall t \in T \tag{8.2}$$

$$\sum_{k=1}^{M_i} \alpha_{i,k,t} = x_{i,t}^{CHP}, \quad 0 \le \alpha_{i,k,t} \le 1, \quad \forall i \in \Phi^{CHP}, \forall t \in T \tag{8.3}$$

For conventional thermal units, the available flexibility only depends on their current power production level, capacity limits, and ramping rate limits, which are described as Eq. (8.5). However, CHP units are primary heat sources, and hence they have to fulfill heat demands while providing reserves for EPSs. Thus, the feasible flexibility domain of CHP units $F^{CHP,flex}$ is modeled by an intersection of the ramping and capacity limit set $F^{EPS,flex}$ in the electricity sector and the heat regulation set $F^{DHS,flex}$ in the heating sector as follows:

$$F^{CHP,flex} = F^{EPS,flex} \cap F^{DHS,flex} \qquad (8.4)$$

$F^{EPS,flex}$

$$= \left\{ \mathbf{R}^{CHP,flex} \in \mathbb{R} \left| \begin{array}{l} R_{i,t}^{CHP,flex} \leq \min(P_{i,t}^{CHP,max} x_{i,t}^{CHP} - P_{i,t}^{CHP}, RU_i^{CHP,flex} x_{i,t}^{CHP}) \\ R_{i,t}^{CHP,flex} \geq -\min(P_{i,t}^{CHP} - P_{i,t}^{CHP,min} x_{i,t}^{CHP}, RD_i^{CHP,flex} x_{i,t}^{CHP}), \end{array} \right. \quad \forall i \in \Phi^{CHP}, \forall t \in T \right\} \qquad (8.5)$$

$F^{DHS,flex}$

$$= \left\{ \mathbf{R}^{CHP,flex} \in \mathbb{R} \left| \begin{array}{l} R_{i,t}^{CHP,flex} = \sum_{k=1}^{M_i} \Delta\alpha_{i,k,t}^{flex} P_{i,k}^{CHP}, \ \Delta H_{i,t}^{CHP\,flex} = \sum_{k=1}^{M_i} \Delta\alpha_{i,k,t}^{flex} H_{i,k}^{CHP} \\ \sum_{k=1}^{M_i} \Delta\alpha_{i,k,t}^{flex} = 0, \ 0 \leq \alpha_{i,k,t} + \Delta\alpha_{i,k,t}^{flex} \leq 1 \\ \mathbf{A}\left(\mathbf{X}^{DA,heat} + \Delta\mathbf{X}^{HEAT,flex}\right) \leq \mathbf{b}, \end{array} \right. \quad \forall i \in \Phi^{CHP}, \forall t \in T \right\} \qquad (8.6)$$

where $\mathbf{X}^{DA,heat}$ and $\Delta\mathbf{X}^{HEAT,flex}$ are the decision variables reflecting the day–ahead and regulated heat operation states in the DHS, respectively. $\Delta\mathbf{X}^{HEAT,flex}$ includes the regulated heat production of CHP units $\Delta H^{CHP,flex}$. $\mathbf{A}$ and $\mathbf{b}$ are parameter matrix and vectors. The detailed descriptions of $\mathbf{X}^{DA,heat}$, $\Delta\mathbf{X}^{HEAT,flex}$, $\mathbf{A}$, and $\mathbf{b}$ are shown in Section 8.4.2.

8.3.2 Primary FRRs and following reserves provision from CHP units and HPs

The flexibility $\mathbf{R}^{flex}$ from CHP units and HPs is exploited to provide both primary FRRs $\mathbf{R}^{FRR}$ for arresting system frequency decline under the outage of the largest generator and the following reserves $\mathbf{R}^{UR}$ and $\mathbf{R}^{DR}$ for offsetting wind power forecast errors.

(1) *Primary FRRs for arresting system frequency decline.* After a generator outage, the system frequency will drop immediately and the primary FRRs are thus required to be delivered fast to arrest the frequency drop. CHP units and HPs are technically capable of providing primary FRRs for EPSs. Generators provide primary frequency responses via governor control, where the primary FRR provided by each generator is proportional to the system frequency deviation. The frequency droop characteristic of a generator is illustrated in Fig. 8.2. $R_{i,t}^{FRR}$ denotes the amount of delivered primary FRR from generator i at time t, RU_i^{FRR} is the upper limit of available primary FRR, and D_i is the droop parameter of generator i.

Mathematically, the primary FRRs provided by conventional thermal units, CHP units, and HPs following a generator outage are modeled

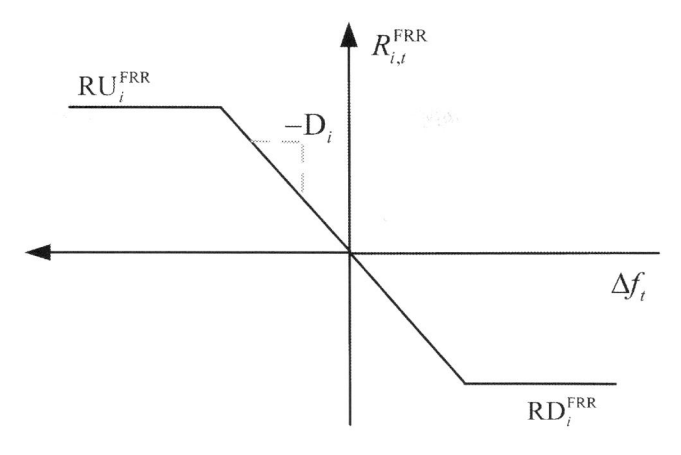

Figure 8.2 Frequency droop characteristic of generators.

as follows [17]:

$$R_{i,t}^{\text{TU/CHP,FRR}} = -\frac{\Delta f_t}{D_i} x_{i,t}^{\text{TU/CHP}}, \quad \forall i \in \Phi^{\text{TU/CHP}}, \forall t \in T \qquad (8.7)$$

$$0 \leq R_{i,t}^{\text{TU/CHP/HP,FRR}} \leq \text{RU}_i^{\text{TU/CHP/HP,FRR}} x_{i,t}^{\text{TU/CHP/HP}},$$
$$\forall i \in \Phi^{\text{TU/CHP/HP}}, \forall t \in T \qquad (8.8)$$

$$\Delta f^{\min} \leq \Delta f_t \leq \Delta f^{\max}, \quad \forall t \in T \qquad (8.9)$$

where Δf_t is the system steady-state frequency deviation after the post-contingency transient, which is set at 1% less than the nominal frequency. Eq. (8.9) denotes the bounds of Δf_t avoiding the activation of under-frequency load shedding relays. Sufficient primary FRRs are required to maintain the power balance all the time and keep system frequency close to its nominal value. The contingency caused by the loss of the largest generator is considered in this chapter, whereas load increases are not taken into account. This implies that there are only negative frequency deviations and only upward FRRs are required following the contingency.

(2) *Following reserves for offsetting wind power forecast errors.* In addition to the primary FRRs, CHP units and HPs also provide the following reserves to cope with the intrinsic uncertainty of wind power. Upward reserves $R_{i,t}^{\text{UR}}$ and downward reserves $R_{i,t}^{\text{DR}}$ from CHP units and HPs are deployed to offset the negative and positive errors of wind power forecasts, respectively. A detailed description is presented in Section 8.4.2.

8.4 Mathematical formulation of decentralized robust energy and reserve co-optimization for multiple IEHSs

8.4.1 Objective function

The objective of the IEHS is to minimize the overall cost of energy generation and provision of both following reserves and primary FRRs over a 24-hour period. It consists of the day–ahead operation cost and expected real-time regulation cost. The day–ahead operation cost includes unit startup cost, minimum load cost, fuel cost, upward/downward following reserve cost, and primary FRR cost. The real-time regulation cost is the expected sum of following reserve deployment cost, wind curtailment cost, and load shedding cost.

$$\min \left\{ C_{\mathrm{TU}}^{\mathrm{DA}} + C_{\mathrm{CHP}}^{\mathrm{DA}} + C_{\mathrm{W}}^{\mathrm{DA}} + E\left[C_{\mathrm{TU}}^{\mathrm{RT}}(\boldsymbol{\xi}) + C_{\mathrm{CHP}}^{\mathrm{RT}}(\boldsymbol{\xi}) + C_{\mathrm{W}}^{\mathrm{RT}}(\boldsymbol{\xi}) + C_{\mathrm{LS}}^{\mathrm{RT}}(\boldsymbol{\xi}) \right] \right\} \tag{8.10}$$

$$C_{\mathrm{TU}}^{\mathrm{DA}} = \sum_{t \in T} \sum_{i \in \Phi^{\mathrm{TU}}} \left\{ c_i^{\mathrm{TU}} y_{i,t}^{\mathrm{TU}} + c_i^{\min,\mathrm{TU}} x_{i,t}^{\mathrm{TU}} + \sum_{j \in \Phi^{\mathrm{J}}} \mathrm{L}_{i,j}^{\mathrm{TU}} P_{i,j,t}^{\mathrm{TU}} \right. $$
$$\left. + c_i^{\mathrm{TU,FRR}} R_{i,t}^{\mathrm{TU,FRR}} + c_i^{\mathrm{TU,UR}} R_{i,t}^{\mathrm{TU,UR}} + c_i^{\mathrm{TU,DR}} R_{i,t}^{\mathrm{TU,DR}} \right\} \tag{8.11}$$

$$C_{\mathrm{CHP}}^{\mathrm{DA}} = \sum_{t \in T} \sum_{i \in \Phi^{\mathrm{CHP}}} \left\{ c_i^{\mathrm{CHP}} y_{i,t}^{\mathrm{CHP}} + \sum_{k=1}^{M_i} \alpha_{i,k,t} c_{i,k}^{\mathrm{CHP}} + c_i^{\mathrm{CHP,FRR}} R_{i,t}^{\mathrm{CHP,FRR}} \right.$$
$$\left. + c_{i,t}^{\mathrm{CHP,UR}} R_{i,t}^{\mathrm{CHP,UR}} + c_{i,t}^{\mathrm{CHP,DR}} R_{i,t}^{\mathrm{CHP,DR}} \right\} \tag{8.12}$$

$$C_{\mathrm{W}}^{\mathrm{DA}} = \sum_{t \in T} \sum_{i \in \Phi^{\mathrm{W}}} c^{\mathrm{W}} P_{i,t}^{\mathrm{DA,WC}} \tag{8.13}$$

$$C_{\mathrm{TU/CHP}}^{\mathrm{RT}}(\boldsymbol{\xi})$$
$$= \sum_{t \in T} \sum_{i \in \Phi^{\mathrm{TU/CHP}}} \left\{ c_i^{\mathrm{TU/CHP},+} r_{i,t}^{\mathrm{TU/CHP,UR}}(\boldsymbol{\xi}) + c_i^{\mathrm{TU/CHP},-} r_{i,t}^{\mathrm{TU/CHP,DR}}(\boldsymbol{\xi}) \right\} \tag{8.14}$$

$$C_{\mathrm{W}}^{\mathrm{RT}}(\boldsymbol{\xi}) = \sum_{t \in T} \sum_{i \in \Phi^{\mathrm{W}}} c^{\mathrm{W}} P_{i,t}^{\mathrm{RT,WC}}(\boldsymbol{\xi}), \quad C_{\mathrm{LS}}^{\mathrm{RT}}(\boldsymbol{\xi}) = \sum_{t \in T} \sum_{i \in \Phi^{\mathrm{LD}}} c^{\mathrm{LS}} P_{i,t}^{\mathrm{LS}}(\boldsymbol{\xi}) \tag{8.15}$$

8.4.2 Day-ahead operation constraints of IEHSs under normal conditions

Day-ahead operation of an IEHS is considered as the first stage, where the energy and following reserve are scheduled with wind power forecasts. The detailed formulations of the operation constraints of both the EPS and DHS are shown as follows.

(1) *Day-ahead operation constraints of the EPS under normal conditions*

Generation constraints of conventional thermal units. The production of conventional thermal units is expressed with piecewise linearization, and the corresponding limits are described next.

$$P_{i,t}^{\mathrm{TU}} = \sum_{j \in \Phi^J} P_{i,j,t}^{\mathrm{TU}}, \quad \forall i \in \Phi^{\mathrm{TU}}, \forall t \in T \tag{8.16}$$

$$0 \le P_{i,j,t}^{\mathrm{TU}} \le \mathrm{P}_{i,j}^{\mathrm{TU,max}}, \quad \forall i \in \Phi^{\mathrm{TU}}, \ \forall j \in \Phi^J, \forall t \in T \tag{8.17}$$

$$\mathrm{P}_i^{\mathrm{TU,min}} x_{i,t}^{\mathrm{TU}} \le P_{i,t}^{\mathrm{TU}} \le \mathrm{P}_i^{\mathrm{TU,max}} x_{i,t}^{\mathrm{TU}}, \quad \forall i \in \Phi^{\mathrm{TU}}, \forall t \in T \tag{8.18}$$

Operating state, minimum startup, and shutdown time constraints of all generators.

$$y_{i,t}^{\mathrm{TU/CHP}} - z_{i,t}^{\mathrm{TU/CHP}} = x_{i,t}^{\mathrm{TU/CHP}} - x_{i,t-1}^{\mathrm{TU/CHP}}, \quad \forall i \in \Phi^{\mathrm{TU/CHP}}, \forall t \in T \tag{8.19}$$

$$y_{i,t}^{\mathrm{TU/CHP}} + z_{i,t}^{\mathrm{TU/CHP}} \le 1, \quad \forall i \in \Phi^{\mathrm{TU/CHP}}, \forall t \in T \tag{8.20}$$

$$\sum_{\tau=t-t_i^{\mathrm{up,min}}+1}^{t} y_{i,\tau}^{\mathrm{TU/CHP}} \le x_{i,t}^{\mathrm{TU/CHP}}, \quad \forall i \in \Phi^{\mathrm{TU/CHP}}, \forall t \in T \tag{8.21}$$

$$\sum_{\tau=t-t_i^{\mathrm{down,min}}+1}^{t} z_{i,\tau}^{\mathrm{TU/CHP}} \le 1 - x_{i,t}^{\mathrm{TU/CHP}}, \quad \forall i \in \Phi^{\mathrm{TU/CHP}}, \forall t \in T \tag{8.22}$$

Ramping constraints of all generators. The continuously generation variations of all generators are limited by upward and downward ramping rate, respectively.

$$P_{i,t}^{\mathrm{TU/CHP}} - P_{i,t-1}^{\mathrm{TU/CHP}} \le \mathrm{RU}_i^{\mathrm{TU/CHP}}(1 - y_{i,\tau}^{\mathrm{TU/CHP}}) + \mathrm{P}_i^{\mathrm{TU,max}} y_{i,\tau}^{\mathrm{TU/CHP}},$$
$$\forall i \in \Phi^{\mathrm{TU/CHP}}, \forall t \in T \tag{8.23}$$

$$P_{i,t-1}^{\mathrm{TU/CHP}} - P_{i,t}^{\mathrm{TU/CHP}} \le \mathrm{RD}_i^{\mathrm{TU/CHP}}(1 - z_{i,\tau}^{\mathrm{TU/CHP}}) + \mathrm{P}_i^{\mathrm{TU,max}} z_{i,\tau}^{\mathrm{TU/CHP}},$$
$$\forall i \in \Phi^{\mathrm{TU/CHP}}, \forall t \in T \tag{8.24}$$

Reserve capacity coupling between primary FRRs and upward following reserves. The primary FRRs and upward following reserves provision from conventional thermal units, CHP units, and HPs are coupled and constrained by their capacity limits.

$$P_{i,t}^{\text{TU/CHP}} + R_{i,t}^{\text{TU/CHP,UR}} + R_{i,t}^{\text{TU/CHP,FRR}} \leq P_{i,t}^{\text{TU/CHP,max}} x_{i,t}^{\text{TU/CHP}},$$
$$\forall i \in \Phi^{\text{TU/CHP}}, \forall t \in T \tag{8.25}$$

$$R_{i,t}^{\text{TU/CHP,UR}} \leq RU_i^{\text{TU/CHP}} x_{i,t}^{\text{TU/CHP}}, \quad \forall i \in \Phi^{\text{TU/CHP}}, \forall t \in T \tag{8.26}$$

$$P_i^{\text{HP,min}} \leq P_{i,t}^{\text{HP}} - R_{i,t}^{\text{HP}} - R_{i,t}^{\text{HP,FRR}}, \quad \forall i \in \Phi^{\text{HP}}, \forall t \in T \tag{8.27}$$

Downward following reserve constraints of all generators and HPs. In addition to upward following reserves, conventional thermal units, CHP units, and HPs provide downward following reserves to offset positive wind power forecast errors.

$$R_{i,t}^{\text{TU/C2,DR}} \leq \min\left\{ P_{i,t}^{\text{TU/CHP}} - P_{i,t}^{\text{TU/CHP,min}} x_{i,t}^{\text{TU/CHP}}, RD_i^{\text{TU/CHP}} x_{i,t}^{\text{TU/CHP}} \right\},$$
$$\forall i \in \Phi^{\text{TU/CHP}}, \forall t \in T \tag{8.28}$$

$$P_{i,t}^{\text{HP}} + R_{i,t}^{\text{HP}} \leq P_i^{\text{HP,max}}, \quad \forall i \in \Phi^{\text{HP}}, \forall t \in T \tag{8.29}$$

Operation constraints of HPs.

$$P_i^{\text{HP,min}} \leq P_{i,t}^{\text{HP}} \leq P_i^{\text{HP,max}}, \quad \forall i \in \Phi^{\text{HP}}, \forall t \in T \tag{8.30}$$

Wind power curtailment in day-ahead operation.

$$0 \leq P_{i,t}^{\text{DA,WC}} \leq P_{i,t}^{\text{W,pre}}, \quad \forall i \in \Phi^{\text{W}}, \forall t \in T \tag{8.31}$$

Power balance constraints. Total electricity production should be equal to demands all the time.

$$\sum_{i \in \Phi^{\text{CHP}}} P_{i,t}^{\text{CHP}} - \sum_{i \in \Phi^{\text{HP}}} P_{i,t}^{\text{HP}} + \sum_{i \in \Phi^{\text{TU}}} P_{i,t}^{\text{TU}} + \sum_{i \in \Phi^{\text{W}}} (P_{i,t}^{\text{W,pre}} - P_{i,t}^{\text{DA,WC}}) - \sum_{i \in \Phi^{\text{LD}}} P_{i,t}^{\text{LD}} = 0,$$
$$\forall t \in T \tag{8.32}$$

Transmission line limits.

$$-\bar{f}_l \leq \sum_{i \in \Omega_n^{\text{CHP}}} g_{l,i}^{\text{CHP}} P_{i,t}^{\text{CHP}} - \sum_{i \in \Omega_n^{\text{HP}}} g_{l,i}^{\text{HP}} P_{i,t}^{\text{HP}} + \sum_{i \in \Omega_n^{\text{TU}}} g_{l,i}^{\text{TU}} P_{i,t}^{\text{TU}}$$

$$+ \sum_{i \in \Omega_n^W} g_{l,i}^W (P_{i,t}^{W,\text{pre}} - P_{i,t}^{DA,WC}) - \sum_{i \in \Omega_n^{LD}} g_{l,i}^{LD} P_{i,t}^{LD} \leq \bar{f}_l,$$

$$\forall l \in \Phi^L, \ \forall n \in \Phi^B, \ \forall t \in T \tag{8.33}$$

Load shedding management with chance constraints. Given the increasing penetration of uncertain wind power, load shedding may occur due to deficient upward following reserves. To achieve a good trade-off between the cost-efficiency and following reserve sufficiency, chance constraints are incorporated into the optimization such that the load shedding is allowed within a small probability. The chance constraints limiting the probability of load shedding caused by wind power uncertainty are formulated as follows.

$$\Pr \left\{ \sum_{i \in \Phi^{CHP}} (P_{i,t}^{CHP} + R_{i,t}^{CHP,UR}) - \sum_{i \in \Phi^{HP}} (P_{i,t}^{HP} - R_{i,t}^{HP}) + \sum_{i \in \Phi^{TU}} (P_{i,t}^{TU} + R_{i,t}^{TU,UR}) \right.$$

$$\left. + \sum_{i \in \Phi^W} \tilde{P}_{i,t}^W(\xi) - \sum_{i \in \Phi^{LD}} P_{i,t}^{LD} \geq 0 \right\} \geq 1 - \varepsilon, \forall t \in T \tag{8.34}$$

(2) *Day-ahead operation constraints of DHS under normal conditions.*

In general, a DHS consists of primary and secondary networks, which are similar to transmission and distribution networks in EPSs. Here, the secondary network in the DHS is not considered, where thermal energy is generated by heat sources and transported to heat-exchange stations through primary pipelines. The EPS and DHS are coupled via CHP plants and HPs.

Heat sources. In this chapter, CHP units and HPs equipped with HAs are considered as heat sources. The heat source model includes the operation model of CHP units (Eqs. (8.2) and (8.3)), the relationship between heat production and the temperature difference of supply and return water (Eq. (8.35)), the thermal energy generated by HPs (Eq. (8.36)), and the thermal model of HAs (Eqs. (8.37)–(8.39)).

$$\sum_{i \in \Omega_j^{CHP}} H_{i,t}^{CHP} + \sum_{i \in \Omega_j^{HP}} H_{i,t}^{HP} - \sum_{i \in \Omega_j^{HA}} \Delta H_{i,t}^{HA} = c \cdot m_{j,t}^{HS} \cdot \left(\tau_{nd,t}^{S,ND} - \tau_{nd,t}^{R,ND} \right),$$

$$\forall j \in \Phi^{HS}, nd = \Omega_j^{HN}, \forall t \in T \tag{8.35}$$

$$H_{i,t}^{HP} = \eta_i P_{i,t}^{HP}, \ \forall i \in \Phi^{HP}, \forall t \in T \tag{8.36}$$

$$-\Delta H_i^{\text{HA,max}} \leq \Delta H_{i,t}^{\text{HA}} \leq \Delta H_i^{\text{HA,max}}, \ \forall i \in \Phi^{\text{HA}}, \forall t \in T \tag{8.37}$$

$$H_{i,t+1}^{\text{HA}} = H_{i,t}^{\text{HA}} + \Delta H_{i,t}^{\text{HA}}, \ \forall i \in \Phi^{\text{HA}}, \forall t \in T \tag{8.38}$$

$$H_i^{\text{HA,min}} \leq H_{i,t}^{\text{HA}} \leq H_i^{\text{HA,max}}, \ \forall i \in \Phi^{\text{HA}}, \forall t \in T \tag{8.39}$$

Heat exchange stations. Heat exchange stations are considered as heat demands in primary networks. The consumed heat demand is expressed as follows.

$$H_{i,t}^{\text{HD}} = c \cdot m_{i,t}^{\text{HD}} \cdot \left(\tau_{nd,t}^{\text{S,ND}} - \tau_{nd,t}^{\text{R,ND}}\right), \ \ \forall i \in \Phi^{\text{HD}}, nd = \Omega_i^{\text{HN}}, \forall t \in T \tag{8.40}$$

Temperature drop. Since the flow rate of thermal mass in pipelines is slow, temperature drop occurs in the supply and return pipelines caused by heat losses.

$$\tau_{p,t}^{\text{S/R,out}} - \tau_t^{a} = \left(\tau_{p,t}^{\text{S/R,in}} - \tau_t^{a}\right) e^{\frac{-\lambda l_p}{c \cdot m_{p,t}^{\text{S/R,pipe}}}}, \ \forall p \in \Phi^{\text{S/R,pipe}}, \forall t \in T \tag{8.41}$$

Temperature mixture. The temperature at each heat node is calculated as the mixture temperature of mass flows flowing into the node, as follows:

$$\sum_{p \in \Omega_{nd}^{\text{S/R,E,pipe}}} \left(m_{p,t}^{\text{S/R,pipe}} \cdot \tau_{p,t}^{\text{S/R,out}}\right) = \tau_{nd,t}^{\text{S/R,ND}} \sum_{p \in \Omega_{nd}^{\text{S/R,B,pipe}}} m_{p,t}^{\text{S/R,pipe}},$$

$$\forall nd \in \Phi^{\text{HN}}, \ \forall t \in T \tag{8.42}$$

$$\tau_{p,t}^{\text{S/R,in}} = \tau_{nd,t}^{\text{S/R,ND}}, \forall nd \in \Phi^{\text{HN}}, \forall p \in \Omega_{nd}^{\text{S/R,B,pipe}}, \forall t \in T \tag{8.43}$$

$$\tau^{\text{S/R,min}} \leq \tau_{nd,t}^{\text{S/R,ND}} \leq \tau^{\text{S/R,max}}, \forall nd \in \Phi^{\text{HN}}, \forall t \in T \tag{8.44}$$

Where Eq. (8.43) means that the temperature of pipelines flowing out of one node is equal to the temperature at this node, and Eq. (8.44) denotes the temperature limits for heat nodes.

To simplify the expression in the following sections, the DHS operation constraints (8.2) and (8.3) and (8.35) through (8.44) are rewritten in a compact form as Eq. (8.45):

$$\mathbf{AX}^{\text{DA,HEAT}} + \mathbf{BX}^{\text{POWER}} \leq \mathbf{b} \tag{8.45}$$

where $\mathbf{X}^{\text{DA,HEAT}}$ are the decision variables reflecting the DHS operation states in the day-ahead stage, which includes heat production of

CHP units $H_{i,t}^{CHP}$ and HPs $H_{i,t}^{HP}$, variables of HAs $H_{i,t}^{HA}$ and $\Delta H_{i,t}^{HA}$, and the temperature variables in the DHS $\tau_{i,t}^{S/R,ND}$, $\tau_{i,t}^{S/R,in}$, and $\tau_{i,t}^{S/R,out}$. $\mathbf{X}^{POWER}$ is the power production and consumption of CHP units and HPs, respectively, namely $P_{i,t}^{CHP}$ and $P_{i,t}^{HP}$. $\mathbf{A}, \mathbf{B}$, and $\mathbf{b}$ are the parameter matrix and vectors, which can be derived from Eqs. (8.2), (8.3), and (8.35) through (8.44).

8.4.3 Contingency operation constraints of IEHSs following the outage of the largest generator

After the outage of the largest generator, CHP units and HPs provide primary FRRs together with those from conventional thermal units to arrest system frequency decline. It should be ensured that the total primary FRR is larger than the lost generator, as represented in Eq. (8.46). In addition, Eqs. (8.7) through (8.9) modeling the primary FRRs delivered by conventional thermal units, CHP units, and HPs are also included in this contingency operation constraint set, which are introduced in Section 8.3.2.

$$\sum_{i \in \Phi^{TU}} R_{i,t}^{TU,FRR} + \sum_{i \in \Phi^{CHP}} R_{i,t}^{CHP,FRR} + \sum_{i \in \Phi^{HP}} R_{i,t}^{HP,FRR} \geq P_t^{OUT}, \quad \forall t \in T$$

$$(8.46)$$

Since the primary function of CHP units is to fulfill heat demands, CHP units are also constrained by the heat regulation set $F^{DHS,FRR}$ while providing $\mathbf{R}^{CHP,FRR}$, which can be found in Section Section 8.4.4.

It is noted that the transmission line constraints in the EPS are not included in the contingency condition, because transmission lines can be overloaded in a short interval.

8.4.4 Real-time regulation constraints of IEHSs accommodating wind power forecast errors

The second-stage problem is to regulate the outputs of non–wind-generating units to accommodate wind power forecast errors in the real–time operation. Once the uncertainty of wind power is revealed, thermal units, CHP units, and HPs will regulate their outputs according to the scheduled following reserves in the first stage.

Reserve deployment constraints. The upward and downward following reserves accommodating forecast errors deployed in the real–time operation should not be larger than the scheduled capacity in the day-ahead stage.

$$0 \leq r_{i,t}^{TU/CHP,UR}(\boldsymbol{\xi}) \leq R_{i,t}^{TU/CHP,UR}, \quad \forall i \in \Phi^{TU/CHP}, \forall t \in T \qquad (8.47)$$

$$0 \leq r_{i,t}^{\text{TU/CHP,DR}}(\boldsymbol{\xi}) \leq R_{i,t}^{\text{TU/CHP,DR}}, \quad \forall i \in \Phi^{\text{TU/CHP}}, \forall t \in T \qquad (8.48)$$

$$-R_{i,t}^{\text{HP}} \leq r_{i,t}^{\text{HP}}(\boldsymbol{\xi}) \leq R_{i,t}^{\text{HP}}, \quad \forall i \in \Phi^{\text{HP}}, \forall t \in T \qquad (8.49)$$

Ramping constraints of thermal units and CHP units in real-time operation.

$$(P_{i,t}^{\text{TU/CHP}} + r_{i,t}^{\text{TU/CHP,UR}}(\boldsymbol{\xi}) - r_{i,t}^{\text{TU/CHP,DR}}(\boldsymbol{\xi})) - (P_{i,t-1}^{\text{TU/CHP}} + r_{i,t-1}^{\text{TU/CHP,UR}}(\boldsymbol{\xi}) - r_{i,t-1}^{\text{TU/CHP,DR}}(\boldsymbol{\xi}))$$
$$\leq \text{RU}_i^{\text{TU/CHP}}(1 - \gamma_{i,\tau}^{\text{TU/CHP}}) + P_i^{\text{TU,max}} \gamma_{i,\tau}^{\text{TU/CHP}}, \quad \forall i \in \Phi^{\text{TU/CHP}}, \forall t \in T$$
$$(8.50)$$

$$(P_{i,t-1}^{\text{TU/CHP}} + r_{i,t-1}^{\text{TU/CHP,UR}}(\boldsymbol{\xi}) - r_{i,t-1}^{\text{TU/CHP,DR}}(\boldsymbol{\xi})) - (P_{i,t}^{\text{TU/CHP}} + r_{i,t}^{\text{TU/CHP,UR}}(\boldsymbol{\xi}) - r_{i,t}^{\text{TU/CHP,DR}}(\boldsymbol{\xi}))$$
$$\leq \text{RD}_i^{\text{TU/CHP}}(1 - z_{i,\tau}^{\text{TU/CHP}}) + P_i^{\text{TU,max}} z_{i,\tau}^{\text{TU/CHP}}, \quad \forall i \in \Phi^{\text{TU/CHP}}, \forall t \in T$$
$$(8.51)$$

Wind power curtailment and load shedding constraints in real-time operation.

$$0 \leq P_{i,t}^{\text{RT,WC}}(\xi) \leq \tilde{P}_{i,t}^{\text{W}}(\xi), \quad \forall i \in \Phi^{\text{W}}, \forall t \in T \qquad (8.52)$$

$$0 \leq P_{i,t}^{\text{RT,LS}}(\boldsymbol{\xi}) \leq P_{i,t}^{\text{LD}}, \quad \forall i \in \Phi^{\text{LD}}, \forall t \in T \qquad (8.53)$$

Transmission line constraints in real-time operation.

$$-\bar{f}_l \leq \sum_{i \in \Omega_n^{\text{CHP}}} g_{l,i}^{\text{CHP}}(P_{i,t}^{\text{CHP}} + r_{i,t}^{\text{CHP,UR}}(\boldsymbol{\xi}) - r_{i,t}^{\text{CHP,DR}}(\boldsymbol{\xi}))$$

$$- \sum_{i \in \Omega_n^{\text{HP}}} g_{l,i}^{\text{HP}}(P_{i,t}^{\text{HP}} - r_{i,t}^{\text{HP}}(\boldsymbol{\xi})) + \sum_{i \in \Omega_n^{\text{TU}}} g_{l,i}^{\text{TU}}(P_{i,t}^{\text{TU}} + r_{i,t}^{\text{TU,UR}}(\boldsymbol{\xi}) - r_{i,t}^{\text{TU,DR}}(\boldsymbol{\xi}))$$

$$+ \sum_{i \in \Omega_n^{\text{W}}} g_{l,i}^{\text{W}}(\tilde{P}_{i,t}^{\text{W}}(\xi) - P_{i,t}^{\text{RT,WC}}(\xi)) - \sum_{i \in \Omega_n^{\text{LD}}} g_{l,i}^{\text{LD}}(P_{i,t}^{\text{LD}} - P_{i,t}^{\text{RT,LS}}(\boldsymbol{\xi})) \leq \bar{f}_l,$$

$$\forall l \in \Phi^{\text{L}}, \ \forall n \in \Phi^{\text{B}}, \ \forall t \in T$$
$$(8.54)$$

Power rebalance constraints in real-time operation.

$$\sum_{i \in \Phi^{\text{CHP}}} (r_{i,t}^{\text{CHP,UR}}(\boldsymbol{\xi}) - r_{i,t}^{\text{CHP,DR}}(\boldsymbol{\xi})) + \sum_{i \in \Phi^{\text{HP}}} r_{i,t}^{\text{HP}}(\boldsymbol{\xi}) + \sum_{i \in \Phi^{\text{TU}}} (r_{i,t}^{\text{TU,UR}}(\boldsymbol{\xi}) - r_{i,t}^{\text{TU,DR}}(\boldsymbol{\xi}))$$

$$+ \sum_{i \in \Phi^{\text{W}}} ((\tilde{P}_{i,t}^{\text{W}}(\xi) - P_{i,t}^{\text{RT,WC}}(\xi)) - (P_{i,t}^{\text{W,pre}} - P_{i,t}^{\text{DA,WC}})) + \sum_{i \in \Phi^{\text{LD}}} P_{i,t}^{\text{RT,LS}}(\boldsymbol{\xi}) = 0, \quad \forall t \in T$$
$$(8.55)$$

Compact-formed heat regulation set in real-time operation. Since CHP units and HPs are coupling devices in the EPS and DHS, CHP units and HPs should be constrained by the heat regulation set as well during the deployment of following reserves and scheduled primary FRRs. The coupling constraint is shown next:

$$\mathbf{A}\left(\mathbf{X}^{\text{DA,HEAT}} + \Delta\mathbf{X}^{\text{HEAT,RT}}(\boldsymbol{\xi})\right) + \mathbf{B}\left(\mathbf{X}^{\text{POWER}}\right.$$
$$+ \mathbf{R}^{\text{CHP,FRR}} + \mathbf{r}^{\text{CHP,UR}}(\boldsymbol{\xi}) - \mathbf{r}^{\text{CHP,DR}}(\boldsymbol{\xi}) - \mathbf{R}^{\text{HP,FRR}} - \mathbf{r}^{\text{HP}}(\boldsymbol{\xi})\right) \leq \mathbf{b} \tag{8.56}$$

where $\Delta\mathbf{X}^{\text{HEAT,RT}}(\boldsymbol{\xi})$ are the decision variables reflecting regulated DHS operation states with reserve deployments, which includes regulated heat production of CHP units and HPs, variables of HAs, and the temperatures in the real-time operation. $\mathbf{R}^{\text{CHP,FRR}}$ and $\mathbf{R}^{\text{HP,FRR}}$ are the decision variables describing the primary FRRs from CHP units and HPs following the outage of the largest generator, namely $R_{i,t}^{\text{CHP,FRR}}$ and $R_{i,t}^{\text{HP,FRR}}$. $\mathbf{r}^{\text{CHP,UR}}(\boldsymbol{\xi})$, $\mathbf{r}^{\text{CHP,DR}}(\boldsymbol{\xi})$, and $\mathbf{r}^{\text{HP}}(\boldsymbol{\xi})$ are the decision variables describing the deployed following reserves from CHP units and HPs accommodating wind power forecast errors, namely $r_{i,t}^{\text{CHP,UR}}(\xi)$, $r_{i,t}^{\text{CHP,DR}}(\xi)$, and $r_{i,t}^{\text{HP}}(\xi)$.

8.5 Reformulation as a mixed-integer linear program

To make the proposed chance-constrained two-stage energy and multi-type reserves optimization computationally tractable, the nonlinear constraints (8.7) are linearized and the nonconvex chance constraints (8.34) are approximated with the contional value at risk (CVaR). The mixed-integer nonlinear programming (MINLP) problem is finally reformulated as a mixed-integer linear programming (MILP) solver.

8.5.1 Linearization of constraint (8.7)

According to Bertsimas and Tsitsiklis [18], the product of a binary and a continuous variable in the nonlinear constraint (8.7) can be equivalently expressed by a pair of linear constraints. Denoting $-\frac{\Delta f_t}{D_i}$ by a new continuous variable $v_{i,t}$, the equivalent linear constraints are shown as follows.

$$v_{i,t}^{\min} x_{i,t}^{\text{TU/CHP}} \leq R_{i,t}^{\text{TU/CHP,FRR}} \leq v_{i,t}^{\max} x_{i,t}^{\text{TU/CHP}} \tag{8.57}$$

$$0 \leq v_{i,t} - R_{i,t}^{\text{TU/CHP,FRR}} \leq v_{i,t}^{\max}(1 - x_{i,t}^{\text{TU/CHP}}) \tag{8.58}$$

8.5.2 Convex approximation of chance constraint (8.34) based on CVaR

The introduced chance constraint (8.34) addressing the uncertain wind power generation $\tilde{P}^{\text{W}}(\xi)$ can be equivalently interpreted as a constraint on

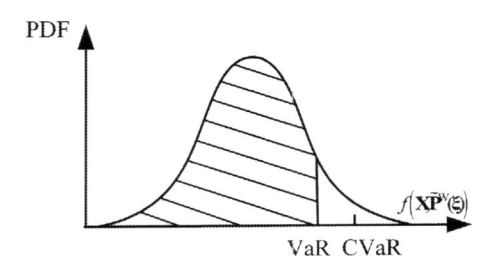

Figure 8.3 Relationship between VaR and CVaR.

the value-at-risk (VaR) at the probability level $1-\varepsilon$ [19], which is expressed as follows:

$$\text{VaR}_{1-\varepsilon}\left(f\left(\mathbf{X},\tilde{\mathbf{P}}^{\text{W}}(\xi)\right)\right) \leq 0 \qquad (8.59)$$

where the VaR is defined as follows, and $\mathbf{X}$ are the decision variables in day-ahead operation without uncertainty.

$$\mathbf{X} = \left[P_{i,t}^{\text{CHP}}, R_{i,t}^{\text{CHP,UR}}, P_{i,t}^{\text{HP}}, R_{i,t}^{\text{HP}}, P_{i,t}^{\text{TU}}, R_{i,t}^{\text{TU,UR}}\right]; \quad \tilde{\mathbf{P}}^{\text{W}}(\xi) = \left[P_{i,t}^{\text{W}}(\xi)\right] \qquad (8.60)$$

CVaR is a conditional expectation of the load shedding taking into account the right tail of the load shedding distribution. The relationship between VaR and CVaR is shown in Fig. 8.3. It is noted that $\text{CVaR}_{1-\varepsilon}\left(f\left(\mathbf{X}, \tilde{\mathbf{P}}^{\text{W}}(\xi)\right)\right) \geq \text{VaR}_{1-\varepsilon}\left(f\left(\mathbf{X}, \tilde{\mathbf{P}}^{\text{W}}(\xi)\right)\right)$, and thus $\text{CVaR}_{1-\varepsilon}\left(f\left(\mathbf{X}, \tilde{\mathbf{P}}^{\text{W}}(\xi)\right)\right) \leq 0$ is a safe and sufficient approximation to the VaR constraint (8.59). Since the VaR is difficult to optimize due to its nonconvexity [20], CVaR is adopted as an alternative to approximate the chance constraint (8.34).

It is hard to derive the analytical expression of CVaR constraints due to the integral operation and unavailable probability distribution function (PDF) of uncertain wind power. Thus, the PDF of wind power is represented using a scenario set $\{\tilde{\text{P}}_{s}^{\text{W}}, s \in \text{S}\}$, and $\text{CVaR}_{1-\varepsilon}\left(f\left(\mathbf{X}, \tilde{\text{P}}^{\text{W}}(\xi)\right)\right)$ is approximately computed in a discrete form by minimizing Eq. (8.61):

$$F_{1-\varepsilon}(\mathbf{X}, \alpha) = \alpha + \frac{1}{\varepsilon}\sum_{s=1}^{S} \pi_{s}\left[f\left(\mathbf{X},\tilde{\mathbf{P}}_{s}^{\text{W}}\right) - \alpha\right]^{+} \qquad (8.61)$$

where π_{s} is the probability of scenarios, and the indicator $[\,a\,]^{+} = \max\{0,\ a\,\}$.

Then, the original nonlinear chance constraint (8.34) is finally reformulated as the following linear constraints (8.62) and (8.63), where δ_{s} is an

Table 8.1 Description of five study cases.

Case	Primary FRR Provision			Following Reserve Provision		
	Thermal Unit + CHP	HP	Method	Thermal Unit	CHP	HP
1	✓	✗	M2	✓	✗	✗
2	✓	✓	M2	✓	✗	✗
3	✓	✓	M1	✓	✗	✗
4	✓	✓	M2	✓	✓	✗
5	✓	✓	M2	✓	✓	✓

introduced auxiliary variable.

$$\text{CVaR}_{1-\varepsilon}\left(f\left(\mathbf{X},\tilde{\mathbf{P}}^{\mathrm{W}}(\boldsymbol{\xi})\right)\right) = \alpha + \frac{1}{\varepsilon}\sum_{s=1}^{S}\pi_s\delta_s \leq 0 \qquad (8.62)$$

$$\delta_s \geq f\left(\mathbf{X},\tilde{\mathbf{P}}^{\mathrm{W}}(\boldsymbol{\xi})\right) - \alpha, \ \delta_s \geq 0 \qquad (8.63)$$

The mathematical formulation of the chance-constrained two-stage energy and reserve optimization in Section 8.4 is finally formulated as an MILP with objective (8.10) subject to the day-ahead operation constraints under normal conditions (Eqs. (8.2) and (8.3), (8.16)–(8.33), (8.35)–(8.44), and (8.62) and (8.63), contingency operation constraints following the outage of the largest generator (Eqs. (8.8) and (8.9), (8.46), (8.57)–(8.58)), and real-time regulation constraints under normal conditions (Eqs. (8.47)–(8.56)).

8.6 Simulation results

This section presents the simulation results on a small-scale IEHS to demonstrate the effectiveness of the proposed two-stage chance-constrained energy and multi-type reserves scheduling scheme. The proposed model is implemented using the YALMIP toolbox in MATLAB and solved by GUROBI 9.0.2 as the MILP solver.

The conventional energy and reserve optimization model M1 without the governor model (8.7) and frequency security requirements (8.9) [21] is adopted as a comparison with respect to the proposed model M2. Five cases are studied to analyze the impact of exploiting the flexibility from CHP units and HPs on the economic and reliable operation of the IEHS, which are listed in Table 8.1. Case 1 is considered as the basic case, where both conventional thermal units and CHP units provide primary FRRs, whereas only thermal units provide following reserve for wind power forecast errors.

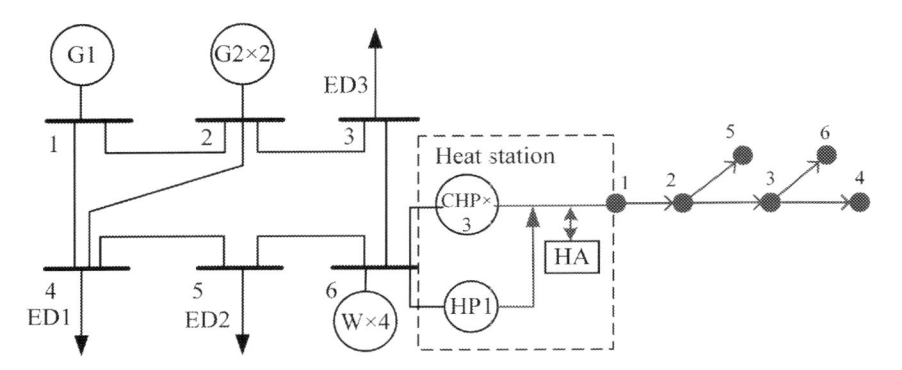

Figure 8.4 Test system with a 6-bus EPS and a 6-node DHS.

In Case 2, HPs are considered to provide primary FRRs. In Case 3, M1 is adopted instead of the proposed model M2 with frequency security requirements. In addition to Case 2, the flexibility from CHP units and HPs are used to provide following reserves in Cases 4 and 5, respectively.

8.6.1 Test system

The first test system consists of a modified 6-bus EPS and a 6-node DHS. The topology is shown in Fig. 8.4. The EPS and DHS are coupled with CHP units and HPs. In the EPS, there is one power plant (G1) with capacity of 85 MW, two 80-MW thermal power units (G2–G3), three 70-MW CHP units (CHP1–CHP3), one 30-MW HP, four wind farms (W1–W4), and three electric loads (ED1, ED2, and ED3). The wind power dataset is from the Wind Integration National Dataset (WIND) toolkit [22] and is enlarged to fit the system capacity. Five hundred wind power scenarios are generated and then reduced to 50 scenarios, which are used to calculate the CVaR value of load shedding. The governor droop parameters of conventional thermal units and CHP units are 4% and 3%, respectively. The electricity-to-heat ratio of HP1 is 3. The energy and reserve prices of thermal units and CHP units are from Zhang et al. [12]. G1 is the largest generator with constant output 85 MW, of which the outage is considered. System nominal frequency is 50 Hz, and system frequency regulation is limited by the maximum frequency deviation of 500 mHz. Since load shedding has worse impacts than wind spillage, the cost of load shedding and wind spillage are set at 500 $/MWh and 350 $/MWh, respectively. The defaulted confidence level controlling the load shedding is set at 0.9 in all cases. In the DHS, there is one HA1 with a capacity of 20 MW and three heat exchange

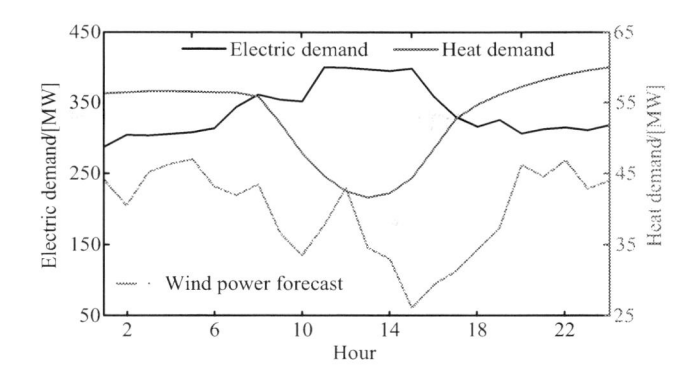

Figure 8.5 Hourly electric demand, heat demand, and wind power forecast in the test system.

stations, where the constant mass flow and variant temperature (CF-VT) control strategy is used. The parameters of the district heating pipelines are from Li et al. [23]. The hourly electric load and heat load are shown in Fig. 8.5.

8.6.2 Impact of exploiting additional reserves from CHP units and HPs on scheduling results

First, the impact of utilizing the flexibility from HPs to provide primary FRRs and the effectiveness of considering frequency security requirements are analyzed through comparing the scheduling results in Cases 1 through 3. Table 8.2 shows the day-ahead operation costs, real-time operation costs, and total costs in Cases 1 through 3 following the outage of G1, as well as the primary FRR scheduling results at the first hour. Table 8.3 shows the day-ahead unit commitment in Cases 1 through 3. Each value in the table indicates the total number of thermal units and CHP units that are ON in these cases. It can be seen from Table 8.2 that at the first hour in Case 1, there are 39.23-MW primary FRRs from thermal units plus 45.77-MW primary FRRs from CHP units compensating the loss of 85 MW, where HP1 does not contribute to the frequency response. In Case 2, HP1 contributes an additional 12.504-MW primary FRRs for the EPS in addition to the 33.46-MW and 39.036-MW primary FRRs provided by thermal units and CHP units, which reduces the system frequency deviation from −0.490 to −0.418 Hz. This demonstrates that exploiting the flexibility from HPs to provide primary FRRs can improve the system frequency regulation.

Table 8.2 Total system cost and primary FRR provision under three different cases.

Case	Costs ($)			Primary FRR at the First Hour (MW)						Δf (Hz)
	C^{DA}	C^{RT}	C^{TOTAL}	G2	G3	CHP1	CHP2	CHP3	HP	
1	8.070×10^5	2.685×10^5	1.076×10^6	19.615	19.615	22.885	0	22.885	0	−0.490
2	7.865×10^5	2.641×10^5	1.051×10^6	16.730	16.730	19.518	0	19.518	12.504	−0.418
3	6.894×10^5	2.181×10^5	9.075×10^5	23.496	24	25	0	0	12.504	/

Table 8.3 Total unit commitment of thermal units and CHP units under three different cases.

Case	Hours (1–24)
1	4 4 4 4 4 4 4 5 5 5 5 5 5 5 5 5 5 5 5 5 4 4 4 4
2	4 4 4 4 4 4 4 4 4 4 5 5 5 5 4 4 4 3 4 3 3 3 3 3
3	3 3 3 3 3 3 4 4 4 4 5 5 5 5 5 5 4 5 4 4 4 3 3 3

Table 8.4 Total system cost and expected wind curtailment under three different cases.

Case	C^{DA} (\$)	C^{RT} (\$)	C^{TOTAL} (\$)	Expected Wind Curtailment (%)
2	7.865×10^5	2.641×10^4	1.051×10^6	13.873
4	6.807×10^5	1.871×10^5	8.678×10^5	8.610
5	6.860×10^5	1.351×10^3	8.211×10^5	5.608

Moreover, with exploiting the additional primary FRRs from HPs, the total system cost in Case 2 decreases by 2.32% compared with that in Case 1. As can be seen from Table 8.3, the dispatched generating units in Case 2 are less than those in Case 1. All thermal units and CHP units in Case 1 are turned on, and some thermal units have to operate at their minimum outputs most of the time to provide FRRs to cover the outage of the largest generator. This leads to a higher operation cost in Case 1. Even though the overall operation cost in Case 3 is lower than that in Case 2, the primary FRR scheduling in Case 3 may be insecure and infeasible because method M1 adopted in Case 3 does not include frequency security requirements. Taking the first hour as an example, Table 8.2 shows that CHP1 is scheduled to provide 25-MW primary FRR. However, according to the governor droop characteristic shown in Eq. (8.7), it can only deliver the full scheduled 25-MW primary FRR when the system frequency deviation is equal to or below −0.536 Hz, which violates the system frequency security limit.

Second, Cases 4 and 5 are studied to illustrate the benefits of utilizing the flexibility from CHP units and HPs to provide following reserves for offsetting wind power forecast errors. Case 2 is considered as a benchmark. The comparison results of system operation cost and expected wind power curtailment in real-time operation are listed in Table 8.4. In Case 2, conventional thermal units are the only providers of following reserves. Operating at their minimum outputs most of the time, thermal units cannot regulate down and the system downward following reserve is thus insufficient. In Case 4, CHP units are considered to provide upward and downward following reserves together with thermal units. Compared with Case 2, the total operation cost and wind power curtailment in Case 4 decreases by

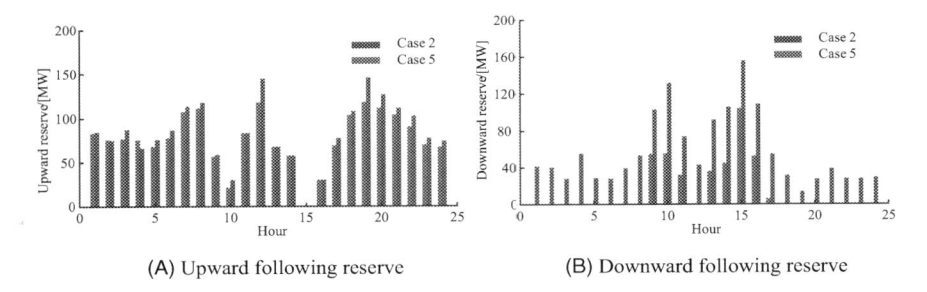

(A) Upward following reserve   (B) Downward following reserve

Figure 8.6 Optimal scheduling of upward and downward following reserves in Cases 2 and 5.

Table 8.5 Optimal results with different risk levels in Case 5.

Risk Level [	Total cost ($)	R^{up} (MW)	Expected Load Shedding (MW)
1%	8.434×10^5	2.417×10^3	0
3%	8.339×10^5	2.324×10^3	1.699
6%	8.299×10^5	2.148×10^3	5.964
10%	8.211×10^5	2.015×10^3	10.924
SP	7.822×10^5	1.785×10^3	29.202

17.43% and 37.94%, respectively. Utilizing the reserves from HPs, the total operation cost and wind power curtailment in Case 5 are reduced further compared to those in Case 4. Fig. 8.6 shows the scheduled upward and downward following reserves in Case 2 and Case 5, respectively. As can be seen, compared with Case 2, both the upward and downward following reserves in Case 5 are increased after exploiting the reserves from CHP units and HPs, and hence more wind power generation is utilized in the real-time stage.

8.6.3 Impact of chance constraints with adjustable risk levels on scheduling results

To demonstrate the effectiveness of the proposed chance-constrained optimization, the energy and reserves are optimized with adjustable risk levels and SP is chosen as a benchmark for comparison. The impact of risk level [on scheduling results is investigated based on Case 5. The optimal scheduling results are listed in Table 8.5. The third column is the total upward following reserve provided over the scheduling horizon (i.e., 24 hours). The fourth column denotes the expected loading shedding in the second stage. The table shows that with reducing the risk level ε from 10% to 1%, the upward following reserve increases and the expected load shedding decreases, which leads to a more conservative scheduling with higher operation cost. It can be seen that the total operation cost in SP is the lowest, whereas the amount of expected load shedding is the highest, since no constraint is imposed on it. In the proposed chance-constrained optimization with risk level [set at 10% and 3%, the expected load shedding decreases by 62.59% and 94.18%, respectively, compared with that in SP.

8.7 Conclusion

This chapter proposed a two-stage chance-constrained energy and multi-type reserves scheduling scheme for IEHSs considering wind forecast errors and the outage of the largest generator. First, the flexibility of CHP units and HPs are modeled to provide two types of ancillary services: following reserves for offsetting wind forecast errors and primary FRRs for arresting system frequency decline after the outage of the largest generator. Then, the probability of load shedding caused by wind uncertainty is managed by chance constraints, which are approximated by the conditional VaR. The proposed scheduling scheme is finally reformulated as an MILP. Simulation results demonstrate that exploiting the primary FRRs from HPs can alleviate the frequency regulation burden on conventional generators and reduce the online generating units operating at their minimum outputs to provide primary FRRs, leading to the improvement of system frequency regulation and economic efficiency. In addition, CHP units and HPs can further increase system economic efficiency and wind power integration by providing following reserves to offset wind forecast errors. The probability of load shedding can also be managed via a varying risk level, which achieves the balance between system economic efficiency and wind power uncertainty.

References

[1] G Strbac, D Pudjianto, M Aunedi, P Djapic, F Teng, X Zhang, et al., Role and value of flexibility in facilitating cost-effective energy system decarbonisation, Prog Energy 2 (2020) 42001.
[2] H Holttinen, M Milligan, S Member, E Ela, N Menemenlis, J Dobschinski, et al., Methodologies to determine operating reserves due to increased wind power, IEEE Trans Sustain Energy 3 (2012) 713–723.
[3] Z Chu, U Markovic, G Hug, F Teng, Towards optimal system scheduling with synthetic inertia provision from wind turbines, IEEE Trans Power Syst 35 (5) (2020) 4056–4066.
[4] A Bloess, WP Schill, A Zerrahn, Power-to-heat for renewable energy integration: a review of technologies, modeling approaches, and flexibility potentials, Appl Energy 212 (2018) 1611–1626.

[5] PD Lund, J Lindgren, J Mikkola, J Salpakari, Review of energy system flexibility measures to enable high levels of variable renewable electricity, Renew Sustain Energy Rev 45 (2015) 785–807.

[6] W Meesenburg, WB Markussen, T Ommen, B Elmegaard, Optimizing control of two-stage ammonia heat pump for fast regulation of power uptake, Appl Energy 271 (2020) 115126.

[7] X Xu, Q Lyu, M Qadrdan, J Wu, Quantification of flexibility of a district heating system for the power grid, IEEE Trans Sustain Energy 11 (4) (2020) 2617–2630.

[8] Z Pan, Q Guo, H Sun, Feasible region method based integrated heat and electricity dispatch considering building thermal inertia. Appl Energy 192 (2017) 395–407.

[9] Y Zhou, W Hu, L Zheng, Y Min, L Chen, Z Lu, L Dong, Power and energy flexibility of district heating system and its application in wide-area power and heat dispatch, Energy 190 (2020) 116426.

[10] F Teng, V Trovato, G Strbac, Stochastic scheduling with inertia-dependent frequency regulation, IEEE Trans Power Syst 31 (2) (2016) 1557–1566.

[11] J Tan, Q Wu, W Wei, F Liu, C Li, B Zhou, Decentralized robust energy and reserve co-optimization for multiple integrated electricity and heating systems, Energy 205 (2020) 118040.

[12] M Zhang, Q Wu, J Wen, B Pan, S Qi, Two-stage stochastic optimal operation of integrated electricity and heat system considering reserve of flexible devices and spatial-temporal correlation of wind power, Appl Energy 275 (2020) 115357.

[13] A Turk, Q Wu, M Zhang, J Østergaard, Day-ahead stochastic scheduling of integrated multi-energy system for flexibility synergy and uncertainty balancing, Energy 196 (2020) 117130.

[14] Y Zhang, J Wang, B Zeng, Z Hu, Chance-constrained two-stage unit commitment under uncertain load and wind power output using bilinear benders decomposition, IEEE Trans Power Syst 32 (2017) 3637–3647.

[15] H Holttinen, M Milligan, S Member, E Ela, N Menemenlis, J Dobschinski, et al., Methodologies to determine operating reserves due to increased wind power, IEEE Trans Sustain. Energy 3 (2012) 713–723.

[16] X Chen, MB McElroy, C Kang, Integrated energy systems for higher wind penetration in China: formulation, implementation, and impacts, IEEE Trans Power Syst 33 (2018) 1309–1319.

[17] JF Restrepo, FD Galiana, Unit commitment with primary frequency regulation constraints, IEEE Trans Power Syst 20 (2005) 1836–1842.

[18] D Bertsimas, JN Tsitsiklis, Introduction to Linear Optimization, Athena Scientific, Belmont, MA, 1997.

[19] X Geng, L Xie, Data-driven decision making in power systems with probabilistic guarantees: theory and applications of chance-constrained optimization, Ann Rev Control 47 (2019) 341–363.

[20] RT Rockafellar, S Uryasev, Optimization of conditional value-at-risk, J Risk 2 (1999) 21–42.

[21] J Wang, M Shahidehpour, Z Li, S Member, Contingency-constrained reserve requirements in joint energy and ancillary services auction, IEEE Trans Power Syst 24 (2009) 1457–1468.

[22] C Draxl, A Clifton, BM Hodge, J McCaa, The Wind Integration National Dataset (WIND) toolkit, Appl Energy 151 (2015) 355–366.

[23] Z Li, W Wu, M Shahidehpour, J Wang, B Zhang, Combined heat and power dispatch considering pipeline energy storage of district heating network, IEEE Trans Sustain Energy 7 (1) (2016) 12–22.

CHAPTER 9

Day-ahead stochastic optimal operation of the integrated electricity and heating system considering reserve of flexible devices

9.1 Introduction

Wind power has been growing rapidly over the past decade. In Denmark, 47% of the power consumption was from the wind in 2019 [1]. The significant uncertainties and variations of wind power may result in the difficulty of scheduling in the power system [2]. Sufficient flexibility is needed to balance the wind power fluctuation and accommodate more wind power in a grid [3].

The integrated electricity and heat system (IEHS), composed of the electric power system (EPS) and district heating system (DHS), has been proposed to improve the flexibility of the EPS [4] and accommodate the increasing penetration of wind power [5]. Extensive research was conducted on the modeling and solution of the optimal operation of the IEHS, including thermal and hydraulic processes of the DHS [6], the time delay of pipelines [7], heat regulation modes [8], thermal inertia of buildings [9], heat storage of the DHS network [10], market framework design of the IEHS [11], and distributed operation of the EPS and DHS [12].

In the IEHS, the EPS and DHS are mainly coupled by the combined heat and power (CHP) units [13,14]. In addition, the coupling between these two energy sectors can be relaxed by the integration of large-scale heat pumps (HPs), electric boilers (EBs), and heat storage tanks (STs). Chen et al. [15] consider both STs and EBs to improve flexibility on the heat side in the economic dispatch. Based on that work [15], a more recent work by Chen et al. [16] considers the unit commitment of CHP units to further improve operational flexibility. Zhang et al. [17] employ EBs to reduce wind curtailment and CO_2 emission in the power system chronological

Optimal operation of integrated multi-energy systems under uncertainty. Copyright © 2022 Elsevier Inc.
DOI: https://doi.org/10.1016/B978-0-12-824114-1.00011-1   All rights reserved.

simulation. Nielsen et al. [18] consider the EBs, HPs, and STs to increase flexibility in the DHS.

In the preceding research, the flexibility from the HPs, EBs, STs, and CHP units is mainly investigated in energy scheduling. Little research exploits the potential and benefits of these flexible devices in providing reserve and heat regulation in the IEHS. Biegel et al. [19] and Wang et al. [20] utilize small-scale HPs on the residential side to provide reserves. However, small-scale HPs are used for individual heating, and they are not connected to the IEHS. Cai et al. [21] model the reserve capacity of the EBs, but it is not incorporated into the scheduling of the IEHS. Verda et al. [22] propose a multiscale model of STs, which is based on the partial differential equation. However, such a model is highly nonlinear, limiting its application in the scheduling of the IEHS. Zugno et al. [23] consider the heat regulation from accumulator tanks in the two-stage unit commitment. However, HPs and EBs are not simultaneously considered to provide reserve.

The CHP units are usually classified into two categories: the back-pressure unit and extraction condensing unit [13]. Tan et al. [24] consider the reserve modeling of back-pressure CHP units in the economic dispatch and utilize the thermal inertia of buildings to balance the heat regulation caused by reserve deployment. Little work has been done on the reserve modeling of condensing CHP units. The operational region of a condensing CHP unit is usually represented by the linear combination of operational region vertices. Lahdelma and Hakonen [25] and Abdolmohammadi and Kazemi [26] model the operational region of condensing CHP units based on convex and nonconvex hull vertices, respectively. However, those authors [25, 26] do not consider the reserve provision in the modeling of condensing CHP units.

In addition, due to the uncertainty of wind power, the decision-making methods under uncertainty are also needed in the scheduling of the IEHS. Robust optimization and stochastic programming are the two popular techniques to cope with uncertainty in the IEHS. The former pursues the feasibility and minimum operational cost of the decision under the worst-case scenario but suffers from the conservativeness [27]. The latter focuses on the expected cost and can reduce the conservativeness [28]. This chapter utilizes stochastic programming to cope with the uncertainty of renewables.

Based on the aforementioned research gaps, this chapter develops a detailed reserve provision and heat regulation model of condensing CHP units, EBs, HPs, and STs in the optimal operation of the IEHS. In addition, two-stage stochastic programming is utilized to cope with uncertainties and co-optimize energy and reserve.

9.2 Two-stage stochastic optimal dispatching scheme of the IEHS

9.2.1 Structure of the IEHS

The schematic representation of the IEHS is depicted in Fig. 9.1. The heat and power sectors are coupled by CHP plants and large-scale HPs and EBs. CHP units co-generate power and thermal energy, whereas HPs and EBs consume power to produce thermal energy. In the power sector, electricity demand is also supplied by conventional thermal power plants and wind power. In the heat sector, the thermal STs are used to release the flexibility of CHP plants, HPs, and EBs.

9.2.2 Framework of the dispatching scheme

The framework of the proposed two-stage stochastic optimal dispatching scheme for the IEHS is illustrated in Fig. 9.2. In the proposed scheme, the system operator aims to achieve the optimal operation of the IEHS and balance the uncertainty of renewables in an economic way by coordinating the reserve deployment and heat regulation from conventional thermal units, CHP units, HPs, EBs, and STs while keeping the balance between supply and demand for both power and heat sectors.

In Fig. 9.2, the optimal operation of the IEHS involves two stages. The first stage corresponds to the day-ahead scheduling, and the second stage relates to the real-time regulation. By coordinating these two stages, the day-ahead decisions are made with the impact of real-time uncertainties incorporated.

As shown in Fig. 9.2, the optimal operation of the IEHS minimizes the operational cost including the day-ahead energy cost and expected real-time regulation cost. The operational constraints of the IEHS can be divided into four parts according to the energy carriers and operational stages: (1) day-ahead operational constraints of the EPS, (2) day-ahead operational constraints of the DHS, (3) real-time operational constraints of the EPS in each uncertainty scenario, and (4) real-time operational constraints of the DHS in each uncertainty scenario.

In the day-ahead stage, the EPS and DHS are coupled by the operational region of the CHP units, HPs, and EBs. In the real-time stage, when the

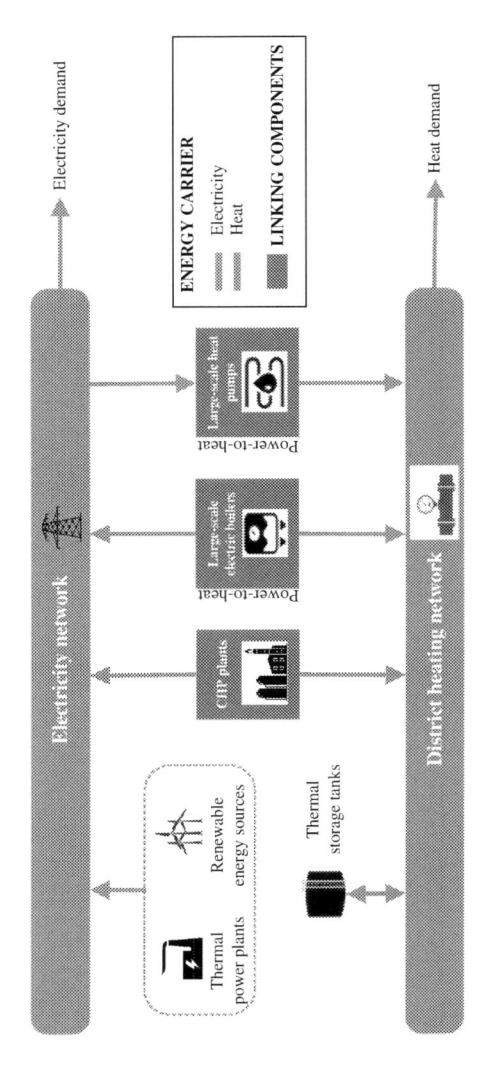

Figure 9.1   Structure of the IEHS.

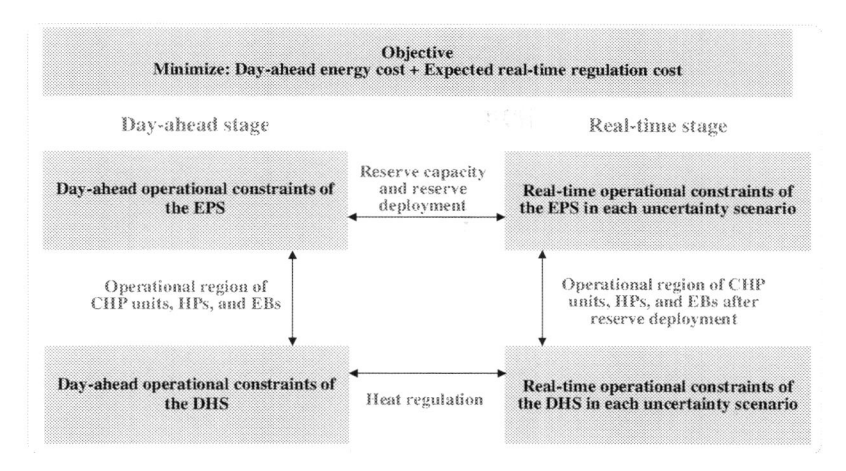

Figure 9.2 Framework of two-stage stochastic optimal dispatching scheme of the IEHS.

reserve deployment and heat regulation are implemented, the two energy sectors are still linked by the operational region of the CHP units, HPs, and EBs. In the EPS, the day–ahead stage and the real–time stage are coupled by the reserve capacity and reserve deployment in each uncertainty scenario of wind power. In the DHS, the two stages are linked by the heat regulation.

In the dispatching scheme, all units and devices can be divided into three categories according to their functions. The first category includes conventional thermal units, which can participate in both the energy and reserve scheduling in the power sector. The second category includes the CHP units, HPs, and EBs, which not only participate in the energy and reserve scheduling in the power sector but also produce heat in the heat sector. The third category includes the STs, which store or release heat energy to balance the heat supply and heat demand.

9.3 Reserve provision and heat regulation from condensing CHP units

The reserve provision and heat regulation model of condensing CHP units is developed in this section, which considers the unit commitment and coupling between reserve deployment and heat regulation.

The operational region of a condensing CHP unit is restricted by multiple boundaries, as shown in Fig. 9.3. The boundaries AB, BC, CD, and DA reflect the maximum limit of power output, maximum limit of

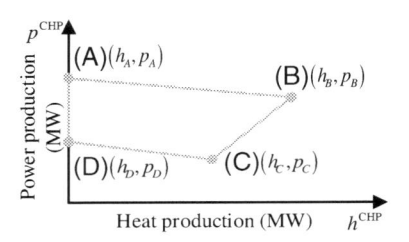

Figure 9.3 Operational region of a condensing CHP unit for heat and power production.

fuel injection, maximum heat rate, and minimum limit of steam injection, respectively [15].

The heat and power production with on/off status of unit commitment is first formulated in Eqs. (9.1) through (9.5) due to their close relation with the reserve capacity in a CHP unit. Any heat and power production levels in the operational region can be represented by the linear combination of the convex–hull vertices [16], as shown in Fig. 9.3. We have the following formulations:

$$p_{i,t}^{CHP} = \alpha_{i,t,A}p_i^{CHP,A} + \alpha_{i,t,B}p_i^{CHP,B} + \alpha_{i,t,C}p_i^{CHP,C} + \alpha_{i,t,D}p_i^{CHP,D},$$
$$\forall t \in \Lambda^T, \forall i \in \Psi^{CHP} \tag{9.1}$$

$$h_{i,t}^{CHP} = \alpha_{i,t,A}h_i^{CHP,A} + \alpha_{i,t,B}h_i^{CHP,B} + \alpha_{i,t,C}h_i^{CHP,C} + \alpha_{i,t,D}h_i^{CHP,D},$$
$$\forall t \in \Lambda^T, \forall i \in \Psi^{CHP} \tag{9.2}$$

$$f_{i,t}^{CHP} = \alpha_{i,t,A}f_i^{CHP,A} + \alpha_{i,t,B}f_i^{CHP,B} + \alpha_{i,t,C}f_i^{CHP,C} + \alpha_{i,t,D}f_i^{CHP,D},$$
$$\forall t \in \Lambda^T, \quad \forall i \in \Psi^{CHP} \tag{9.3}$$

$$\alpha_{i,t,A} + \alpha_{i,t,B} + \alpha_{i,t,C} + \alpha_{i,t,D} = x_{i,t}^{CHP}, \forall t \in \Lambda^T, \quad \forall i \in \Psi^{CHP} \tag{9.4}$$

$$0 \leq \alpha_{i,t,A}, \alpha_{i,t,B}, \alpha_{i,t,C}, \alpha_{i,t,D} \leq 1, \forall t \in \Lambda^T, \quad \forall i \in \Psi^{CHP} \tag{9.5}$$

where Ψ^{CHP} represents the set of indices of CHP units, Λ^T represents the set of indices of time periods, $p_i^{CHP,A/B/C/D}$ represents power production of vertices A/B/C/D for CHP unit i, $h_i^{CHP,A/B/C/D}$ represents heat production of vertices A/B/C/D for CHP unit i, $f_i^{CHP,A/B/C/D}$ represents fuel cost of vertices A/B/C/D for CHP unit i, $x_{i,t}^{CHP}$ represents the on/off status of thermal/CHP unit i in period t, and $\alpha_{i,t,A/B/C/D}$ represents the coefficient of vertices A/B/C/D of CHP unit i in period t.

In Eqs. (9.1) through (9.5), Eqs. (9.1) and (9.2) represent the power production and heat production, respectively; Eq. (9.3) is the simplified linearized fuel cost function; Eq. (9.4) combines the coefficients of vertices with the on/off status of unit commitment; and Eq. (9.5) limits the range of coefficients.

On the basis of Eqs. (9.1) through (9.5), a detailed reserve provision and heat regulation model of the condensing CHP units is proposed, which includes the reserve capacity in day–ahead and reserve deployment and heat regulation in real time. Define variables $\Delta\alpha_{i,t,s,A}^{\mathrm{UR/DR}}$, $\Delta\alpha_{i,t,s,B}^{\mathrm{UR/DR}}$, $\Delta\alpha_{i,t,s,C}^{\mathrm{UR/DR}}$, and $\Delta\alpha_{i,t,s,D}^{\mathrm{UR/DR}}$, and the linkage between reserve deployment and heat regulation in real time can be established as follows:

$$r_{i,t}^{\mathrm{CHP,UR}} \leq \min\{\mathrm{U}_i^{\mathrm{CHP}}x_{i,t}^{\mathrm{CHP}}, \bar{\mathrm{P}}_i^{\mathrm{CHP}}x_{i,t}^{\mathrm{CHP}} - p_{i,t}^{\mathrm{CHP}}\}, \forall t \in \Lambda^{\mathrm{T}}, \forall i \in \Psi^{\mathrm{CHP}}$$
$$(9.6)$$

$$r_{i,t}^{\mathrm{CHP,DR}} \leq \min\{\mathrm{D}_i^{\mathrm{CHP}}x_{i,t}^{\mathrm{CHP}}, p_{i,t}^{\mathrm{CHP}} - \underline{\mathrm{P}}_i^{\mathrm{CHP}}x_{i,t}^{\mathrm{CHP}}\}, \forall t \in \Lambda^{\mathrm{T}}, \forall i \in \Psi^{\mathrm{CHP}}$$
$$(9.7)$$

$$0 \leq \Delta r_{i,t,s}^{\mathrm{CHP,UR}} \leq r_{i,t}^{\mathrm{CHP,UR}}, \forall t \in \Lambda^{\mathrm{T}}, \forall i \in \Psi^{\mathrm{CHP}}, \forall s \in \Psi^{\mathrm{S}} \qquad (9.8)$$

$$0 \leq \Delta r_{i,t,s}^{\mathrm{CHP,DR}} \leq r_{i,t}^{\mathrm{CHP,DR}}, \forall t \in \Lambda^{\mathrm{T}}, \forall i \in \Psi^{\mathrm{CHP}}, \forall s \in \Psi^{\mathrm{S}} \qquad (9.9)$$

$$\Delta r_{i,t,s}^{\mathrm{CHP,UR}} = \sum_j \Delta\alpha_{i,t,s,j}^{\mathrm{UP}}\mathrm{p}_i^{\mathrm{CHP},j}, \; j \in \{A, B, C, D\},$$

$$\forall t \in \Lambda^{\mathrm{T}}, \forall i \in \Psi^{\mathrm{CHP}}, \forall s \in \Psi^{\mathrm{S}} \qquad (9.10)$$

$$\Delta r_{i,t,s}^{\mathrm{CHP,DR}} = \sum_j \Delta\alpha_{i,t,s,j}^{\mathrm{DN}}\mathrm{p}_i^{\mathrm{CHP},j}, \; j \in \{A, B, C, D\},$$

$$\forall t \in \Lambda^{\mathrm{T}}, \; \forall i \in \Psi^{\mathrm{CHP}}, \; \forall s \in \Psi^{\mathrm{S}} \qquad (9.11)$$

$$\Delta\alpha_{i,t,s,A}^{\mathrm{UP}}+\Delta\alpha_{i,t,s,B}^{\mathrm{UP}}+\Delta\alpha_{i,t,s,C}^{\mathrm{UP}}+\Delta\alpha_{i,t,s,D}^{\mathrm{UP}} = 0, \forall t \in \Lambda^{\mathrm{T}}, \forall i \in \Psi^{\mathrm{CHP}}, \forall s \in \Psi^{\mathrm{S}}$$
$$(9.12)$$

$$\Delta\alpha_{i,t,s,A}^{\mathrm{DN}}+\Delta\alpha_{i,t,s,B}^{\mathrm{DN}}+\Delta\alpha_{i,t,s,C}^{\mathrm{DN}}+\Delta\alpha_{i,t,s,D}^{\mathrm{DN}} = 0, \forall t \in \Lambda^{\mathrm{T}}, \forall i \in \Psi^{\mathrm{CHP}}, \forall s \in \Psi^{\mathrm{S}}$$
$$(9.13)$$

$$0 \leq \alpha_{i,t,j} + \Delta\alpha_{i,t,s,j}^{\mathrm{UP}} \leq 1, j \in \{A, B, C, D\}, \forall t \in \Lambda^{\mathrm{T}}, \forall i \in \Psi^{\mathrm{CHP}}, \forall s \in \Psi^{\mathrm{S}}$$
$$(9.14)$$

$$0 \leq \alpha_{i,t,j} - \Delta\alpha_{i,t,s,j}^{\mathrm{DN}} \leq 1, j \in \{A, B, C, D\}, \forall t \in \Lambda^{\mathrm{T}}, \forall i \in \Psi^{\mathrm{CHP}}, \forall s \in \Psi^{\mathrm{S}}$$
$$(9.15)$$

$$0 \leq \alpha_{i,t,j} + \Delta\alpha_{i,t,s,j}^{\mathrm{UP}} - \Delta\alpha_{i,t,s,j}^{\mathrm{DN}} \leq 1, \, j \in \{A, B, C, D\}, \forall t \in \Lambda^{\mathrm{T}},$$
$$\forall i \in \Psi^{\mathrm{CHP}}, \quad \forall s \in \Psi^{\mathrm{S}} \tag{9.16}$$

$$\Delta h_{i,t,s}^{\mathrm{CHP,UP}} = \sum_{j} \Delta\alpha_{i,t,s,j}^{\mathrm{UP}} \mathrm{h}_{i,t}^{\mathrm{CHP},j}, \forall t \in \Lambda^{\mathrm{T}}, \forall i \in \Psi^{\mathrm{CHP}}, \forall s \in \Psi^{\mathrm{S}} \tag{9.17}$$

$$\Delta h_{i,t,s}^{\mathrm{CHP,DN}} = \sum_{j} \Delta\alpha_{i,t,s,j}^{\mathrm{DN}} \mathrm{h}_{i,t}^{\mathrm{CHP},j}, \forall t \in \Lambda^{\mathrm{T}}, \forall i \in \Psi^{\mathrm{CHP}}, \forall s \in \Psi^{\mathrm{S}} \tag{9.18}$$

where Ψ^{S} represents the set of indices of wind power scenarios, $\Delta\alpha_{i,t,s,j}^{\mathrm{UP/DN}}$ represents coefficient regulation of vertice j for CHP unit i in period t under scenario s relating to upward/downward reserve, $\Delta h_{i,t,s}^{\mathrm{CHP,UP/DN}}$ represents heat regulation of CHP unit i in period t under scenario s relating to $\Delta\alpha_{i,t,s,j}^{\mathrm{UP/DN}}$, $r_{i,t}^{\mathrm{CHP,UR/DR}}$ represents upward/downward reserve of CHP unit i in period t, $\bar{\mathrm{P}}_i^{\mathrm{CHP}}/\underline{\mathrm{P}}_i^{\mathrm{CHP}}$ represents maximum/minimum output of CHP unit i, and $\Delta r_{i,t,s}^{\mathrm{CHP,UR/DR}}$ represents upward/downward reserve deployment for CHP unit i in period t under scenario s.

In Eqs. (9.6) through (9.18), Eqs. (9.6) and (9.7) represent the upward and downward reserve capacity in day–ahead, and Eqs. (9.8) through (9.18) represent the reserve deployment and heat regulation in each scenario in real time. Among Eqs. (9.8) through (9.18), Eqs. (9.8) and (9.9) ensure the upward and downward reserve deployments in each scenario no more than the reserve capacity; Eqs. (9.10) and (9.11) utilize the linear combination of convex-hull vertices to represent the reserve deployment in each scenario; Eqs. (9.12) and (9.13) represent the relation of the regulation coefficient of four vertices, which can be deduced from Eq. (9.4) and $(\alpha_{i,t,A} + \Delta\alpha_{i,t,s,A}^{\mathrm{DR/UR}}) + (\alpha_{i,t,B} + \Delta\alpha_{i,t,s,B}^{\mathrm{DR/UR}}) + (\alpha_{i,t,C} + \Delta\alpha_{i,t,s,C}^{\mathrm{DR/UR}}) + (\alpha_{i,t,D} + \Delta\alpha_{i,t,s,D}^{\mathrm{DR/UR}}) = x_{i,t}^{\mathrm{CHP}}$; Eqs. (9.14) through (9.16) limit the range of regulation coefficient; and Eqs. (9.17) and (9.18) represent how the heat regulation relates to reserve deployment.

9.4 Mathematical formulation of stochastic optimal operation of the IEHS

According to the framework in Section 9.2, the mathematical formulation of the two–stage stochastic optimal operation of the IEHS is detailed in this section.

9.4.1 Objective function of the IEHS

The objective function of the proposed dispatching scheme can be expressed as Eq. (9.19) and Eq. (9.20). It minimizes the total operational cost in two stages (i.e., the day–ahead operational cost and expected real-time regulation cost):

$$\min f = f_1 + \sum_{s=1}^{S} p_s f_{2,s} \tag{9.19}$$

$$f_1 = f_1^{CHP} + f_1^{TU} + f_1^{W}$$

$$f_{2,s} = f_{2,s}^{CHP} + f_{2,s}^{TU} + f_{2,s}^{W} + f_{2,s}^{LD}$$

$$f_1^{TU} = \sum_{t \in \Lambda^T} \sum_{i \in \Psi^{TU}} \left\{ C_i^{TU} u_{i,t}^{TU} + f_i^{\min,TU} x_{i,t}^{TU} \right.$$

$$\left. + \sum_{k \in \Phi^{TU}} K_{i,k}^{TU} p_{i,t,k}^{TU} + C_i^{TU,UR} r_{i,t}^{TU,UR} + C_i^{TU,DR} r_{i,t}^{TU,DR} \right\}$$

$$f_1^{CHP} = \sum_{t \in \Lambda^T} \sum_{i \in \Psi^{CHP}} \left\{ C_i^{CHP} u_{i,t}^{CHP} + f_{i,t}^{CHP} + C_{i,t}^{CHP,UR} r_{i,t}^{CHP,UR} \right.$$

$$\left. + C_{i,t}^{CHP,DR} r_{i,t}^{CHP,DR} \right\}$$

$$f_1^{W} = \sum_{t \in \Lambda^T} \sum_{i \in \Psi^{W}} \left\{ C^{W} \Delta w_{i,t} \right\}$$

$$f_{2,s}^{TU} = \sum_{t \in \Lambda^T} \sum_{i \in \Psi^{TU}} \left\{ C_i^{TU,UR,dep} \Delta r_{i,t,s}^{TU,UR} + C_i^{TU,DR,dep} \Delta r_{i,t,s}^{TU,DR} \right\}$$

$$f_{2,s}^{CHP} = \sum_{t \in \Lambda^T} \sum_{i \in \Psi^{CHP}} \left\{ C_i^{CHP,UR,dep} \Delta r_{i,t,s}^{CHP,UR} + C_i^{CHP,DR,dep} \Delta r_{i,t,s}^{CHP,DR} \right\}$$

$$f_{2,s}^{W} = \sum_{t \in \Lambda^T} \sum_{i \in \Psi^{W}} \left\{ C^{W} \Delta w_{i,t,s} \right\}$$

$$f_{2,s}^{LD} = \sum_{t \in \Lambda^T} \sum_{i \in \Psi^{B}} \left\{ C^{LD} \Delta L_{i,t,s} \right\} \tag{9.20}$$

where f represents the total operational cost, f_1 represents the day–ahead operational cost, $f_{2,s}$ represents the real-time regulation cost under scenario s, $f_1^{TU/CHP}$ represents the total operational cost of all thermal/CHP units in day–ahead, $f_{2,s}^{TU/CHP}$ represents the real-time regulation cost of all thermal/CHP units under scenario s, f_1^{W} represents the day–ahead penalty cost of wind power curtailment for all wind farms, $f_{2,s}^{W}$ represents the real-time penalty cost of wind power curtailment for all wind farms under scenario s, $f_{2,s}^{\mathrm{LD}}$ represents the real-time penalty cost of load shedding under scenario s, $f_{i,t}^{\mathrm{CHP}}$ represents the day-ahead energy cost of CHP unit i in period t, Ψ^{TU} represents the set of indices of thermal power units, Ψ^{W} represents the set of indices of wind farms, Ψ^{B} represents the set of indices of buses in the EPS, $C_i^{\mathrm{TU/CHP}}$ represents the startup cost of thermal/CHP unit i, C^{W} represents the penalty price of wind power curtailment, C^{LD} represents the penalty price of load shedding, $C_i^{\mathrm{TU,UR/DR}}$ represents the capacity price of upward/downward reserve of thermal unit i, $C_i^{\mathrm{TU,UR/DR,dep}}$ represents the price of upward/downward reserve deployment for thermal unit i, $C_i^{\mathrm{CHP,UR/DR,dep}}$ represents the price of upward/downward reserve deployment for CHP unit i, $C_i^{\mathrm{CHP,UR/DR}}$ represents the capacity price of upward/downward reserve of CHP unit i, $f_i^{\mathrm{min,TU}}$ represents the fuel cost corresponding to the minimum output level of thermal unit i, $p_{i,t}^{\mathrm{TU/CHP}}$ represents the power production of thermal/CHP unit i in period t, $p_{i,t,k}^{\mathrm{TU}}$ represents the scheduled value of segment k for thermal unit i in period t, $\Delta w_{i,t}$ represents wind power curtailment in the base case for wind farm i in period t, and $\Delta w_{i,t,s}$ represents wind power curtailment for wind farm i in period t under scenario s.

9.4.2 Day-ahead operational constraints of the EPS

Power balance constraint.

$$\sum_{i\in\Psi^{\mathrm{TU}}} p_{i,t}^{\mathrm{TU}} + \sum_{i\in\Psi^{\mathrm{CHP}}} \left(p_{i,t}^{\mathrm{CHP}} - p_{i,t}^{\mathrm{WP}}\right) - \sum_{i\in\Psi^{\mathrm{EB}}} p_{i,t}^{\mathrm{EB}} - \sum_{i\in\Psi^{\mathrm{HP}}} p_{i,t}^{\mathrm{HP}}$$
$$+ \sum_{i\in\Psi^{\mathrm{W}}} \left(\mathrm{W}_{i,t} - \Delta w_{i,t}\right) = \sum_{i\in\Psi^{\mathrm{B}}} \mathrm{L}_{i,t}, t \in \Lambda^{\mathrm{T}} \tag{9.21}$$

where $\mathrm{L}_{i,t}$ represents load at bus i in period t, $\mathrm{W}_{i,t}$ represents forecast value of wind farm i in period t, $p_{i,t}^{\mathrm{HP/EB}}$ represents electricity consumption of HP/EB i in period t, and $p_{i,t}^{\mathrm{WP}}$ represents electricity consumption of water pump (WP) i in period t.

Constraints of thermal units and CHP units.

$$\underline{\mathrm{P}}_i^{\mathrm{TU}} x_{i,t} \le p_{i,t}^{\mathrm{TU}} \le \bar{\mathrm{P}}_i^{\mathrm{TU}} x_{i,t}, \forall t \in \Lambda^{\mathrm{T}}, \forall i \in \Psi^{\mathrm{TU}} \tag{9.22}$$

$$p_{i,t}^{\mathrm{TU}} = \sum_{k\in\Phi^{\mathrm{TU/CHP}}} p_{i,t,k}^{\mathrm{TU}}, \forall t \in \Lambda^{\mathrm{T}}, \forall i \in \Psi^{\mathrm{TU}}, \forall k \in \Phi^{\mathrm{TU}} \tag{9.23}$$

$$0 \le p_{i,t,k}^{\mathrm{TU}} \le \bar{\mathrm{P}}_{i,k}^{\mathrm{TU}}, \forall t \in \Lambda^{\mathrm{T}}, \forall i \in \Psi^{\mathrm{TU}}, \forall k \in \Phi^{\mathrm{TU}} \tag{9.24}$$

$$p_{i,t}^{TU/CHP} - p_{i,t-1}^{TU/CHP} \leq U_i^{TU/CHP}(1 - u_{i,t}^{TU/CHP}) + \bar{P}_i^{TU/CHP}u_{i,t}^{TU/CHP},$$
$$\forall t \in \Lambda^T, \forall i \in \Psi^{TU/CHP}$$
$$(9.25)$$

$$p_{i,t-1}^{TU/CHP} - p_{i,t}^{TU/CHP} \leq D_i^{TU/CHP}(1 - v_{i,t}^{TU/CHP}) + \bar{P}_i^{TU/CHP}v_{i,t}^{TU/CHP},$$
$$\forall t \in \Lambda^T, \forall i \in \Psi^{TU/CHP}$$
$$(9.26)$$

$$u_{i,t}^{TU/CHP} - v_{i,t}^{TU/CHP} = x_{i,t}^{TU/CHP} - x_{i,t-1}^{TU/CHP}, \forall t \in \Lambda^T, \forall i \in \Psi^{TU/CHP}$$
$$(9.27)$$

$$u_{i,t}^{TU/CHP} + v_{i,t}^{TU/CHP} \leq 1, \forall t \in \Lambda^T, \forall i \in \Psi^{TU/CHP}$$
$$(9.28)$$

$$x_{i,t}^{TU/CHP} - x_{i,t-1}^{TU/CHP} \leq x_{i,\tau}^{TU/CHP}, \forall \tau \in \left[t + 1, \min\{N_T, t + t_{st,i}^{TU/CHP} - 1\}\right]$$
$$(9.29)$$

$$x_{i,t-1}^{TU/CHP} - x_{i,t}^{TU/CHP} \leq 1 - x_{i,\tau}^{TU/CHP}, \forall \tau \in \left[t + 1, \min\{N_T, t + t_{dn,i}^{TU/CHP} - 1\}\right]$$
$$(9.30)$$

$$r_{i,t}^{TU,UR} \leq \min\{U_i^{TU}x_{i,t}^{TU}, \bar{P}_i^{TU}x_{i,t}^{TU} - p_{i,t}^{TU}\}, \forall t \in \Lambda^T, \forall i \in \Psi^{TU} \quad (9.31)$$

$$r_{i,t}^{TU,DR} \leq \min\{D_i^{TU}x_{i,t}^{TU}, p_{i,t}^{TU} - \underline{P}_i^{TU}x_{i,t}^{TU}\}, \forall t \in \Lambda^T, \forall i \in \Psi^{TU} \quad (9.32)$$

where Φ^{TU} represents the set of indices of segments of the linearized fuel cost function of thermal units, $K_{i,k}^{TU}$ represents the slope of segment k of the linearized fuel cost function of thermal unit i, $t_{st,i}^{TU/CHP}$ represents the minimum startup time of thermal/CHP unit i, $t_{dn,i}^{TU/CHP}$ represents the minimum shutdown time of thermal/CHP unit i, $U_i^{TU/CHP}/D_i^{TU/CHP}$ represents the upward/downward ramping rate of thermal/CHP unit i, $\bar{P}_i^{TU}/\underline{P}_i^{TU}$ represents the maximum/minimum output of thermal unit i, $\bar{P}_i^{CHP}/\underline{P}_i^{CHP}$ represents the maximum/minimum output of CHP unit i, $\bar{P}_{i,k}^{TU}$ represents the maximum value of segment k of thermal unit i, $r_{i,t}^{TU,UR/DR}$ represents the upward/downward reserve of thermal unit i in period t, $r_{i,t}^{CHP,UR/DR}$ represents the upward/downward reserve of CHP unit i in period t, $u_{i,t}^{TU/CHP}$ represents the startup indicator of thermal/CHP unit i in period t, $v_{i,t}^{TU/CHP}$ represents the shutoff indicator of thermal/CHP unit i in period t, and $x_{i,t}^{TU}$ represents the on/off status of thermal unit i in period t.

In Eqs. (9.22) through (9.32), Eqs. (9.22) through (9.24) represent the output limits of thermal units with piece-wise linearization, Eqs. (9.25) and

(9.26) describe the upward and downward ramping rate limits of the thermal units and CHP units, Eqs. (9.27) through (9.30) model the changes of the on/off status of the thermal units and CHP units and their minimum up- and downtime requirements [29], and Eqs. (9.31) and (9.32) limit the reserve capacity of thermal units.

Constraints of HPs/EBs/WPs.

$$\underline{p}_i^{HP} x_{i,t}^{HP} \leq p_{i,t}^{HP} \leq \bar{p}_i^{HP} x_{i,t}^{HP}, \forall t \in \Lambda^T, \forall i \in \Psi^{HP} \tag{9.33}$$

$$0 \leq p_{i,t}^{EB} \leq \bar{p}_i^{EB}, \forall t \in \Lambda^T, \forall i \in \Psi^{EB} \tag{9.34}$$

$$\underline{p}_i^{WP} \leq p_{i,t}^{WP} \leq \bar{p}_i^{WP}, \forall t \in \Lambda^T, \forall i \in \Psi^{WP} \tag{9.35}$$

where Ψ^{WP} represents the set of indices of WPs, Ψ^{HP} represents the set of indices of HPs, Ψ^{EB} represents the set of indices of EBs, $\bar{p}_i^{HP}/\bar{p}_i^{EB}$ represents the maximum power consumption of HP/EB i, $\bar{p}_i^{WP}/\underline{p}_i^{WP}$ represents maximum/minimum power consumption of WP i, and $x_{i,t}^{HP}$ represents the on/off status of HPs.

Eq. (9.33) represents the power consumption constraint of HPs with the on/off status, Eq. (9.34) is the power consumption constraint of the EBs [18], and Eq. (9.35) represents the power consumption limit of WPs to sustain a certain level of pressure for the water cycle.

Constraints of wind power curtailment.

$$0 \leq \Delta w_{i,t} \leq W_{i,t}, \forall t \in \Lambda^T, \forall i \in \Psi^W \tag{9.36}$$

Constraints of transmission lines.

$$\left| \sum_{i \in \Theta^{TU,d}} g_{d,l}^{TU} p_{i,t}^{TU} + \sum_{i \in \Theta^{CHP,d}} g_{d,l}^{CHP} \left(p_{i,t}^{CHP} - p_{i,t}^{WP} \right) + \sum_{i \in \Theta^{EB,d}} g_{d,l}^{EB} p_{i,t}^{EB} \right.$$
$$\left. + \sum_{i \in \Theta^{HP,d}} g_{d,l}^{HP} p_{i,t}^{HP} \sum_{i \in \Theta^{W,d}} g_{d,l}^{W} \left(W_{i,t} - \Delta w_{i,t} \right) - \sum_{i \in \Psi^B} g_{d,l}^{B} L_{i,t} \right|$$
$$\leq \bar{f}_l, \forall t \in \Lambda^T, \forall d \in \Psi^B, \forall l \in \Psi^L \tag{9.37}$$

where Ψ^L represents the set of indices of transmission lines in the EPS, $\Theta^{TU/CHP/W/EB/HP,d}$ represents the set of indices of thermal units/CHP units/wind farms/EB/HP located on bus d, $\bar{f}_l$ represents capacity of transmission line l, and $g_{d,l}^{TU/CHP/W/EB/HP/B}$ represents distribution factor of the thermal power unit/CHP unit/wind farm/HP/EB/load at bus d on line l.

Apart from the preceding constraints, the constraints of the condensing CHP units in Eq. (9.1) and Eqs. (9.4) through (9.7) should also be included in this part.

9.4.3 Day-ahead operational constraints of the DHS

Heat output and heat exchange.

$$\sum_{i\in\Psi^{CHP}h_{i,t}^{CHP}} + \sum_{i\in\Psi^{EB}} h_{i,t}^{EB} + \sum_{i\in\Psi^{HP}} h_{i,t}^{HP} - \sum_{i\in\Psi^{ST}} \Delta h_{i,t}^{ST}$$
$$= C^{wa}m_t^{HS}(T_{n,t}^S - T_{n,t}^R), \forall t \in \Lambda^T, n \in \Psi^{HS} \tag{9.38}$$

$$C^{wa}m_t^{HES}(T_{n,t}^S - T_{n,t}^R) = H_{e,t}^{HES}, \forall t \in \Lambda^T, n \in \Psi^{HES}, e \in \Psi^{HD} \tag{9.39}$$

Eq. (9.38) reflects the heat balance of the heat network. Eq. (9.39) describes that the heat exchange in the heat station is equal to the heat demand, where Ψ^{HS} represents the set of indices of nodes with heat sources, Ψ^{HES} represents the set of indices of nodes with heat exchange stations, Ψ^{HD} represents the set of indices of nodes with heat loads, C^{wa} represents the specific heat capacity of water, $m_t^{HS/HES}$ represents the mass flow rate in heat sources/heat exchange stations, $h_{i,t}^{CHP}$ represents the heat production of CHP unit i in period t, $h_{i,t}^{HP/EB}$ represents the heat production of HP/EB i in period t, and $T_{n,t}^{S/R}$ represents the supply/return temperature of node n in period t.

Constraints of the thermal process.

$$\underline{T}^{S/R} \le T_{n,t}^{S/R} \le \bar{T}^{S/R}, \forall t \in \Lambda^T, \forall n \in \Psi^{node} \tag{9.40}$$

$$TO_{b,t}^{S/R} - T_t^{Am} = (TI_{b,t}^{S/R} - T_t^{Am})e^{-\frac{\lambda l_b}{C^{wa}m_{b,t}^{S/R}}}, \forall t \in \Lambda^T, \forall b \in \Psi^{pipe} \tag{9.41}$$

$$\left(\sum_{b\in\Omega_n^{pipe+}} m_{b,t}^{S/R}\right) T_{n,t}^{S/R} = \sum_{b\in\Omega_n^{pipe-}} m_{b,t}^{S/R} TI_{b,t}^{S/R}, \forall t \in \Lambda^T, \forall b \in \Psi^{pipe} \tag{9.42}$$

where Eq. (9.40) limits the temperature range of supply and return pipes, Eq. (9.41) represents the temperature drop along the pipe, Eq. (9.42) presents the water mixture process [30], Ψ^{pipe} represents the set of indices of pipes in the DHS, Ψ^{node} represents the set of indices of nodes in the DHS, $\Omega_n^{pipe,+}/\Omega_n^{pipe,-}$ represents the set of indices of pipes to/from node n, L_b represents the length of pipeline b, $m_{b,t}^{S/R}$ represents the mass flow rate of pipeline b in period t in the supply/return network; λ represents the heat transfer coefficient of pipes, $\bar{T}^{S/R}/\underline{T}^{S/R}$ represents the maximum/minimum temperature of supply/return pipes, T_t^{Am} represents the ambient temperature in period t, $TI_{b,t}^{S/R}$ represents the inlet supply/return temperature of pipe b in period t, and $TO_{b,t}^{S/R}$ represents the outlet supply/return temperature of pipe b in period t.

Constraints of the hydraulic process.

$$pr_{n,t}^{S} - pr_{n,t}^{R} \geq pr_{e,t}^{HES}, \forall t \in \Lambda^{T}, \forall n \in \Psi^{node} \tag{9.43}$$

$$p_{i,t}^{WP} = \frac{m_t^{HS}\left(pr_{n,t}^{S} - pr_{n,t}^{R}\right)}{\eta_i^{WP} \cdot \rho}, \forall t \in \Lambda^{T}, \forall n \in \Psi^{HS} \tag{9.44}$$

$$pr_{n_1,t}^{S/R} - pr_{n_2,t}^{S/R} = \mu_b\left(m_{b,t}^{S/R}\right)^2, \forall t \in \Lambda^{T}, \forall b \in \Psi^{pipe}, n_1 \in \Omega^{b+}, n_2 \in \Omega^{b-} \tag{9.45}$$

where Eq. (9.43) represents that the pressure difference between supply and return water in a heat exchange station is required to be larger than a specified level to sustain the mass flow, Eq. (9.44) describes that the power consumption of WPs is proportional to the pressure difference, Eq. (9.45) represents that the pressure difference between the two ends of a pipe is proportional to the square of mass flow rate [31], Ω^{b+}/Ω^{b-} represents the set of indices of start nodes and end nodes of pipe b, ρ represents the density of water, η_i^{WP} represents the efficiency of water pump i, and $pr_{n,t}^{S/R}$ represents pressure of node n in period t for the supply/return network.

Heat production constraints of HPs/EBs.

$$h_{i,t}^{HP} = cop^{HP} p_{i,t}^{HP}, \forall t \in \Lambda^{T}, \forall i \in \Psi^{HP} \tag{9.46}$$

$$h_{i,t}^{EB} = \eta^{EB} p_{i,t}^{EB}, \forall t \in \Lambda^{T}, \forall i \in \Psi^{EB} \tag{9.47}$$

where cop^{HP} represents the coefficient of performance of the HP and η^{EB} represents the energy conversion efficiency of the EB.

Constraints of STs.

$$-\bar{h}_i^{ST} \leq \Delta h_{i,t}^{ST} \leq \bar{h}_i^{ST}, \forall t \in \Lambda^{T}, \forall i \in \Psi^{ST} \tag{9.48}$$

$$H_{i,t+1}^{ST} = H_{i,t}^{ST} + \Delta h_{i,t}^{ST}, \forall t \in \Lambda^{T}, \forall i \in \Psi^{ST} \tag{9.49}$$

$$\underline{H}_i^{ST} \leq H_{i,t}^{ST} \leq \bar{H}_i^{ST}, \forall t \in \Lambda^{T}, \forall i \in \Psi^{ST} \tag{9.50}$$

where Eq. (9.48) limits the maximum charging/discharging rate, Eq. (9.49) reflects the changes in energy level, and Eq. (9.50) constrains the capacity of heat storage; Ψ^{ST} represents the set of indices of STs, $\bar{H}_i^{ST}/\underline{H}_i^{ST}$ represents the maximum/minimum energy storage of STs, $\bar{h}_i^{ST}$ represents the maximum charging and discharging rate of STs, $H_{i,t}^{ST}$ represents the heat energy level of ST i in period t, and $\Delta h_{i,t}^{ST}$ represents the charging/discharging heat energy of ST i in period t.

Apart from the preceding constraints, constraint (9.2) for the heat production of CHP units should also be included in this part.

9.4.4 Real-time operational constraints of the EPS

In this section, the reserve from the thermal units, CHP units, HPs, and EBs is deployed to balance the wind power deviations in scenarios. As the deployed reserve from those devices is tightly coupled with the base plan in the first stage, the linkage constraints between the two stages are modeled for each device.

Power rebalance constraints.

$$
\sum_{i\in\Psi^{TU}}\left(\Delta r_{i,t,s}^{TU,UR}-\Delta r_{i,t,s}^{TU,DR}\right)+\sum_{i\in\Psi^{CHP}}\left(\Delta r_{i,t,s}^{CHP,UR}-\Delta r_{i,t,s}^{CHP,DR}\right)
$$

$$
-\sum_{i\in\Psi^{EB}}\Delta p_{i,t,s}^{EB}-\sum_{i\in\Psi^{HP}}\Delta p_{i,t,s}^{HP}+\sum_{i\in\Psi^{W}}\left(W_{i,t,s}-W_{i,t}-\Delta w_{i,t,s}+\Delta w_{i,t}\right)
$$

$$
=\sum_{i\in\Psi^{B}}-\Delta L_{i,t,s}, \forall t\in\Lambda^{T}, \forall s\in\Psi^{S} \tag{9.51}
$$

where $W_{i,t,s}$ represents the wind power output for wind farm i in scenario s in period t, $\Delta r_{i,t,s}^{TU,UR/DR}$ represents the upward/downward reserve deployment for thermal unit i in period t under scenario s, $\Delta L_{i,t,s}$ represents the load shedding of bus i in period t under scenario s, and $\Delta p_{i,t,s}^{HP/EB}$ represents the regulation of electricity consumption of HP/EB i in period t under scenario s.

The linkage constraints of thermal units/HPs/EBs between two stages.

$$
0\leq\Delta r_{i,t,s}^{TU,UR/DR}\leq r_{i,t}^{TU,UR/DR}, \forall t\in\Lambda^{T}, \forall i\in\Psi^{TU}, \forall s\in\Psi^{S} \tag{9.52}
$$

$$
\underline{p}_i^{HP}x_{i,t}^{HP}-p_{i,t}^{HP}\leq\Delta p_{i,t,s}^{HP}\leq\bar{p}_i^{HP}x_{i,t}^{HP}-p_{i,t}^{HP}, \forall t\in\Lambda^{T}, \forall i\in\Psi^{HP}, \forall s\in\Psi^{S} \tag{9.53}
$$

$$
\underline{p}_i^{EB}-p_{i,t}^{EB}\leq\Delta p_{i,t,s}^{EB}\leq\bar{p}_i^{EB}-p_{i,t}^{EB}, \forall t\in\Lambda^{T}, \forall i\in\Psi^{EB}, \forall s\in\Psi^{S} \tag{9.54}
$$

where Eq. (9.52) represents that the deployed upward/downward reserve of thermal units in each scenario is no more than the scheduled capacity in the first stage, and Eq. (9.53) and Eq. (9.54) describe that the reserve deployment of the HPs and EBs is mainly by changing the power consumption.

Wind power curtailment and load shedding under each scenario.

$$0 \leq \Delta w_{i,t,s} \leq \mathrm{W}_{i,t,s}, \forall t \in \Lambda^{\mathrm{T}}, \forall i \in \Psi^{\mathrm{W}} \tag{9.55}$$

$$0 \leq \Delta L_{i,t,s} \leq \mathrm{L}_{i,t}, \forall t \in \Lambda^{\mathrm{T}}, \forall i \in \Psi^{\mathrm{B}} \tag{9.56}$$

Constraints of transmission lines under each scenario.

$$\left| \begin{array}{l} \sum_{i \in \Theta^{\mathrm{TU},d}} g_{d,l}^{\mathrm{TU}} \left(p_{i,t}^{\mathrm{TU}} + \Delta r_{i,t,s}^{\mathrm{TU,UR}} - \Delta r_{i,t,s}^{\mathrm{TU,DR}} \right) + \sum_{i \in \Theta^{\mathrm{CHP},d}} g_{d,l}^{\mathrm{CHP}} \left(p_{i,t}^{\mathrm{CHP}} + \Delta r_{i,t,s}^{\mathrm{CHP,UR}} - \Delta r_{i,t,s}^{\mathrm{CHP,DR}} - p_{i,t}^{\mathrm{WP}} \right) + \\ \sum_{i \in \Theta^{\mathrm{W},d}} g_{d,l}^{\mathrm{W}} \left(\mathrm{W}_{i,t,s} - \Delta w_{i,t,s} \right) + \sum_{i \in \Theta^{\mathrm{EB},d}} g_{d,l}^{\mathrm{EB}} \left(p_{i,t}^{\mathrm{EB}} + \Delta p_{i,t,s}^{\mathrm{EB}} \right) + \sum_{i \in \Theta^{\mathrm{HP},d}} g_{d,l}^{\mathrm{HP}} \left(p_{i,t}^{\mathrm{HP}} + \Delta p_{i,t,s}^{\mathrm{HP}} \right) - \sum_{i \in \Psi^{\mathrm{B}}} g_{d,l}^{\mathrm{B}} \left(\mathrm{L}_{i,t} - \Delta L_{i,t,s} \right) \end{array} \right| \leq f_l^{\max} \tag{9.57}$$

Ramping constraints after reserve deployment under each scenario.

$$\left(p_{i,t}^{\mathrm{TU/CHP}} + \Delta r_{i,t,s}^{\mathrm{TU/CHP,UR}} - \Delta r_{i,t,s}^{\mathrm{TU/CHP,DR}} \right)$$
$$- \left(p_{i,t-1}^{\mathrm{TU/CHP}} + \Delta r_{i,t-1,s}^{\mathrm{TU/CHP,UR}} - \Delta r_{i,t-1,s}^{\mathrm{TU/CHP,DR}} \right)$$
$$\leq \mathrm{U}_i^{\mathrm{TU/CHP}} (1 - u_{i,t}^{\mathrm{TU/CHP}}) + \bar{\mathrm{P}}_i^{\mathrm{TU/CHP}} u_{i,t}^{\mathrm{TU/CHP}} \tag{9.58}$$

$$\left(p_{i,t-1}^{\mathrm{TU/CHP}} + \Delta r_{i,t-1,s}^{\mathrm{TU/CHP,UR}} - \Delta r_{i,t-1,s}^{\mathrm{TU/CHP,DR}} \right)$$
$$- \left(p_{i,t}^{\mathrm{TU/CHP}} + \Delta r_{i,t,s}^{\mathrm{TU/CHP,UR}} - \Delta r_{i,t,s}^{\mathrm{TU/CHP,DR}} \right)$$
$$\leq \mathrm{D}_i^{\mathrm{TU/CHP}} (1 - v_{i,t}^{\mathrm{TU/CHP}}) + \bar{\mathrm{P}}_i^{\mathrm{TU/CHP}} v_{i,t}^{\mathrm{TU/CHP}} \tag{9.59}$$

In addition, constraint Eqs. (9.8) through (9.15) for the reserve deployment of the condensing CHP units should be included in this part.

9.4.5 Real-time operational constraints of the DHS

Heat rebalance constraints under each scenario.

$$\sum_{i \in \Psi^{\mathrm{CHP}}} \left(\Delta h_{i,t,s}^{\mathrm{CHP,DR}} - \Delta h_{i,t,s}^{\mathrm{CHP,UR}} \right) + \sum_{i \in \Psi^{\mathrm{EB}}} \Delta h_{i,t,s}^{\mathrm{EB}}$$
$$+ \sum_{i \in \Psi^{\mathrm{HP}}} \Delta h_{i,t,s}^{\mathrm{HP}} - \sum_{i \in \Psi^{\mathrm{ST}}} \left(\Delta h_{i,t,s}^{\mathrm{ST}} - \Delta h_{i,t}^{\mathrm{ST}} \right)$$
$$= \mathrm{C}^{\mathrm{wa}} \mathrm{m}_t^{\mathrm{HS}} (T_{n,t,s}^{\mathrm{S}} - T_{n,t,s}^{\mathrm{R}} - T_{n,t}^{\mathrm{S}} + T_{n,t}^{\mathrm{R}}), \forall t \in \Lambda^{\mathrm{T}}, \forall s \in \Psi^{\mathrm{S}} \tag{9.60}$$

$$C^{wa} m_t^{HES} (T_{n,t,s}^S - T_{n,t,s}^R) = H_{e,t}^{HES}, \forall t \in \Lambda^T, n \in \Psi^{HES}, e \in \Psi^{HD}, \forall s \in \Psi^S$$

$$(9.61)$$

where $TI_{b,t,s}^{S/R}$ represents the inlet temperature of pipe b in period t under scenario s, $\Delta h_{i,t,s}^{ST}$ represents the charging/discharging heat energy of ST i in period t under scenario s, and $\Delta h_{i,t,s}^{HP/EB}$ represents the heat regulation of HP/EB i in period t under scenario s.

Thermal process under each scenario.

$$\underline{T}^{S/R} \leq T_{n,t,s}^{S/R} \leq \bar{T}^{S/R}, \forall t \in \Lambda^T, \forall n \in \Psi^{node}, \forall s \in \Psi^S \qquad (9.62)$$

$$TO_{b,t,s}^{S/R} - T_t^{Am} = (TI_{b,t,s}^{S/R} - T_t^{Am})e^{-\frac{\lambda L_b}{C^{wa} m_{b,t}^S}}, \forall t \in \Lambda^T, \forall b \in \Psi^{pipe}, \forall s \in \Psi^S$$

$$(9.63)$$

$$\left(\sum_{b \in \Omega_n^{pipe+}} m_{b,t}^{S/R}\right) T_{n,t,s}^{S/R} = \sum_{b \in \Omega_n^{pipe-}} m_{b,t}^{S/R} TI_{b,t,s}^{S/R}, \forall t \in \Lambda^T, \forall b \in \Psi^{pipe}, \forall s \in \Psi^S$$

$$(9.64)$$

where Eqs. (9.62) through (9.64) are similar to Eqs. (9.40) through (9.42), which represent the thermal process in each scenario; $TO_{b,t,s}^{S/R}$ represents the outlet temperature of pipe b in period t under scenario s and $T_{n,t,s}^{S/R}$ represents the temperature of node n in period t under scenario s.

Heat regulation of HPs/EBs.

$$\Delta h_{i,t,s}^{HP} = \eta^{HP} \Delta p_{i,t,s}^{HP}, \forall t \in \Lambda^T, \forall i \in \Psi^{HP} \qquad (9.65)$$

$$\Delta h_{i,t,s}^{EB} = \eta^{EB} \Delta p_{i,t,s}^{EB}, \forall t \in \Lambda^T, \forall i \in \Psi^{EB} \qquad (9.66)$$

where Eqs. (9.65) and (9.66) reflect that the heat regulation of HPs and EBs is linearly related to their power regulation in each scenario.

Heat regulation of STs under each scenario.

$$-\bar{h}_i^{ST} \leq \Delta h_{i,t,s}^{ST} \leq \bar{h}_i^{ST}, \forall t \in \Lambda^T, \forall i \in \Psi^{ST} \qquad (9.67)$$

$$H_{i,t+1,s}^{ST} = H_{i,t,s}^{ST} + \Delta h_{i,t,s}^{ST}, \forall t \in \Lambda^T, \forall i \in \Psi^{ST} \qquad (9.68)$$

$$\underline{H}_i^{ST} \leq H_{i,t,s}^{ST} \leq \bar{H}_i^{ST}, \forall t \in \Lambda^T, \forall i \in \Psi^{ST} \qquad (9.69)$$

where Eqs. (9.67) through (9.69) are similar to Eqs. (9.48) through (9.50), representing the charging/discharging rate, changes of heat energy level, and heat storage capacity under each scenario; $H_{i,t,s}^{ST}$ represents the heat energy level of ST i in period t under scenario s.

In addition, constraint Eqs. (9.17) and (9.18) for the heat regulation of the condensing CHP units should be included in this part.

9.4.6 Compact form of two-stage stochastic optimal operation

In summary, the compact form of the mathematical formulation for the two-stage stochastic optimal operation of the IEHS can be described as follows.

$$\min f_1(\mathbf{x}) + \mathrm{E}\big(f_{2,s}(\mathbf{y}_s)\big)$$
$$s.t. \begin{cases} A_1\mathbf{x} \le b_1 \\ A_2\mathbf{x} + B_2\mathbf{y}_s \le b_{2s}, \forall s \end{cases} \tag{9.70}$$

The decision variables include the day–ahead scheduling variable $\mathbf{x}$ and real–time regulation variable $\mathbf{y}_s$, where $\mathbf{x}=\{u_{i,t}^{TU/CHP}, v_{i,t}^{TU/CHP}, r_{i,t}^{TU,UR/DR}, r_{i,t}^{CHP,UR/DR}, h_{i,t}^{CHP}, p_{i,t}^{TU/CHP}, \alpha_{i,t,j}p_{i,t}^{WP}, \Delta w_{i,t}, x_{i,t}^{HP}, p_{i,t}^{HP/EB}, h_{i,t}^{HP/EB}, H_{i,t}^{ST}, \Delta h_{i,t}^{ST}, pr_{n,t}^{S/R}, TI_{b,t}^{S/R}, TO_{b,t}^{S/R}, T_{n,t}^{S/R}\}$ and $\mathbf{y}_s=\{\Delta r_{i,t,s}^{TU,UR/DR}, \Delta r_{i,t,s}^{CHP,UR/DR}, \Delta h_{i,t,s}^{CHP,UR/DR}, \Delta\alpha_{i,t,s,j}^{UR/DR}, \Delta w_{i,t,s}, \Delta L_{i,t,s}, \Delta p_{i,t,s}^{HP/EB}, \Delta h_{i,t,s}^{HP/EB}, H_{i,t,s}^{ST}, \Delta h_{i,t,s}^{ST}, TI_{b,t,s}^{S/R}, TO_{b,t,s}^{S/R}, T_{n,t,s}^{S/R}\}$, respectively.

9.5 Case study

9.5.1 System description

The test system consists of a 6–bus EPS and a 6–node DHS, which is used to demonstrate the efficacy of the proposed two-stage stochastic optimal dispatch scheme of the IEHS. Four wind farms with spatial and temporal correlations are integrated into this system. The scenario generation of these four wind farms is based on the method in Chapter 3. Fig. 9.4 shows the single–line diagram of the test system. Fig. 9.5 shows the hourly power demand, heat demand, and wind power aggregate forecast values of the test system.

In the test system, the EPS sector includes two thermal units (G1 and G2), one condensing CHP unit (CHP1), four wind farms (W1–W4), two EBs, two HPs, one WP (WP1), and seven transmission lines. The data of transmission lines is from Li et al. [32]. Each EB has a power capacity of 2.5 MW, and the conversion efficiency is 0.9. Each HP has a capacity of 2.5 MW and its conversion efficiency is set as 3. The DHS sector includes five

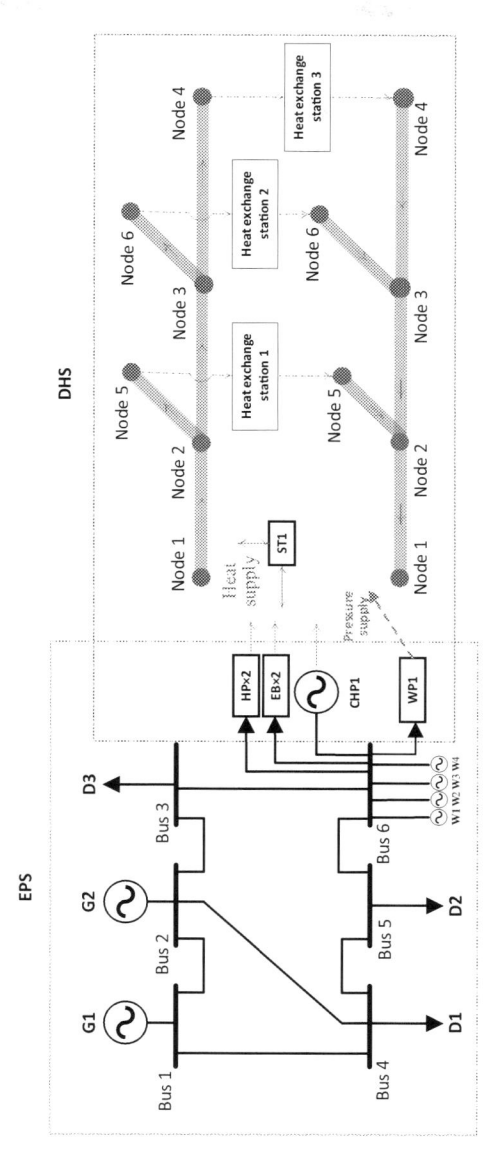

Figure 9.4 Configuration of the 6-bus and 6-node integration system.

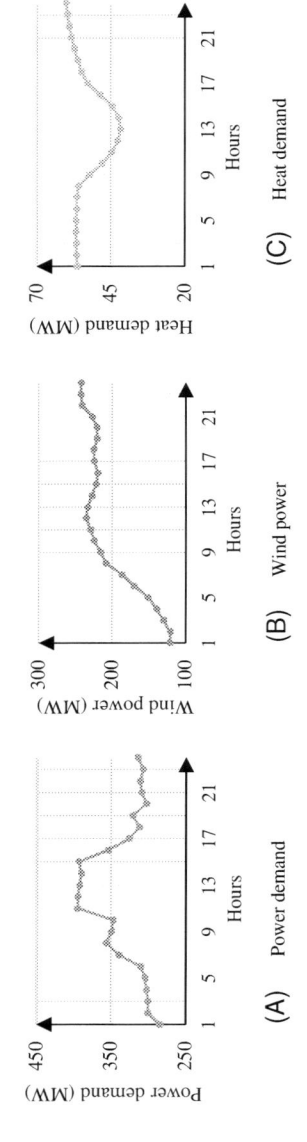

Figure 9.5 Power demand, heat demand, and wind power forecast values in the test system.

pipelines, three heat exchange stations, and one ST (ST1) with a capacity of 20 MW.

In the simulation, the cost parameters are set as follows. The compensation for wind curtailment and load shedding is 80 $/MWh and 1500 $/MWh, respectively. The reserve capacity price of conventional units is 40% of their highest incremental cost of producing energy [33]. The reserve capacity price of CHP1 is 25 $/MWh. The reserve deployment prices of conventional units and CHP1 are 40 $/MWh.

Five cases are set for analyzing the impact and effect of different flexible devices on reducing the wind curtailment rate and operational cost, which relates to the energy and economic efficiency of the IEHS. The aim of the comparison is to identify which measure is more beneficial to achieve a high-share renewable-based energy system while guaranteeing a cost-effective solution. A description for case setting is presented in Table 9.1.

In Case 1, the conventional thermal units are the only reserve resource to balance wind power uncertainty. This case is the base case for comparison. In Case 2, the condensing CHP units and STs are included as additional power reserve and heat regulation sources. In Case 3, the CHP units, EBs, and STs are integrated to provide power reserve and heat regulation. In Case 4, the CHP units, HPs, and STs are integrated to provide power reserve and heat regulation. In Case 5, the CHP units, HPs, EBs, and STs are all taken into account to provide power reserve and heat regulation.

9.5.2 Improved flexibility from condensing CHP units and flexible devices

The energy scheduling of the test system for five cases is shown in Fig. 9.6. The impact and effect of utilizing different flexible devices are analyzed based on comparisons of operational cost and wind power utilization. The operational cost and wind power curtailment for five cases are compared in Table 9.2. In this table, the improvement of economic efficiency of Cases 2 through 5 are calculated with Case 1 as the base case.

In Case 1, only the reserve of conventional thermal units is considered. Due to the limited flexibility, Case 1 has the highest wind curtailment and operational cost. In Fig. 9.6(A), the wind power is mainly curtailed at periods when the load is in its valley while the wind power has high availability. During those time periods, the heat demand is at a high level due to the lower ambient temperature at night. In Case 2, after considering the reserve

Table 9.1 Case setting.

Cases	Energy Scheduling					Reserve Provision				Heat Regulation			
	Thermal Units	CHP Units	EBs	HPs	STs	Thermal Units	CHP Units	EBs	HPs	CHP Units	EBs	HPs	STs
1	√	√	×	×	√	√	×	×	×	×	×	×	×
2	√	√	×	×	√	√	√	×	×	√	×	×	√
3	√	√	√	×	√	√	√	√	×	√	√	×	√
4	√	√	×	√	√	√	√	×	√	√	×	√	√
5	√	√	√	√	√	√	√	√	√	√	√	√	√

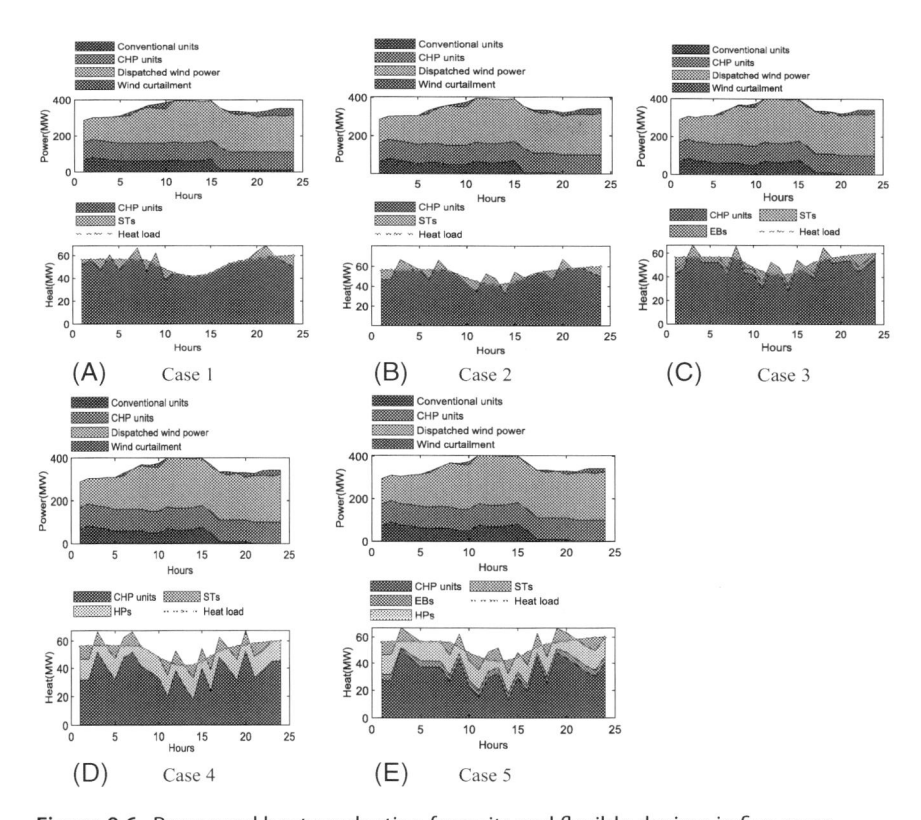

Figure 9.6 Power and heat production for units and flexible devices in five cases.

of CHP1 and heat regulation of ST1, the total cost decreases by 6.31% over that of Case 1. The wind curtailment rate decreases by 2.01% over that of Case 1.

In Case 3 and Case 4, when the EBs and HPs are introduced, the heat supply originally from CHP units can be replaced by the EBs and HPs, then the fuel cost of units can be saved and more wind power can be used for power consumption of the EBs and HPs. Even with the same integrated power capacity, the EBs and HPs present different impacts on reducing the operational cost and wind curtailment. Compared with Case 1, EBs and HPs can reduce the wind power curtailment by 3.07% and 2.93%, respectively. However, the HPs can bring higher economic efficiency than the EBs with the same capacity. Compared with Case 1, EBs and HPs can reduce the operational cost by 11.60% and 12.54%, respectively.

Table 9.2 Operational cost and wind curtailment of five cases.

Case	1	2	3	4	5
Day–Ahead Cost/$	1.8078×105	1.7485×105	1.6986×105	1.6720×105	1.5873×105
Expected Real–Time Cost/$	6.5291×104	5.5688×104	4.7654×104	4.8017×104	4.4428×104
Total Cost/$	2.4607×105	2.3054×105	2.1752×105	2.1521×105	2.0316×105
Improvement of Economic Efficiency (%)	—	6.31	11.60	12.54	17.44
Wind Curtailment Rate (%)	10.23	8.21	7.16	7.30	6.09

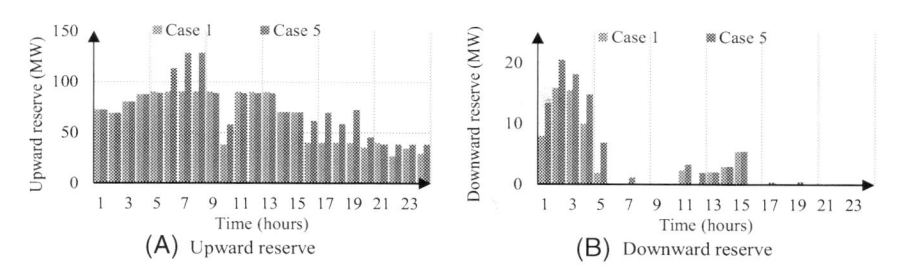

Figure 9.7 Comparison between Case 1 and Case 5 for deployed upward and downward reserve.

In Case 5, when the reserve and heat regulation of CHP units, STs, EBs, and HPs are all considered, the test system has the lowest operational cost and wind curtailment, reducing the wind curtailment and operational cost by 4.14% and 17.44%, respectively. As can be seen from Fig. 9.6(E), the heat production from CHP1 has been reduced compared with other cases due to the simultaneous integration of EBs, HPs, and STs, releasing more flexibility for the system to accommodate wind power.

Fig. 9.7 shows the comparison between Case 1 and Case 5 for the expected value of deployed upward and downward reserve under all wind power scenarios. Compared with Case 1, Case 5 shows a great improvement in available reserve after considering the condensing CHP units, HPs, EBs, and STs. It should be noted that the downward reserve is in low availability from period 16 to period 24 when the power load is in its valley while the wind power and heat demand is high. At these periods, both the thermal power units and CHP units are in their minimum outputs, leading them hard to release the downward reserve and accommodate the wind power.

In addition, to validate the effectiveness of reserve provision and heat regulation model of the condensing CHP units, the feasibility of operating points of CHP1 after reserve deployment is tested. Fig. 9.8 shows that the

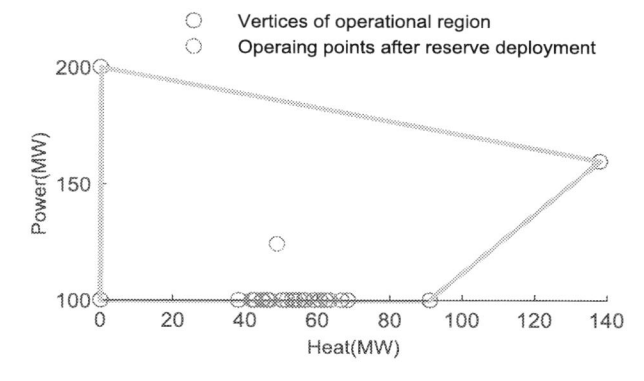

Figure 9.8 Operating points of CHP1 after reserve deployment.

operating points in different time periods under a certain wind power scenario are still within the operational region of CHP1 after reserve deployment, which validates the effectiveness of the modeling of condensing CHP units in the feasibility guarantee.

9.6 Conclusion

In this chapter, a two-stage stochastic optimal dispatching scheme was proposed for the IEHS, which considers detailed modeling of reserve provision and heat regulation for condensing CHP units, HPs, EBs, and heat STs. A 6-bus energy system was simulated under different cases. By coordinating different flexible devices, the dispatching scheme of the IEHS can improve the upward and downward reserve capacity, reduce wind curtailment, and save operational cost. The reserve provision and heat regulation model of condensing CHP units is effective at guaranteeing a feasible operating point after reserve deployment. Compared with the EBs, the HPs with the same capacity can save more operational cost due to their high energy conversion efficiency.

References

[1] Reuters, Denmark sources record 47% of power from wind in 2019, 2020. https://www.reuters.com/article/us-climate-change-denmark-windpower/denmark-sources-record-47-of-power-from-wind-in-2019-idUSKBN1Z10KE. (Accessed 19 April 2021).
[2] P Pinson, L Mitridati, C Ordoudis, J Ostergaard, Towards fully renewable energy systems: experience and trends in Denmark, CSEE J. Power Energy Syst 3 (1) (2017) 26–35.
[3] E Cesena, N Good, M Panteli, J Mutale, P Mancarella, Flexibility in sustainable electricity systems: multivector and multisector nexus perspectives, IEEE Electrification Mag 7 (2) (2019) 12–21.
[4] A Bloess, W Schill, A Zerrahn, Power-to-heat for renewable energy integration: a review of technologies, modeling approaches, and flexibility potentials, Appl Energy 212 (2018) 1611–1626.
[5] I Dimoulkas, M Amelin, F Levihn, District heating system operation in power systems with high share of wind power, J. Mod. Power Syst Clean Energy 5 (6) (2017) 850–862.
[6] X Liu, J Wu, N Jenkins, A Bagdanavicius, Combined analysis of electricity and heat networks, Appl Energy 162 (2016) 1238–1250.
[7] C Wu, W Gu, P Jiang, Z Li, H Cai, B Li, Combined economic dispatch considering the time-delay of district heating network and multi-regional indoor temperature control, IEEE Trans. Sustain Energy 9 (1) (2018) 118–127.
[8] D Wang, Y Zhi, H Jia, K Hou, S Zhang, W Du, et al., Optimal scheduling strategy of district integrated heat and power system with wind power and multiple energy stations considering thermal inertia of buildings under different heating regulation modes, Appl. Energy 240 (2019) 341–358.
[9] W Gu, J Wang, S Lu, Z Luo, C Wu, Optimal operation for integrated energy system considering thermal inertia of district heating network and buildings, Appl. Energy 199 (2017) 234–246.

[10] Z Li, W Wu, J Wang, B Zhang, T Zheng, Transmission-constrained unit commitment considering combined electricity and district heating networks, IEEE Trans. Sustain. Energy 7 (2016) 480–492.

[11] Y Cao, W Wei, L Wu, S Mei, M Shahidehpour, Z Li, Decentralized operation of interdependent power distribution network and district heating network: a market-driven approach, IEEE Trans. Smart Grid 10 (2019) 5374–5385.

[12] C Lin, W Wu, B Zhang, Y Sun, Decentralized solution for combined heat and power dispatch through benders decomposition, IEEE Trans. Sustain. Energy 8 (2017) 1361–1372.

[13] J Huang, Z Li, Q Wu, Coordinated dispatch of electric power and district heating networks: a decentralized solution using optimality condition decomposition, Appl. Energy 206 (2017) 1508–1522.

[14] H Lu, C Wang, Q Li, R Wiser, K Porter, Reducing wind power curtailment in China: comparing the roles of coal power flexibility and improved dispatch, Clim. Policy 19 (5) (2019) 623–635.

[15] X Chen, C Kang, M O'Malley, Q Xia, J Bai, C Liu, et al., Increasing the flexibility of combined heat and power for wind power integration in China: modeling and implications, IEEE Trans. Power Syst. 30 (4) (2015) 1848–1857.

[16] X Chen, M.B McElroy, C Kang, Integrated energy systems for higher wind penetration in China: formulation, implementation, and impacts, IEEE Trans. Power Syst. 33 (2) (2018) 1309–1319.

[17] N Zhang, X Lu, MB McElroy, C Nielsen, X Chen, Y Deng, et al., Reducing curtailment of wind electricity in China by employing electric boilers for heat and pumped hydro for energy storage, Appl. Energy 184 (2016) 987–994.

[18] M Nielsen, J Morales, M Zugno, T Pedersen, H Madsen, Economic valuation of heat pumps and electric boilers in the Danish energy system, Appl. Energy 167 (2016) 189–200.

[19] B Biegel, P Andersen, TS Pedersen, K Nielsen, J Stoustrup, L Hanse, Electricity market optimization of heat pump portfolio, IEEE International Conference on Control Applications, 2013 in: Proc2013.

[20] D Wang, S Parkinson, W Miao, H Jia, C Crawford, N Djilali, Hierarchical market integration of responsive loads as spinning reserve, Appl. Energy 104 (2013) 229–238.

[21] L Cai, Z Zhang, D Xie, X Song, Y Lai, Evaluation and simulation on the reserve capacity of thermal storage electric boiler clusters, E3S Web Conf. 143 (2020) 1–7.

[22] V Verda, F Colella, Primary energy savings through thermal storage in district heating networks, Energy 36 (2011) 4278–4286.

[23] M Zugno, JM Morales, H Madsen, Commitment and dispatch of heat and power units via affinely adjustable robust optimization, Comput. Oper. Res. 75 (2016) 191–201.

[24] J Tan, Q Wu, Q Hu, W Wei, F Liu, Adaptive robust energy and reserve co-optimization of integrated electricity and heating system considering wind uncertainty, Appl. Energy 260 (2020) 114230.

[25] R Lahdelma, H Hakonen, An efficient linear programming algorithm for combined heat and power production, Eur. J. Oper. Res. 148 (2003) 141–151.

[26] HR Abdolmohammadi, A Kazemi, A., Benders decomposition approach for a combined heat and power economic dispatch, Energy Convers. Manag. 71 (2013) 21–31.

[27] H Zhou, Z Li, J Zheng, Q Wu, H Zhang, Robust scheduling of integrated electricity and heating system hedging heating network uncertainties, IEEE Trans. Smart Grid 11 (2) (2020) 1543–1555.

[28] J Wang, Z Wei, B Yang, Y Yong, M Xue, G Sun, et al., Two-stage integrated electricity and heat market clearing with energy stations, IEEE Access 7 (2019) 44928–44938.

[29] M Zhang, J Fang, X Ai, H Shuai, W Yao, H He, et al., Feasibility identification and computational efficiency improvement for two-stage RUC with multiple wind farms, IEEE Trans. Sustain. Energy 11 (3) (2020) 1669–1678.

[30] Z Pan, Q Guo, H Sun, Interactions of district electricity and heating systems considering time-scale characteristics based on quasi-steady multi-energy flow, Appl. Energy 167 (2016) 230–243.

[31] S Huang, W Tang, Q Wu, C Li, Network constrained economic dispatch of integrated heat and electricity systems through mixed integer conic programming, Energy 179 (2019) 464–474.

[32] Z Li, W Wu, M Shahidehpour, J Wang, B Zhang, Combined heat and power dispatch considering pipeline energy storage of district heating network, IEEE Trans. Sustain. Energy 7 (1) (2016) 12–22.

[33] E Heydarian-Forushani, MEH Golshan, MP Moghaddam, M Shafie-Khah, JPS Catalão, Robust scheduling of variable wind generation by coordination of bulk energy storages and demand response, Energy Convers. Manage 106 (2015) 941–950.

CHAPTER 10

Two-stage stochastic optimal operation of integrated energy systems

10.1 Introduction

The installed capacity of renewable energy sources (RESs) is increasing at a high rate. In the future RES-based power system, challenges in power system stability and security are more likely to happen due to fluctuations and variability of RESs. Hence, measures on how to accommodate RESs should be introduced. One of the prominent solutions is the integration of different energy sectors to provide flexibility and increase efficiency of the multi-energy systems. Furthermore, due to variability of wind power, a mismatch between the day-ahead (DA) schedule and real-time operation occurs. To achieve the most optimal and efficient operating points for the next day while taking into account uncertainties and hence minimize the deviation between the DA schedule and real-time operation, a stochastic programming approach is proposed. This chapter introduces a two-stage stochastic DA scheduling of integrated energy systems (IESs). The scheduling of generation is performed in the first stage, whereas the second stage accommodates the wind power uncertainties. Wind power uncertainties are represented as set of scenarios that is generated by a proposed method that takes into account temporal correlation and presents a reliable input to the second stage.

Section 10.2 provides an introduction to the two-stage DA scheduling approach. Section 10.3 demonstrates the mathematical model of the two-stage DA scheduling approach for IESs. Then, the stochasticity of the uncertainty through realistic scenarios obtained by scenario generation method is described in Section 10.4. Finally, the example of the case study and summary are given in Sections 10.5 and 10.6, respectively.

10.2 Background and DA scheduling

This section introduces the background and a two-stage DA scheduling for IESs and gives an overview of the stochastic programming approaches.

Optimal operation of integrated multi-energy systems under uncertainty. Copyright © 2022 Elsevier Inc.
DOI: https://doi.org/10.1016/B978-0-12-824114-1.00008-1 All rights reserved.

10.2.1 Background

Today, a high increase in installed capacity of RESs and replacement of conventional generators by RESs is taking place. Due to the fluctuating nature of RESs, mismatch between generation and demand can occur, leading to an unstable system. To realize the long-term goal of RES-based systems operating in a secure, reliable, and economic manner, new solutions for accommodation of RESs have be explored [1]. The technologies can enable bi-directional energy conversion, such as the power-to-gas (P2G) unit [2].

Studies for optimal power flow of the electric power system (EPS) have been performed for a long time with the main aim of supplying the load reliably and as economically as possible [3]. Moreover, much research today has been focused on reliable and stable operation of the power system with high penetration of RESs. As a promising solution for reliable and stable operation, integration of the natural gas system (NGS) and EPS is a considerably researched area [4, 1, 2]. In addition, interactions between the NGS and EPS through P2G have been investigated in the work of Yang et al. [5] and Fang et al. [6]. The mentioned work shows that excess wind power can be used to produce hydrogen through electrolysis. Therefore, P2G is a prominent solution for handling electricity overproduction and underproduction. It has shown a balancing potential in peak regulation due to its fast response to changes. P2G can convert surplus electricity into gas while decarbonizing the gas sector and optimizing network investments, and can be available as an energy storage, sustainable feedstock, and sustainable fuel [7, 1]. In addition, co-generation and conversion units in the EPS can be a linkage to the district heating system (DHS) [8]. Therefore, the synergy between electricity, heat, and gas systems must be investigated.

In respect to the mentioned work, interactions between multi-energy systems through P2G and combined heat and power (CHP) can accommodate RESs and provide flexibility required by the future RES-based system. Therefore, integration of EPS, DHS, and NGS is a promising solution to balance RES production and increase flexibility [9, 10]. The interactions among different energy sectors will provide more flexibility required by the future RES-based system.

The scheduling of generation and demand is performed in three markets: the DA market, intra-day market, and real-time (RT) market [11]. The DA market balances the production and consumption based on the forecasted data. The forecasted data usually includes the demand and renewable energy generation. It is traditionally based on a single deterministic forecast that fails to capture the uncertainties in the energy systems. Moreover, a specific amount of reserves is not set aside and the expected value of the cost for providing reserves is not included. The DA market closes the day before the delivery day and provides a schedule for each hour of the next day. The intra-day market is the market closer to the actual delivery hour. It closes 1 hour before the actual delivery. The RT market, also called the *balancing market*, is a single-period market that is balancing the generation and demand using updated data. Conventionally, traditional generators usually provide the regulating power in the balancing market. It can be noted that during operational hours in a system with integrated RESs, the scheduling provided in DA based on single forecasted data is no longer economically efficient. Therefore, a mismatch between the DA schedule and RT operation occurs. Moreover, inaccurate DA forecasts can lead to wrong and unsuitable dispatch and therefore increase the costs of RT operation. To achieve the most optimal and efficient operating points for the next day while taking uncertainties into account, and hence minimize the deviation between the DA schedule and RT operation, a stochastic programming approach is proposed in this chapter.

The stochastic programming approach has the following benefits and characteristics. First, it is based on stochastic forecast of the uncertain parameters. Second, a clear interaction between the DA stage and RT stage is taken into account. Third, it achieves lower expected system costs and reserves are optimized. Finally, it takes a realistic range of scenarios and probabilities into account. Therefore, a stochastic programming approach has two stages. The first stage is the DA stage that interacts with the second stage (the RT stage). In the first stage, operating points of the generating units are dispatched, whereas the second stage creates realizations of the wind power output and accommodates the uncertainties through reserves.

10.2.2 Overview of two-stage stochastic programming approaches

The stochastic programming approach has gained a lot of attention in recent years in the field of optimization under uncertainty. A two-stage stochastic programming approach for the IES taking into account wind power uncertainties is presented by Zeng et al. [12]. The focus is on DA scheduling, with the first stage representing the scheduling stage and the second stage representing RT operation based on wind realization. The aim is to minimize the total expected operational cost of the EPS, DHS, and NGS while assuming constant water temperature and gas pressure. Schulzea and McKinnon [13] compared the deterministic and multistage stochastic programming approaches showing minimum operational costs by using the stochastic approach.

Several approaches exist to consider uncertainty in stochastic scenario–based programming. RES uncertainties have been considered by Wang et al. [14], Ghasemi et al. [15], and Su et al. [16]. In the work of Wang et al. [14], wind power uncertainty is considered in a security-constrained unit commitment. To capture the volatility of wind power output, the authors generated a large number of scenarios using Monte Carlo simulation, as the wind is assumed to be subjected to a normal distribution. Each scenario in that case is assigned a probability that is one divided by the total number of generated scenarios. In the work of Ghasemi et al. [15] and Su et al. [16], a set of scenarios is generated by Monte Carlo simulation as well. In the work of Su et al. [16], the solar and wind power outputs are used as uncertainty sources where a correlation between wind and solar output power is ignored. To decrease the variance of the performed simulation, the authors used Latin hypercube sampling.

However, to present all scenarios with the corresponding probability of occurrence, there is a large number of scenarios of possible future states to be included in the optimization problem. Consequently, the computation time rises and the approach is no longer suitable for RT operation [17]. Hence, SR techniques are used. Su et al. [16] eliminate scenarios with low probability and aggregate similar scenarios. The applied technique is measuring the distance between scenarios based on probability metrics. The purpose of scenario reduction methods is to decrease the computational burden. In the work of Growe–Kuska et al. [18], scenario reduction methods and procedures are described. The basic idea behind scenario reduction algorithms is creating a new initial scenario set consisting of the preserved scenarios and assigning new probabilities to the preserved scenarios. The scenarios might be constructed in a scenario tree where the number of nodes is deduced compared with a fan structure. Two algorithms are accounted for: simultaneous backward reduction and the fast forward selection algorithm. A thorough description and the steps for mentioned algorithms can be found in the work of Wu [19], Growe–Kuska et al. [18], and Wang [20]. Wang [20] formulates the importance of the stochastic programming approach and gives an overview of scenario-based approaches in stochastic programming. In the work of Wu [19], it is implied that a large number of scenarios will cause computational burden that will limit the tractability of the solution. Therefore, the author suggests relevant methodologies for solving a two-stage stochastic problem by using a moment matching method based on historical data as the statistical specifications are used to approximate the original distribution without the real probability distribution known.

Based on the literature review conducted, the most common methods to obtain the appropriate scenario set are summarized as follows. First, Rayleigh probability density function (PDF) of the wind speed and wind turbine power curve can be used to generate scenarios as given by Rabiee and Soroudi [21] and Atwa and El-Saadany [22]. Second, the Weibull PDF of wind speed and Monte Carlo have been used by Shi et al. [23] and Bornapour et al. [24]. Last, a normal PDF can be used to generate random variables. However, based on the work of Zhang et al. [25], Yıldız et al. [26], and Ma et al. [27], the following conclusions can be summarized. The PDF of wind speed cannot be assumed to follow a theoretical distribution due to the fluctuating nature of wind speed. Hence, using the wind speed to generate scenarios for wind power is not accurate enough, as a significant error can appear due to inaccuracies in the wind speed to wind power conversion. Moreover, forecast error of wind power is dynamic, and a temporal correlation and seasonal effect due to wind variability are important. Hence, approaches like the autoregressive moving average model lack a description of the nature of RESs.

The more suitable method to generate scenarios in this chapter is based on the empirical cumulative distribution function (ECDF) of measured historical values (MHVs) based on Ma et al. [27]. The scenario generation method is proven to have the best performance according to evaluation indices of Zhang et al. [25] and comparison performed by Ma et al. [27]. The stochasticity is described as generated scenarios based on historical data and represented through realistic scenarios and assigned probabilities generated by the scenario generation method. The scenario generation algorithm is based on historical observations and provides a more advanced approach to characterize the stochasticity of wind power.

However, not many mentioned algorithms have been applied to the operation of the IES. In regard to stochastic DA scheduling, the research has been focused on the EPS field [28–30]. Therefore, this chapter introduces two aspects. The first aspect is focused on flexibility provision through synergy of multiple energy systems to accommodate wind power uncertainty. The second aspect introduces a reliable method to generate scenarios to achieve higher flexibility and more reliable and economical decisions. Finally, three expectations are summarized as follows. First, it is expected that a two–stage stochastic programming approach for the IES shows improvement in efficiency, flexibility, and security of the IES. Moreover, the synergy of different energy subsystems should reduce system costs and increase wind power utilization. Second, linking components, such as P2G, CHP, (electric boiler) (EB), and heat pump (HP), as well as linepack, thermal energy storage, and gas storages, are expected to increase the flexibility of the entire IES. Third, a reliable method for scenario generation is used as an input to the second stage of the stochastic approach. The proposed method ensures temporal correlation. As an expectation, the total expected cost is minimized and reserves are optimized. Additionally, scenario reduction is performed to decrease the computational burden.

The mathematical model is presented in the next section, whereas the detailed procedure for the proposed scenario generation method is given in Section 10.4.

10.3 Mathematical model of the IES for two-stage DA scheduling

In this section, a mathematical model of the integrated and coordinated multi-energy system is designed. The IES combines the EPS, DHS, and NGS along with the linking components and considers the network constraints for each of the energy systems. The DA scheduling model for the IES is formulated as a two–stage optimization problem. In the first stage, the DA stage, decisions are made before the wind realization. In the second stage, the RT stage, the realization of wind power is known and decisions are made dependent on the first-stage decisions and wind realization. In the second stage, the reserves are deployed accordingly. The wind uncertainties are represented through a number of scenarios for a period of 24 hours. Each scenario has its corresponding probability.

The proposed model is exploring the synergy of different energy subsystems simultaneously. The model utilizes coordinated flexibility to balance wind power fluctuations in both stages. Besides the conventional generation and demand, the model considers the thermal energy storage and gas storage,

P2G, EB, and HP. The thermal energy storage and gas storage are commonly found in the DHS and NGS, respectively [31, 32]. The thermal energy storage is usually a hot water tank (HWT) in the vicinity of the CHP plant and provides flexibility in moments of high- and low-demand periods. In the case of high electricity demand and lower heat demand, the excess heat power produced by CHP is stored in the HWT. On the contrary, in the case when heat demand is higher than heat generated by CHP, the heat is provided by the HWT. Furthermore, P2G has gained a lot of attention in recent years due to its flexibility to convert excess electricity from RESs to gas. The proposed coordination of the EPS, DHS, and NGS with integrated P2G, EB, HP, and storages provides higher flexibility and increases efficiency of the entire IES.

To sum up, two sets of constraints for DA and RT stages are given. Each set is divided into four subsets of constraints containing the EPS, DHS, NGS, and linking constraints. The objective of the entire model is to minimize total expected costs for the IES while being subjected to security constraints.

To simplify the notation, the DA and RT superscript is ignored in the description of parameters and variables throughout the chapter. The notation for sets is as described next and will be used in the following section.

Symbol	Description	Symbol	Description
$c \in \Omega^{GC}$	Gas compressors	$n, m \in \Omega^{EPS/DHS/NGS}$	Set of nodes in the EPS/DHS/NGS
$d \in \Omega^{GD}$	Gas demand	$p \in \Omega^{P2G}$	P2G
$e \in \Omega^{ED}$	Electric demand	$s \in \Omega^{HS}$	Heat storages
$eb \in \Omega^{EB}$	Electric boilers	$st \in \Omega^{ST}$	Gas storages
$g \in \Omega^{GS}$	Gas sources	$t \in T$	Set of timesteps in the scheduling horizon
$hp \in \Omega^{HP}$	Heat pumps	$w \in \Omega^{WF}$	Set of wind farms at node n
$j \in \Omega^{CHP}$	CHP	$m \in \Omega^{DHS,mix, S/R}$	Set of nodes with the mixing temperatures in the supply and return system
$l \in \Omega^{HL}$	Heat demand		
$lp \in \Omega^{LP}$	Linepack		

10.3.1 Objective function

The objective of the two-stage stochastic model is to minimize total expected costs for the entire IES. The total expected costs are split into DA scheduling operation cost and expected system cost due to additional balancing in RT operation. The objective is presented in Eq. (10.1).

$$
\begin{aligned}
\min \ & \sum_{t=1}^{T} \left\{ \begin{array}{l} \displaystyle\sum_{j=1}^{n^{CHP}} C_j^{CHP,P} P_{j,t}^{CHP,DA} - \sum_{p=1}^{n^{P2G}} C_p^{P2G} D_{p,t}^{P2G,DA} - \sum_{eb=1}^{n^{EB}} C_{eb}^{EB} D_{eb,t}^{EB,DA} - \sum_{hp=1}^{n^{HP}} C_{hp}^{HP} D_{hp,t}^{HP,DA} \\ \displaystyle + \sum_{g=1}^{n^{GS}} C_g^{GS} Q_{g,t}^{GS,DA} + \sum_{j=1}^{n^{CHP}} C_j^{CHP,H} H_{j,t}^{CHP,DA} \end{array} \right\} \\
& + \sum_{s \in S} \pi_s \sum_{t=1}^{T} \left\{ \begin{array}{l} \displaystyle\sum_{j=1}^{n^{CHP}} \left(C_j^{UR} R_{j,t,s}^{CHP,U} - C_j^{DR} R_{j,t,s}^{CHP,D} \right) + \sum_{j=1}^{n^{CHP}} C_j^{CHP,H} \left(H_{j,t,s}^{CHP,RT} - H_{j,t}^{CHP,DA} \right) \\ \displaystyle - \sum_{p=1}^{n^{P2G}} C_p^{P2G} \left(D_{p,t,s}^{P2G,RT} - D_{p,t}^{P2G,DA} \right) - \sum_{eb=1}^{n^{EB}} C_{eb}^{EB} \left(D_{eb,t,s}^{EB,RT} - D_{eb,t}^{EB,DA} \right) \\ \displaystyle - \sum_{hp=1}^{n^{HP}} C_{hp}^{HP} \left(D_{hp,t,s}^{HP,RT} - D_{hp,t}^{HP,DA} \right) + \sum_{g=1}^{n^{GS}} \left(C_g^{UR} R_{g,t,s}^{GS,U} - C_g^{DR} R_{g,t,s}^{GS,D} \right) \\ \displaystyle + \sum_{e=1}^{n^{ED}} \left(C_e^{VOLL} D_{e,t,s}^{ED,shed} \right) + \sum_{w=1}^{n^{WF}} \left(C_w^{spill} W_{w,t,s}^{spill} \right) \end{array} \right\}
\end{aligned}
$$

(10.1)

where parameters and variables are denoted as follows:

Parameters		
$C_j^{CHP,H/P}$	=	Marginal cost of heat/electricity production by CHP (DKK/MWh),
C_g^{DR}	=	Marginal cost of downward regulating power of the gas source unit g (DKK/MWh),
C_j^{DR}	=	Marginal cost of downward regulating power of the CHP unit j (DKK/MWh),
C_{eb}^{EB}	=	Marginal cost of electricity for producing heat by the EB (DKK/MWh),
C_{hp}^{HP}	=	Marginal cost of electricity for producing heat by the HP (DKK/MWh),
C_g^{GS}	=	Marginal cost of producing gas by the gas source (DKK/MWh),
C_p^{P2G}	=	Cost of electricity for producing gas and heat by P2G (DKK/MWh),
C_g^{UR}	=	Marginal cost of upward regulating power of the gas source unit g (DKK/MWh),
C_j^{UR}	=	Marginal cost of upward regulating power of the CHP unit j (DKK/MWh),
C_e^{VOLL}	=	Value of the lost load (DKK/MWh), and
C_w^{spill}	=	Cost of wind spillage (DKK/MWh).
Variables		
D_{eb}^{EB}	=	Electric consumption of the EB (MW),
$D_e^{ED,shed}$	=	Electric load shedding (MW),
D_{hp}^{HP}	=	Electric consumption of the HP (MW),

(continued on next page)

Parameters

D_p^{P2G}	$=$	Electric consumption of P2G (MW),
H_j^{CHP}	$=$	Heat power output from CHP (MW),
P_j^{CHP}	$=$	Power output of CHP (MW),
Q_g^{GS}	$=$	Gas output from the gas source (MW),
$R_j^{CHP,U/D}$	$=$	Upward/downward regulation power from CHP (MW),
$R_g^{GS,U/D}$	$=$	Upward/downward regulation power from the gas source (MW), and
W_w^{spill}	$=$	Wind spillage of wind farms (MW).

The DA stage includes the cost of producing electricity and heat power by CHP. Furthermore, it considers flexible demand. The flexible demand includes EB, HP, and P2G. Further on, the gas supply cost is included in the last row of the DA stage. The RT stage is dependent on the realization of the scenario and the probability of the scenario to occur. Expected system balancing costs in the RT stage are multiplied with the probability of each scenario and are represented in the last four rows. The first term describes the cost of upward regulation power (URP) and downward regulation power (DRP) from CHP, whereas the second term presents the cost of heat power regulation from CHP. Flexible demand from P2G and the EB are accounted for in the third and fourth term, respectively, as a difference in power between the RT stage and DA stage. Furthermore, in the fifth term, flexible demand provided by the HP is accounted for as a difference in power between the RT stage and DA stage. Moreover, the regulation cost of gas supply is presented in the sixth term. Finally, in the last two terms, the value of lost load and wind power spillage are included, respectively.

The following sections list the first and second stages. As mentioned, each stage is divided into four subsets.

10.3.2 First-stage constraints

In this section, the first-stage constraints for the EPS, NGS, and DHS are presented. At the end, the constraints for linking components are given.

10.3.2.1 Electric power system

The first-stage constraints for the EPS are summarized in Eqs. (10.2) through (10.11). The nodal power balance equation is demonstrated in Eq. (10.2). It presents the balance between generation and demand. Generation includes power output from CHP noted as P_j^{CHP} and wind power output in the first stage noted as W_w^{WF}. The consumption includes P2G, the EB, the HP; the electric consumption of the gas compressor (GC) noted as D_c^{GC}; and electric demand noted as D_e^{ED}. B_{nm} notes susceptance of the transmission line, whereas δ_n denotes the phase angle of bus n. The generated power of CHP is limited by maximum and minimum operational output power as shown in Eq. (10.3). The consumption of P2G, the HP, and the EB are limited by its minimum and maximum consumption levels and shown in Eqs. (10.4) through (10.6), respectively. Furthermore, Eq. (10.7) describes the limitation of transmission capacity by its maximum capacity, denoted as P_{nm}^{rate}. The ramping rate limit for CHP is illustrated in Eq. (10.8), where P_j^{RRL} denotes the downward and upward ramp rate limits expressed in %/h. Eq. (10.9) gives the angle constraint on the reference bus, whereas Eq. (10.10) illustrates the non-negative variable for the electric consumption of the GC. The wind power in the first stage cannot exceed the maximum capacity of the wind power, denoted as $W_w^{WF,max}$, as presented in Eq. (10.11).

$$\sum_{j\in\Omega_n^{CHP}} P_{j,t}^{CHP,DA} + \sum_{w\in\Omega_n^{WF}} W_{w,t}^{WF,DA} - \sum_{p\in\Omega_n^{P2G}} D_{p,t}^{P2G,DA} - \sum_{eb\in\Omega_n^{EB}} D_{eb,t}^{EB,DA}$$
$$- \sum_{hp\in\Omega_n^{HP}} D_{hp,t}^{HP,DA} - \sum_{e\in\Omega_n^{ED}} D_{e,t}^{ED,DA} - \sum_{c\in\Omega_n^{GC}} D_{c,t}^{GC,DA}$$
$$= \sum_{m\in\Omega_n^{EPS}} B_{nm}\left(\delta_{n,t}^{DA} - \delta_{m,t}^{DA}\right), \quad \forall n \in \Omega^{EPS}, \; \forall t \in T$$
(10.2)

$$P_j^{CHP,min} \leq P_{j,t}^{CHP,DA} \leq P_j^{CHP,max}, \quad \forall j \in \Omega^{CHP}, \; \forall t \in T \tag{10.3}$$

$$D_p^{P2G,min} \leq D_{p,t}^{P2G,DA} \leq D_p^{P2G,max}, \quad \forall p \in \Omega^{P2G}, \; \forall t \in T \tag{10.4}$$

$$D_{hp}^{HP,min} \leq D_{hp,t}^{HP,DA} \leq D_{hp}^{HP,max}, \quad \forall hp \in \Omega^{HP}, \; \forall t \in T \tag{10.5}$$

$$D_{eb}^{EB,min} \leq D_{eb,t}^{EB,DA} \leq D_{eb}^{EB,max}, \quad \forall eb \in \Omega^{EB}, \; \forall t \in T \tag{10.6}$$

$$-P_{nm}^{rate} \leq B_{nm}\left(\delta_{n,t}^{DA} - \delta_{m,t}^{DA}\right) \leq P_{nm}^{rate}, \quad \forall n, m \in \Omega^{EPS}, \; \forall t \in T \tag{10.7}$$

$$\left| P_{j,t-1}^{CHP,DA} - P_{j,t}^{CHP,DA} \right| \leq P_j^{RRL}, \quad \forall j \in \Omega^{CHP}, \; \forall t \in T, \; t > 1 \tag{10.8}$$

$$\delta_{ref,t}^{DA} = 0, \quad \forall t \in T \tag{10.9}$$

$$D_{c,t}^{GC,DA} \geq 0, \quad \forall c \in \Omega^{GC}, \; \forall t \in T \tag{10.10}$$

$$0 \leq W_{w,t}^{WF,DA} \leq W_w^{WF,max}, \quad \forall w \in \Omega^{WF}, \; \forall t \in T \tag{10.11}$$

10.3.2.2 District heating system

DHS constraints are split into the hydraulic model, thermal model, and scheduling operating constraints. The hydraulic model is expressed by continuity of flow. Continuity of flow is a constraint indicating that the mass flow entering the node is equal to the mass flow leaving the node and consumption at that node. The continuity of flow is often expressed as follows:

$$A \, m_{mn,t}^{DA} = A_q \circ mq_{m,t}^{DA} \, , \quad \forall m \in \Omega^{DHS}, \, \forall t \in T \tag{10.12}$$

where m_{mn} is the mass flow entering or leaving the node. mq_m is mass flow at the consumption or production node. A is the network incidence matrix. The number of rows and columns correspond to the number of nodes and pipes, respectively. The elements of the matrix can have three values. Value $(+1)$ corresponds to the mass flow entering the node, (-1) represents the mass flow leaving the node, and (0) represents no flow. A_q is a vector of similar values with the length equal to the number of nodes. The value in the vector is $(+1)$ if the mass flow leaves the node, and it corresponds to (-1) if the mass flow is entering the node. As a note, an element-wise operation on the right-hand side is performed. The unit of mass flow is kg/h.

The DHS consists of supply and return pipelines. The supply pipeline provides the heat from the heat source to the demand, whereas the return pipeline includes pipelines returning to the heat source after the heat has been extracted by the heat demand. Therefore, continuity of flow is considered for both supply and return pipelines in Eqs. (10.13) and (10.14), respectively. Moreover, the final equation for the hydraulic model is continuity of flow for supply and return pipelines as illustrated in Eq. (10.15). The superscript S stands for supply pipelines, whereas superscript R denotes the return pipelines.

$$A^S \, m_{mn,t}^{S, \, DA} = A_q^S \circ mq_{m,t}^{DA} \, , \quad \forall m \in \Omega^{DHS}, \, \forall t \in T \tag{10.13}$$

$$A^R \, m_{mn,t}^{R, \, DA} = A_q^R \circ mq_{m,t}^{DA} \, , \quad \forall m \in \Omega^{DHS}, \, \forall t \in T \tag{10.14}$$

$$m_{mn,t}^{S, \, DA} = m_{mn,t}^{R, \, DA}, \quad \forall (mn) \in \Omega^{DHS}, \, \forall t \in T \tag{10.15}$$

The thermal model is given in Eqs. (10.16) through (10.26). The nodal heat balance equation is presented in Eq. (10.16). The balance equation correlates the temperature and heat energy at each node. The difference in the temperature decides if the energy is produced or consumed. The temperature difference is calculated as the difference between the final temperature and initial temperature. If energy from a heat source is calculated, the final temperature is higher than the initial temperature, resulting in a positive temperature difference. The supply temperature is the final temperature, and the return temperature is the initial temperature. On the contrary, when the heat is extracted by a load, the final temperature is lower than the initial temperature. The return temperature is the final temperature, and the supply temperature is the initial temperature. The entire nodal balance equation considers CHP, the HP, P2G, and the EB as the providers of heat power. Additionally, thermal energy storage is integrated into the system to provide additional flexibility. H_l^{HL} denotes heat demand in MW, whereas H_{eb}^{EB}, H_{hp}^{HP}, and H_p^{P2G} denote heat power output from the EB, the HP, and P2G in MW, respectively. $H_s^{HS,in}$ stands for the heat power input to the thermal energy storage in MW, whereas $H_s^{HS,out}$ stands for the heat power output from the thermal energy storage in MW.

$$
\begin{aligned}
&\sum_{j\in\Omega_n^{CHP}} H_{j,t}^{CHP,DA} + \sum_{hp\in\Omega_n^{HP}} H_{hp,t}^{HP,DA} \\
&+ \sum_{eb\in\Omega_n^{EB}} H_{eb,t}^{EB,DA} + \sum_{p\in\Omega_n^{P2G}} H_{p,t}^{P2G,DA} \\
&+ \sum_{s\in\Omega_n^{HS}} \left(H_{s,t}^{HS,out,DA} - H_{s,t}^{HS,in,DA} \right) - \sum_{l\in\Omega_n^{HL}} H_{l,t}^{HL,DA} = \left\{ \begin{array}{l} T_{m,t}^{final} > T_{m,t}^{initial} \ mq_{m,t}^{DA} c \left(T_{m,t}^{S,DA} - T_{m,t}^{R,DA} \right) \\ T_{m,t}^{initial} > T_{m,t}^{final} \ mq_{m,t}^{DA} c \left(T_{m,t}^{R,DA} - T_{m,t}^{S,DA} \right) \end{array} \right\}, \\
&\hspace{8cm} \forall m \in \Omega^{DHS}, \ \forall t \in T
\end{aligned}
\tag{10.16}
$$

In the DHS, the water temperature is considered as the important parameter. In the supply pipelines, the temperature of water is much higher due to heat being transported from a heat source to the heat demand. The return pipelines have lower water temperature, as the heat is being extracted from the heat demand and is being returned back to the heat source. The second equation in the thermal model is the temperature mixing equation at the node. In situations where there is more than one incoming mass flow to the node, the temperature of the mass flow going out of the node needs to be obtained. The temperature mix equations for supply and return pipelines are shown in Eqs. (10.17) and (10.18), respectively. $T_m^{S/R}$ stands for the temperature at a node in the supply or return pipelines. $T_{mn}^{S,OUT}$ and $T_{mn}^{S,IN}$ denote the outlet and inlet temperature of the supply pipelines, respectively, whereas $T_{mn}^{R,OUT}$ and $T_{mn}^{R,IN}$ denote the outlet and inlet temperature of the return pipelines, respectively.

$$
\sum \left(m_{nm,t}^{S,\ DA} T_{nm,t}^{S,OUT,DA} \right) = T_{mn,t}^{S,IN,DA} \sum \left(m_{mn,t}^{S,\ DA} \right), \ \forall m \in \Omega^{DHS,mix,S}, \ \forall t \in T
\tag{10.17}
$$

$$\sum \left(m_{mn,t}^{R,\ DA}\, T_{mn,t}^{R,OUT,DA} \right) + \sum \left(mq_{m,t}^{DA}\, T_{m,t}^{R,DA} \right) = T_{mn,t}^{R,IN,DA} \sum \left(m_{mn,t}^{R,\ DA} \right),$$
$$\forall m \in \Omega^{DHS,mix,R},\ \forall t \in T$$
$$(10.18)$$

The third part of the thermal model is the temperature drop equation. The temperature drop equation expresses the heat loss due to heat transportation due to the difference in high water temperature and ambient temperature. The drop depends on the parameters of the pipeline. Parameter λ depends on the diameter and heat transfer coefficient. Both supply and return pipelines have a drop in the temperature along the pipelines, as shown in Eqs. (10.19) and (10.20), respectively. T^A denotes the ambient temperature in °C, c denotes the specific heat capacity of water in Wh/kgK, λ_{mn} denotes the thermal conductivity of the DHS pipeline in W/mK, and L_{mn} denotes the length of pipeline in meters.

$$T_{mn,t}^{S,OUT,DA} - T^A = \left(T_{mn,t}^{S,IN,DA} - T^A \right) e^{-\frac{\lambda_{mn} L_{mn}}{cm_{mn,t}^{S,\ DA}}},\quad \forall (mn) \in \Omega^{DHS},\ \forall t \in T$$
$$(10.19)$$

$$T_{mn,t}^{R,OUT,DA} - T^A = \left(T_{mn,t}^{R,IN,DA} - T^A \right) e^{-\frac{\lambda_{mn} L_{mn}}{cm_{mn,t}^{R,\ DA}}},\quad \forall (mn) \in \Omega^{DHS},\ \forall t \in T$$
$$(10.20)$$

The last part of the thermal model is illustrated in Eqs. (10.21) through (10.26). Eq. (10.21) assures constant temperature input from the heat source, denoted as $T_m^{S,fixed}$. In addition, for the first node in the supply pipeline, the inlet temperature of the inlet for the first pipeline is equal to the supply temperature at that node. Eq. (10.22) illustrates that for the last node in the return system, the inlet temperature of the pipeline equals the return temperature at that node. Moreover, for the nonmixing nodes, the following constraints should be considered. The outlet temperature of the first pipeline equals the inlet temperature of the following pipeline and supply temperature at that node as illustrated in Eqs. (10.23) and (10.24), respectively. Similarly, the constraints are considered for the return pipelines in Eqs. (10.25) and (10.26).

$$T_{m,t}^{S,DA} = T_{mn,t}^{S,IN,DA} = T_{m,t}^{S,fixed},\quad m = 1, \forall t \in T \tag{10.21}$$

$$T_{m,t}^{R,DA} = T_{mn,t}^{R,IN,DA},\quad m = max(m), \forall t \in T \tag{10.22}$$

$$T_{mn,t}^{S,OUT,DA} = T_{mn,t}^{S,IN,DA},\quad \forall m \notin \Omega^{DHS,mix,S},\ \forall t \in T \tag{10.23}$$

$$T_{mn,t}^{S,OUT,DA} = T_{m,t}^{S,DA},\quad \forall m \notin \Omega^{DHS,mix,S},\ \forall t \in T \tag{10.24}$$

$$T_{mn,t}^{R,OUT,DA} = T_{m,t}^{R,DA}, \quad \forall m \notin \Omega^{DHS,mix,R}, \quad \forall t \in T \tag{10.25}$$

$$T_{mn,t}^{R,OUT,DA} = T_{mn,t}^{R,IN,DA}, \quad \forall m \notin \Omega^{DHS,mix,R}, \quad \forall t \in T \tag{10.26}$$

The last constraints in the DHS are scheduling operating constraints. The mass flow limits in the supply and return pipelines, and mass flow at the consumption or production node, are presented in Eqs. (10.27), (10.28), and (10.29), respectively. The inflow and outflow temperature in the pipelines are limited by its operational temperatures as illustrated in Eqs. (10.30) through (10.33). Eqs. (10.34) and (10.35) limit the supply and return temperatures, respectively. The thermal energy storage balance equation is shown in Eqs. (10.36) and (10.37). H_s^{SOE} denotes the heat stock in the storage in MWh. The state of energy of the thermal energy storage is limited by its maximum and minimum allowed capacity in Eq. (10.38). Operational limits for maximum injection and withdrawal rate are presented in Eqs. (10.39) and (10.40), respectively. Operational constraints of thermal energy storage ensure that only one process, injection or withdrawal, is available at a time as demonstrated in Eqs. (10.39) through (10.43). $x_s^{HS,out}$ is a binary variable and is equal to 1 if active, meaning that extraction from the storage is active, and otherwise the variable is equal to 0. $x_s^{HS,in}$ denotes the binary variable and is equal to 1 if the injection in the storage is active, and otherwise it is equal to 0.

$$m_{mn}^{S,min} \leq m_{mn,t}^{S,\ DA} \leq m_{mn}^{S,max}, \quad \forall (mn) \in \Omega^{DHS}, \forall t \in T \tag{10.27}$$

$$m_{mn}^{R,min} \leq m_{mn,t}^{R,\ DA} \leq m_{mn}^{R,max}, \quad \forall (mn) \in \Omega^{DHS}, \forall t \in T \tag{10.28}$$

$$mq_m^{min} \leq mq_{m,t}^{DA} \leq mq_m^{max}, \quad \forall m \in \Omega^{DHS}, \forall t \in T \tag{10.29}$$

$$T_{mn}^{S,IN,min} \leq T_{mn,t}^{S,IN,DA} \leq T_{mn}^{S,IN,max}, \quad \forall (mn) \in \Omega^{DHS}, \forall t \in T \tag{10.30}$$

$$T_{mn}^{S,OUT,min} \leq T_{mn,t}^{S,OUT,DA} \leq T_{mn}^{S,OUT,max}, \quad \forall (mn) \in \Omega^{DHS}, \forall t \in T \tag{10.31}$$

$$T_{mn}^{R,IN,min} \leq T_{mn,t}^{R,IN,DA} \leq T_{mn}^{R,IN,max}, \quad \forall (mn) \in \Omega^{DHS}, \forall t \in T \tag{10.32}$$

$$T_{mn}^{R,OUT,min} \leq T_{mn,t}^{R,OUT,DA} \leq T_{mn}^{R,OUT,max}, \quad \forall (mn) \in \Omega^{DHS}, \forall t \in T \tag{10.33}$$

$$T_m^{S,min} \leq T_{m,t}^{S,DA} \leq T_m^{S,max}, \quad \forall m \in \Omega^{DHS}, \forall t \in T \tag{10.34}$$

$$T_m^{R,min} \leq T_{m,t}^{R,DA} \leq T_m^{R,max}, \quad \forall m \in \Omega^{DHS}, \forall t \in T \tag{10.35}$$

$$H_{s,t}^{SOE,DA} = H_{s,t-1}^{SOE,DA} + H_{s,t}^{HS,in,DA} - H_{s,t}^{HS,out,DA}, \quad \forall s \in \Omega^{HS}, \forall t \in T, t > 1 \tag{10.36}$$

$$H_{s,t}^{SOE,DA} = H_s^{SOE0,DA} + H_{s,t}^{HS,in,DA} - H_{s,t}^{HS,out,DA}, \quad \forall s \in \Omega^{HS}, \ t = 1 \tag{10.37}$$

$$H_s^{SOE,min} \leq H_{s,t}^{SOE,DA} \leq H_s^{SOE,max}, \quad \forall s \in \Omega^{HS}, \forall t \in T, t > 1 \tag{10.38}$$

$$0 \leq H_{s,t}^{HS,in,DA} \leq H_s^{HS,in,max} x_{s,t}^{HS,in,DA}, \quad \forall s \in \Omega^{HS}, \forall t \in T \tag{10.39}$$

$$0 \leq H_{s,t}^{HS,out,DA} \leq H_s^{HS,out,max} x_{s,t}^{HS,out,DA}, \quad \forall s \in \Omega^{HS}, \forall t \in T \tag{10.40}$$

$$x_{s,t}^{HS,out,DA} + x_{s,t}^{HS,in,DA} \leq 1, \quad \forall s \in \Omega^{HS}, \forall t \in T \tag{10.41}$$

$$x_{s,t}^{HS,out,DA} \in \{0, 1\}, \quad \forall s \in \Omega^{HS}, \forall t \in T \tag{10.42}$$

$$x_{s,t}^{HS,in,DA} \in \{0, 1\}, \quad \forall s \in \Omega^{HS}, \forall t \in T \tag{10.43}$$

10.3.2.3 Natural gas system

The NGS is represented by steady–state flow in Eqs. (10.44) through (10.65). For modeling purposes, isothermal flow is considered as explained in Chapter 2. Due to the pressure difference between two ends of the pipeline, the gas can flow. The gas flow equation is shown in Eq. (10.44). It is assumed that there is no elevation between two ends of the pipeline. $p_{n,t}$ and $p_{m,t}$ are inlet and outlet pressures in the pipeline at a certain time t, respectively, expressed in MPa. Q_{nm} represents a gas flow in the pipeline n-m in m^3/h. The pipeline parameters are expressed through parameter C_{nm}. C_{nm} illustrates the relationship between gas flow, gas pressures, and pipeline parameters as shown in Eq. (10.45) ([33], Shashi [34]). The unit for C_{nm} is (m^3/h)/MPa.

$$\left(\left(p_{n,t}^{DA} \right)^2 - \left(p_{m,t}^{DA} \right)^2 \right) C_{nm}^2 = \left(Q_{nm,t}^{DA} \right)^2, \quad \forall(nm) \in \Omega^{NGS}, \forall t \in T \tag{10.44}$$

$$C_{nm} = C \frac{T_b}{p_b} D_{nm}^{2.5} \left(\frac{1}{L_{nm} \gamma_g T_a Z_a f} \right)^{0.5} \eta_{p,nm}, \quad \forall(nm) \in \Omega^{NGS} \tag{10.45}$$

where

C = constant for calculation of C_{nm} dependent on system unit (47.8917×10^{-6}),

T_b = temperature based on normal cubic meter conditions (K),

p_b = pressure based on normal cubic meter conditions (MPa),

D_{nm} = pipeline diameter (mm),

L_{nm} = pipeline length (km),

γ_g = relative density of natural gas,

T_a = average absolute gas temperature (K),

Z_a = average compressibility factor,

f = friction constant, and

$\eta_{p,nm}$ = efficiency of the pipeline.

The nodal gas balance equation is demonstrated in Eq. (10.46). Gas consumption of CHP expressed in MW is denoted by Q_j^{DCHP}. $Q_{lp}^{LP,IN/OUT}$ denotes gas injection or extraction to or from the linepack in MW, Q_p^{P2G} denotes gas output from P2G in MW, $Q_{st}^{ST,IN/OUT}$ denotes gas injection or extraction to or from gas storage in MW, Q_{nn} denotes the gas flow expressed in MW, and Q_d^{GD} denotes gas demand in MW. It consists of P2G and the gas source as units injecting gas, whereas consumers of gas are gas demand and CHP. It should be noted that gas consumption of the CHP plant is non-negative variable. A gas storage and linepack are integrated into the system as shown.

$$
\begin{aligned}
&\sum_{g \in \Omega_n^{GS}} Q_{g,t}^{GS,DA} + \sum_{st \in \Omega_n^{ST}} \left(Q_{st,t}^{ST,OUT,DA} - Q_{st,t}^{ST,IN,DA} \right) + \sum_{p \in \Omega_n^{P2G}} Q_{p,t}^{P2G,DA} \\
&- \sum_{d \in \Omega_n^{GD}} Q_{d,t}^{GD,DA} - \sum_{j \in \Omega_n^{CHP}} Q_{j,t}^{DCHP,DA} + \sum_{lp \in \Omega_n^{LP}} \left(Q_{lp,t}^{LP,OUT,DA} - Q_{lp,t}^{LP,IN,DA} \right) \\
&= \sum_{m \in \Omega_n^{NGS}} Q_{nm,t}^{DA}, \quad \forall n, m \in \Omega^{NGS}, \forall t \in T
\end{aligned}
$$

(10.46)

The operation limit for injection of the gas from the gas source is limited by the maximum and minimum output of the gas source and is shown in Eq. (10.47). The pressure limits and pressure at the reference node are expressed in Eqs. (10.48) and (10.49), respectively. The gas flow is limited by the maximum allowed flow rate, Q_{nm}^{max}, in Eq. (10.50).

$$
Q_g^{GS,min} \leq Q_{g,t}^{GS,DA} \leq Q_g^{GS,max}, \quad \forall g \in \Omega^{GS}, \forall t \in T \tag{10.47}
$$

$$
\left(p_m^{min} \right)^2 \leq \left(p_{m,t}^{DA} \right)^2 \leq \left(p_m^{max} \right)^2, \quad \forall m \in \Omega^{NGS}, \forall t \in T \tag{10.48}
$$

$$
\left(p_{ref,t}^{DA} \right)^2 = p_{ref}, \quad \forall t \in T \tag{10.49}
$$

$$
-Q_{nm}^{max} \leq Q_{nm,t}^{DA} \leq Q_{nm}^{max}, \quad \forall (nm) \in \Omega^{NGS}, \forall t \in T \tag{10.50}
$$

The GCs are integrated in the pipelines to control the flow and compensate for the energy losses [33, 35]. The flow is controlled by the compressor ratio, CR, as shown in Eq. (10.51). The electricity consumption from the

GC is expressed as presented in Eq. (10.52). C^{GC} represents the constant for the GC dependent on units, Z_a is the average compressibility factor, T_s is the suction temperature in K, E^{GC} is the parasitic efficiency, c_k is the specific heat constant for the GC, Q_c^{GC} is the gas flow rate through the compressor, and η_g^{GC} represents compression efficiency.

$$\left(p_{m,t}^{DA}\right)^2 \leq CR^2 \left(p_{n,t}^{DA}\right)^2, \quad \forall (nm) \in \Omega^{NGS}, \; \forall t \in T \tag{10.51}$$

$$D_{c,t}^{GC,DA} = C^{GC} Z_a \frac{T_s}{E^{GC}\eta_g^{GC}} \frac{c_k}{c_k-1}\left(CR^{\frac{c_k-1}{c_k}} - 1\right)Q_{c,t}^{GC,DA}, \quad \forall c \in \Omega^{GC}, \forall t \in T \tag{10.52}$$

The gas storage state of energy denoted as Q_{st}^{SOE} is calculated based on the injection or withdrawal to and from the gas storage as presented in Eqs. (10.53) and (10.54). Eq. (10.55) limits the gas storage capacity by its minimum and maximum available capacity. The maximum injection and withdrawal rates to and from the gas storage are expressed in Eqs. (10.56) and (10.57), respectively. The binary variables $x_{st}^{ST,IN}$ and $x_{st}^{ST,OUT}$ are introduced as presented in Eqs. (10.56) through (10.60) to provide either injection or extraction to and from the gas storage.

$$Q_{st,t}^{SOE,DA} = Q_{st,t-1}^{SOE,DA} + Q_{st,t}^{ST,IN,DA} - Q_{st,t}^{ST,OUT,DA}, \quad \forall st \in \Omega^{ST}, \forall t \in T, t > 1 \tag{10.53}$$

$$Q_{st,t}^{SOE,DA} = Q_{st}^{SOE0,DA} + Q_{st,t}^{ST,IN,DA} - Q_{st,t}^{ST,OUT,DA}, \quad \forall st \in \Omega^{ST}, t = 1 \tag{10.54}$$

$$Q_{st}^{SOE,min} \leq Q_{st,t}^{SOE,DA} \leq Q_{st}^{SOE,max}, \quad \forall st \in \Omega^{ST}, \; \forall t \in T \tag{10.55}$$

$$Q_{st}^{ST,IN,min} \leq Q_{st,t}^{ST,IN,DA} \leq Q_{st}^{ST,IN,max} x_{st,t}^{ST,IN,DA}, \quad \forall st \in \Omega^{ST}, \forall t \in T \tag{10.56}$$

$$Q_{st}^{ST,OUT,min} \leq Q_{st,t}^{ST,OUT,DA} \leq Q_{st}^{ST,OUT,max} x_{st,t}^{ST,OUT,DA}, \quad \forall st \in \Omega^{ST}, \forall t \in T \tag{10.57}$$

$$x_{st,t}^{ST,IN,DA} + x_{st,t}^{ST,OUT,DA} \leq 1, \quad \forall st \in \Omega^{ST}, \forall t \in T \tag{10.58}$$

$$x_{st,t}^{ST,IN,DA} \in \{0, 1\}, \quad \forall st \in \Omega^{ST}, \forall t \in T \tag{10.59}$$

$$x_{st,t}^{ST,OUT,DA} \in \{0, 1\}, \quad \forall st \in \Omega^{ST}, \forall t \in T \tag{10.60}$$

The linepack balance should be maintained as presented in Eqs. (10.61) and (10.62); the linepack capacity is denoted as QLP_{lp}, and the initial linepack from Eq. (10.62) is calculated as shown in Eq. (10.63):

$$QLP_{lp,t}^{DA} = QLP_{lp,t-1}^{DA} + Q_{lp,t}^{LP,IN,DA} - Q_{lp,t}^{LP,OUT,DA}, \quad \forall lp \in \Omega^{LP}, \forall t \in T, t > 1 \tag{10.61}$$

$$QLP_{lp,t}^{DA} = QLP_{lp}^{0} + Q_{lp,t}^{LP,IN,DA} - Q_{lp,t}^{LP,OUT,DA}, \quad \forall lp \in \Omega^{LP}, t = 1 \tag{10.62}$$

$$QLP_{lp}^{0} = C^{LP} \frac{T_b}{p_b} \frac{1}{Z_a T_a} D_{nm}^2 L_{nm} p_{ave,nm}^{LP}, \quad \forall lp \in \Omega^{LP} \tag{10.63}$$

Where C^{LP} denotes the linepack constant dependent on the system unit, QLP_{lp}^{0} denotes the initial linepack capacity (MWh), and $p_{ave,nm}^{LP}$ denotes the average absolute gas pressure in the pipeline (MPa). The most accurate calculation for the average gas pressure in the pipelines is presented in Eq. (10.64). Note that the majority of the symbols were explained and defined earlier. To provide flexibility in the next optimization period, the linepack constraint in Eq. (10.65) is introduced.

$$p_{ave,nm}^{LP} = \frac{2}{3} \left(p_n + p_m - \frac{p_n p_m}{p_n + p_m} \right), \quad \forall lp \in \Omega^{LP} \tag{10.64}$$

$$QLP_{lp,t}^{DA} = QLP_{lp}^{0}, \quad \forall lp \in \Omega^{LP}, \; t = \max(T) \tag{10.65}$$

10.3.2.4 Linking components

Linking constraints couple several subsystems of the IES. The linking units include P2G, EB, HP, and CHP. The electricity and heat generated from CHP are based on the consumption of the gas in the CHP unit and generating of electricity and heat efficiencies as shown in Eqs. (10.66) and (10.67), respectively. $\eta^{CHP,e}$ and $\eta^{CHP,h}$ denote the generated electricity efficiency and generated heat efficiency of CHP, respectively. The conversion of excess electricity to gas and heat by the P2G unit are illustrated in Eqs. (10.68) and (10.69), respectively. $\eta^{P2G,Q}$ and $\eta^{P2G,H}$ represent the P2G conversion efficiency from electricity to gas and heat, respectively. P2G is considered as one direction energy conversion unit that consumes excess electricity from a wind turbine and produces hydrogen. It is considered that methane only can be injected in the NGS. This means that by using electrolysis, the conversion from electricity to hydrogen happens and, through the methanation process, methane is obtained. Hence, a lower efficiency is considered, as more energy is needed to obtain methane. The relationship between heat generated and electricity consumed by the HP is shown in

Eq. (10.70), where COP denotes the coefficient of performance of the HP. Eq. (10.71) represents the ratio between heat output and electricity consumption of the EB through efficiency of the EB, η^{EB}.

$$P_{j,t}^{CHP,DA} = \eta^{CHP,e} Q_{j,t}^{DCHP,DA}, \quad \forall j \in \Omega^{CHP}, \ \forall t \in T \tag{10.66}$$

$$H_{j,t}^{CHP,DA} = \eta^{CHP,h} Q_{j,t}^{DCHP,DA}, \quad \forall j \in \Omega^{CHP}, \ \forall t \in T \tag{10.67}$$

$$Q_{p,t}^{P2G,DA} = \eta^{P2G,Q} D_{p,t}^{P2G,DA}, \quad \forall p \in \Omega^{P2G}, \ \forall t \in T \tag{10.68}$$

$$H_{p,t}^{P2G,DA} = \eta^{P2G,H} D_{p,t}^{P2G,DA}, \quad \forall p \in \Omega^{P2G}, \ \forall t \in T \tag{10.69}$$

$$H_{hp,t}^{HP,DA} = COP \, D_{hp,t}^{HP,DA}, \quad \forall hp \in \Omega^{HP}, \ \forall t \in T \tag{10.70}$$

$$H_{eb,t}^{EB,DA} = \eta^{EB} D_{eb,t}^{EB,DA}, \quad \forall eb \in \Omega^{EB}, \ \forall t \in T \tag{10.71}$$

10.3.3 Second-stage constraints

The second-stage constraints are listed in this section. Due to similarity of the second-stage constraints and first-stage constraints, not all constraints are listed. Hence, additional explanations will be given for supplementary constraints only.

10.3.3.1 Electric power system

The constraints for the RT stage are represented in Eqs. (10.72) through (10.80). The nodal power balance in Eq. (10.72) considers the difference between the power consumed and generated in RT and DA. In addition, it includes the upward and downward regulation from CHP, load shedding, and wind spillage. Several scenarios represent the wind power in RT, denoted as $W_w^{W,RT}$. Note that the majority of the symbols were explained and defined earlier, except $R_p^{P2G,U/D}$, which represents URP and DRP from P2G. Due to regulation power deployed in the RT stage, the output power of CHP is calculated as illustrated in Eq. (10.73). Moreover, capacity limits for regulation power are demonstrated in Eqs. (10.74) through (10.78). The wind spillage and load shedding limitations are shown in Eqs. (10.79) and (10.80), respectively. The remaining part of the equations in the RT stage is similar to Eqs. (10.3) through (10.10) with the difference of substituting DA with the RT acronym.

$$
\sum_{j\in\Omega_n^{CHP}}\left(R_{j,t,s}^{CHP,U}-R_{j,t,s}^{CHP,D}\right)+\sum_{w\in\Omega_n^{WF}}\left(W_{w,t,s}^{WF,RT}-W_{w,t}^{WF,DA}-W_{w,t,s}^{spill}\right)
$$
$$
-\sum_{p\in\Omega_n^{P2G}}\left(D_{p,t,s}^{P2G,RT}-D_{p,t}^{P2G,DA}\right)-\sum_{eb\in\Omega_n^{EB}}\left(D_{eb,t,s}^{EB,RT}-D_{eb,t}^{EB,DA}\right)
$$
$$
-\sum_{hp\in\Omega_n^{HP}}\left(D_{hp,t,s}^{HP,RT}-D_{hp,t}^{HP,DA}\right)-\sum_{c\in\Omega_n^{GC}}\left(D_{c,t,s}^{GC,RT}-D_{c,t}^{GC,DA}\right)+\sum_{e\in\Omega_n^{ED}}D_{e,t,s}^{ED,shed}
$$
$$
=\sum_{m\in\Omega_n^{EPS}}B_{nm}\left(\delta_{n,t,s}^{RT}-\delta_{m,t,s}^{RT}-\delta_{n,t}^{DA}+\delta_{m,t}^{DA}\right),\quad\forall n\in\Omega^{EPS},\ \forall t\in T,\forall s\in\Omega^S
$$

(10.72)

$$
P_{j,t,s}^{CHP,RT}=P_{j,t}^{CHP,DA}+\left(R_{j,t,s}^{CHP,U}-R_{j,t,s}^{CHP,D}\right),\quad\forall j\in\Omega^{CHP},\ \forall t\in T,\forall s\in\Omega^S
$$

(10.73)

$$
0\le R_{j,t,s}^{CHP,U}\le R_j^{CHP,U,max}x_{j,t,s}^{CHP,U},\quad\forall j\in\Omega^{CHP},\ \forall t\in T,\forall s\in\Omega^S
$$

(10.74)

$$
0\le R_{j,t,s}^{CHP,D}\le R_j^{CHP,D,max}x_{j,t,s}^{CHP,D},\quad\forall j\in\Omega^{CHP},\ \forall t\in T,\forall s\in\Omega^S
$$

(10.75)

$$
x_{j,t,s}^{CHP,U}+x_{j,t,s}^{CHP,D}\le 1,\quad\forall j\in\Omega^{CHP},\ \forall t\in T,\forall s\in\Omega^S
$$

(10.76)

$$
x_{j,t,s}^{CHP,U}\in\{0,1\},\quad\forall j\in\Omega^{CHP},\ \forall t\in T,\forall s\in\Omega^S
$$

(10.77)

$$
x_{j,t,s}^{CHP,D}\in\{0,1\},\quad\forall j\in\Omega^{CHP},\ \forall t\in T,\forall s\in\Omega^S
$$

(10.78)

$$
0\le W_{w,t,s}^{spill}\le W_{w,t,s}^{WF,RT},\quad\forall w\in\Omega^{WF},\ \forall t\in T,\forall s\in\Omega^S
$$

(10.79)

$$
0\le D_{e,t,s}^{ED,shed}\le D_{e,t}^{ED,DA},\quad\forall e\in\Omega^{ED},\ \forall t\in T,\forall s\in\Omega^S
$$

(10.80)

10.3.3.2 District heating system

Due to similarity of the RT stage constraints in the DHS compared with the DA stage, the remaining part of the equations is not shown. However, Eqs. (10.13) through (10.15) and (10.17) through (10.43) are included in the RT stage by substituting the DA acronym with the RT acronym. The slight difference in equations in the RT and DA stage in seen in the nodal heat balance equation. The nodal heat balance equation for the RT stage is shown in Eq. (10.81) as a difference between the values obtained in RT and DA. The wind uncertainty influences all three subsystems. The CHP production varies based on the wind uncertainty and demand, and therefore influences the heat generation in the DHS.

$$\sum_{j \in \Omega_n^{CHP}} \left(H_{j,t,s}^{CHP,RT} - H_{j,t}^{CHP,DA} \right) + \sum_{hp \in \Omega_n^{HP}} \left(H_{hp,t,s}^{HP,RT} - H_{hp,t}^{HP,DA} \right)$$

$$+ \sum_{eb \in \Omega_n^{EB}} \left(H_{eb,t,s}^{EB,RT} - H_{eb,t}^{EB,DA} \right)$$

$$+ \sum_{p \in \Omega_n^{P2G}} \left(H_{p,t,s}^{P2G,RT} - H_{p,t}^{P2G,DA} \right)$$

$$+ \sum_{s \in \Omega_n^{HS}} \left(\begin{matrix} H_{s,t,s}^{HS,out,RT} - H_{s,t,s}^{HS,in,RT} \\ -H_{s,t}^{HS,out,DA} + H_{s,t}^{HS,in,DA} \end{matrix} \right) = \left\{ \begin{matrix} T_{m,t}^{final} > T_{m,t}^{initial} : \\ \left[mq_{m,t,s}^{RT} c(T_{m,t,s}^{S,RT} - T_{m,t,s}^{R,RT}) \right] - \left[mq_{m,t}^{DA} c(T_{m,t}^{S,DA} - T_{m,t}^{R,DA}) \right] \\ T_{m,t}^{initial} > T_{m,t}^{final} : \\ \left[mq_{m,t,s}^{RT} c(T_{m,t,s}^{R,RT} - T_{m,t,s}^{S,RT}) \right] - \left[mq_{m,t}^{DA} c(T_{m,t}^{R,DA} - T_{m,t}^{S,DA}) \right] \end{matrix} \right\},$$

$$\forall m \in \Omega^{DHS}, \ \forall t \in T, \forall s \in \Omega^S \qquad (10.81)$$

10.3.3.3 Natural gas system

The nodal gas balance equation is the difference between the RT stage and DA stage values as shown in Eq. (10.82). The additional variables are upward and downward regulation gas provided by the gas source and P2G. Due to regulation injection of gas deployed by the gas source in the RT stage, the total injection gas is calculated as illustrated in Eq. (10.83). Moreover, the gas capacity limits for injection of gas in RT are depicted in Eqs. (10.84) through (10.88). Similarly, P2G regulation is activated in the RT stage and the total gas provided by P2G in RT is calculated as shown in Eq. (10.89). Similarly as well, the capacities of regulation gas in RT provided by P2G are limited in Eqs. (10.90) through (10.94). The remaining part of the equations in the RT stage is similar to Eqs. (10.44, 10.45), and (10.47) through (10.65) with the difference of substituting DA with the RT acronym.

$$\sum_{g \in \Omega_n^{GS}} \left(R_{g,t,s}^{GS,U} - R_{g,t,s}^{GS,D} \right) + \sum_{st \in \Omega_n^{ST}} \left(Q_{st,t,s}^{ST,OUT,RT} - Q_{st,t,s}^{ST,IN,RT} - Q_{st,t}^{ST,OUT,DA} + Q_{st,t}^{ST,IN,DA} \right)$$

$$+ \sum_{p \in \Omega_n^{P2G}} \left(R_{p,t,s}^{P2G,U} - R_{p,t,s}^{P2G,D} \right) - \sum_{j \in \Omega_n^{CHP}} \left(Q_{j,t,s}^{DCHP,RT} - Q_{j,t}^{DCHP,DA} \right)$$

$$+ \sum_{lp \in \Omega_n^{LP}} \left(Q_{lp,t,s}^{LP,OUT,RT} - Q_{lp,t,s}^{LP,IN,RT} - Q_{lp,t}^{LP,OUT,DA} + Q_{lp,t}^{LP,IN,DA} \right)$$

$$= \sum_{m \in \Omega_n^{NGS}} \left(Q_{mm,t,s}^{RT} - Q_{mm,t}^{DA} \right), \quad \forall n, m \in \Omega^{NGS}, \forall t \in T, \forall s \in \Omega^S$$

$$(10.82)$$

$$Q_{gs,t,s}^{GS,RT} = Q_{gs,t}^{GS,DA} + R_{gs,t,s}^{GS,U} - R_{gs,t,s}^{GS,D}, \quad \forall gs \in \Omega^{GS}, \forall t \in T, \forall s \in \Omega^S \qquad (10.83)$$

$$0 \le R_{gs,t,s}^{GS,U} \le R_{gs}^{GS,U,max} x_{gs,t,s}^{GS,U}, \quad \forall gs \in \Omega^{GS}, \forall t \in T, \forall s \in \Omega^S \qquad (10.84)$$

$$0 \le R_{gs,t,s}^{GS,D} \le R_{gs}^{GS,D,max} x_{gs,t,s}^{GS,D}, \quad \forall gs \in \Omega^{GS}, \forall t \in T, \forall s \in \Omega^S \qquad (10.85)$$

$$x_{gs,t,s}^{GS,U} + x_{gs,t,s}^{GS,D} \leq 1, \quad \forall gs \in \Omega^{GS}, \ \forall t \in T, \forall s \in \Omega^{S} \tag{10.86}$$

$$x_{gs,t,s}^{GS,U} \in \{0, 1\}, \quad \forall gs \in \Omega^{GS}, \ \forall t \in T, \forall s \in \Omega^{S} \tag{10.87}$$

$$x_{gs,t,s}^{GS,D} \in \{0, 1\}, \quad \forall gs \in \Omega^{GS}, \ \forall t \in T, \forall s \in \Omega^{S} \tag{10.88}$$

$$Q_{p,t,s}^{P2G,RT} = Q_{p,t}^{P2G,DA} + R_{p,t,s}^{P2G,U} - R_{p,t,s}^{P2G,D}, \quad \forall p \in \Omega^{P2G}, \ \forall t \in T, \forall s \in \Omega^{S} \tag{10.89}$$

$$0 \leq R_{p,t,s}^{P2G,U} \leq R_{p}^{P2G,U,max} x_{p,t,s}^{P2G,U}, \quad \forall p \in \Omega^{P2G}, \ \forall t \in T, \forall s \in \Omega^{S} \tag{10.90}$$

$$0 \leq R_{p,t,s}^{P2G,D} \leq R_{p}^{P2G,D,max} x_{p,t,s}^{P2G,D}, \quad \forall p \in \Omega^{P2G}, \ \forall t \in T, \forall s \in \Omega^{S} \tag{10.91}$$

$$x_{p,t,s}^{P2G,U} + x_{p,t,s}^{P2G,D} \leq 1, \quad \forall p \in \Omega^{P2G}, \ \forall t \in T, \forall s \in \Omega^{S} \tag{10.92}$$

$$x_{p,t,s}^{P2G,U} \in \{0, 1\}, \quad \forall p \in \Omega^{P2G}, \ \forall t \in T, \forall s \in \Omega^{S} \tag{10.93}$$

$$x_{p,t,s}^{P2G,D} \in \{0, 1\}, \quad \forall p \in \Omega^{P2G}, \ \forall t \in T, \forall s \in \Omega^{S} \tag{10.94}$$

10.3.3.4 Linking components

The second-stage constraints for the linking units are similar to Eqs. (10.66) through (10.71) with the substitution of the DA superscript by the RT superscript.

10.4 Scenario generation and reduction method

As mentioned in Section 10.2, a suitable method for scenario generation is based on the ECDF of MHVs [27, 36, 25]. Based on the work of Ma et al. [27] and Zhang et al. [25], the proposed scenario generation method has shown the best performance according to evaluation indices. Thus, the evaluation of the proposed scenario generation method is not performed here. The stochasticity is described through realistic scenarios and assigned probabilities that are generated by the proposed scenario generation method. The scenario generation algorithm is based on historical observations and provides a more advanced approach to characterize stochasticity of wind power. The scenario generation method consists of four steps, as can be depicted from Fig. 10.1. Each of the steps is elaborated on further in the sections. At the end of this section, the scenario reduction technique is presented.

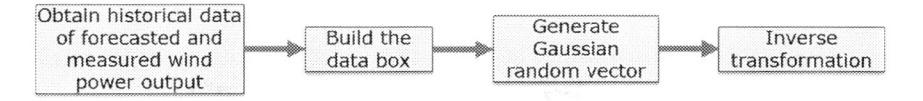

Figure 10.1 Schematic overview for the proposed scenario generation method.

10.4.1 Historical data of forecasted and measured wind power output

As a first step of the proposed scenario generation method, measured and forecasted values of wind power production have to be obtained. One year of data has been collected from a European transmission system operator, found on the European Network of Transmission System Operators for Electricity (ENTSO-E) platform [37]. The resolution of the data is 15 minutes. A matrix is created where a number of rows is equal to a number of timesteps for the entire year. The inputs to the first and second columns of the matrix are data pairs of MHVs and forecasted historical values (FHVs), respectively. The entire matrix is normalized by division with the maximum value of the dataset. Once a matrix is created and normalized, the construction of the data box can be elaborated on further.

10.4.2 Constructing the data box using the historical data

To obtain the distribution for a given point forecast of wind power output, the data box is introduced. The data box contains equally separated power bins. Each power bin is associated with a certain range of forecasted wind power output. The number of power bins in the data box is dependent on the bin width and the number of data used. Since the data are normalized as described earlier, the data box consists of 50 power bins of width 0.02 pu. To build such a data box, the historical data pairs are sorted by FHVs, and FHVs are assigned to the matching power bins with their corresponding MHVs. Each power bin contains a number of the assigned FHV with the corresponding MHV. An illustration for data box construction is presented in Fig. 10.2.

As mentioned, the data box is used to give an approximation of the distribution for a specific point forecast of wind power output. Hence, various distribution fits for the MHV are shown in Fig. 10.3. It can be observed that MHVs are not fitting appropriately to the theoretical distributions. Hence, a nonparametric estimate in the form of an ECDF is created for the MHV

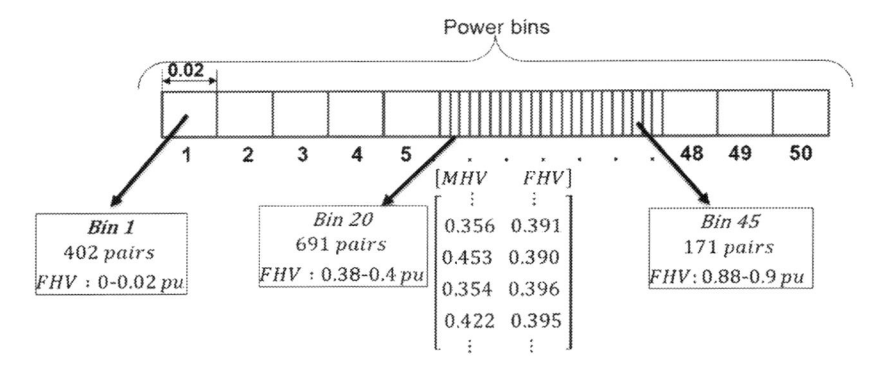

Figure 10.2 Illustration of the data box.

for each power bin. The ECDF is based on historical observations and is not characterized by any theoretical distribution.

10.4.3 Generating Gaussian random vector

The third step is generation of a Gaussian random vector. To generate the Gaussian random vector, two steps are required. The first step is calculation of the parameter representing the correlation between the two random time variables as described in Section 10.4.3.1, whereas the second step is calculation of the covariance matrix based on a mentioned parameter as shown in Section 10.4.3.2. Finally, the Gaussian random vector is generated based on the normal distribution with zero mean and covariance matrix, $\sim N(0, \Sigma)$.

10.4.3.1 Calculation of the parameter representing the correlation between the two random time variables

The parameter epsilon, ε, gives the correlation between the two random time variables for generated scenarios to follow wind variability. To obtain ε, the MHV and generated scenarios are used in the following manner. First, ramping of wind power output between two time intervals of the MHV is calculated as a difference between the wind power output at time t and at time $t+1$. Second, a distribution fit of the ramping values of wind power is obtained. Fig. 10.4 demonstrates the distribution histogram for the ramping values of measured wind power and fitting distributions. The best approximation for distribution fit for the ramping values of measured wind power output is realized by t–location scale distribution [27, 25]. Thus, values of PDF are calculated for a t–locations scale distribution object. The

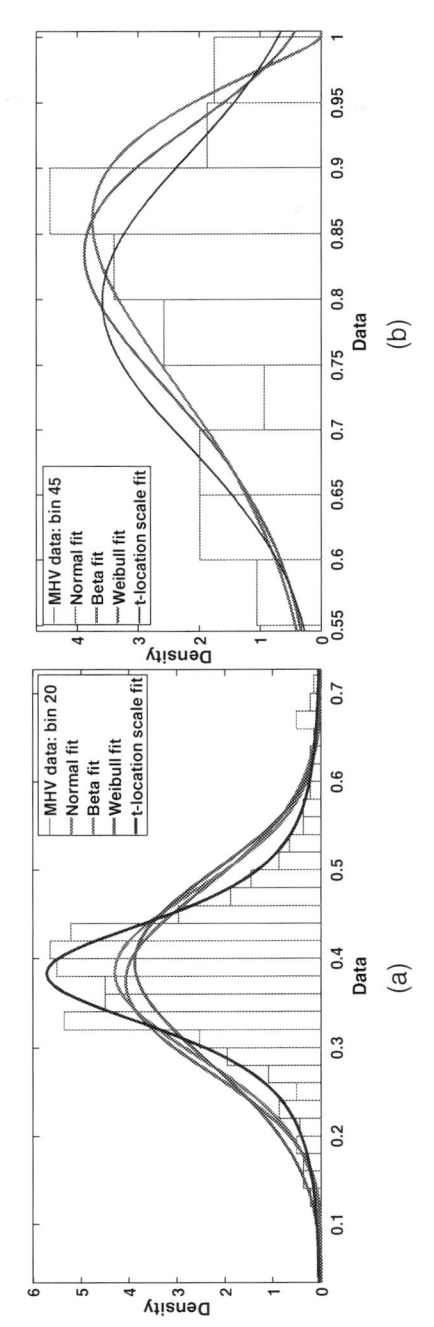

Figure 10.3 Density (PDF) for HMVs and various fits for power (a) bin 20 and (b) bin 45.

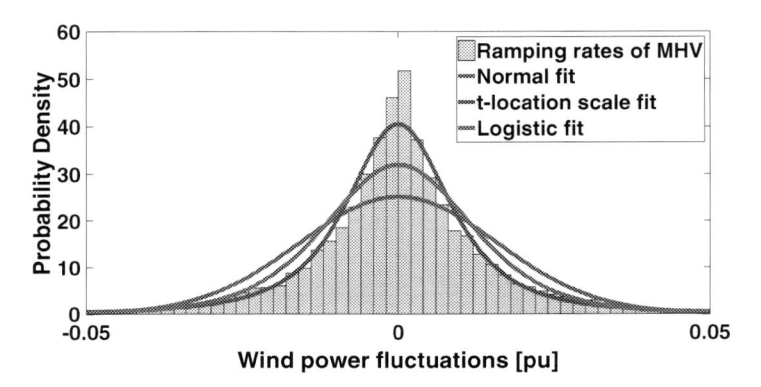

Figure 10.4 Distribution histogram of ramping rates for measured historical wind power output and distribution fits.

sampling size of 1000 in the interval from −0.15 to 0.15 is used according to the existing research of Zhang et al. [25]. A similar procedure is applied to generated scenarios in a few steps as follows. First, the scenarios are generated based on the proposed method with the difference of having the variable ε parameter within a certain range. Second, the ramping values of wind power output between two time intervals are calculated. Third, a t-location scale distribution fitting is used and values of PDF are calculated at equal sampling size and interval range as mentioned earlier. Finally, values of the PDF of wind power output ramping for MHVs and generated scenarios are obtained and used in the calculation of the parameter ε.

The parameter ε is based on the calculation for the indicator epsilon. The indicator epsilon, I_ε, is calculated for each of the variable values of parameters ε. Eq. (10.95) describes the calculation for the indicator epsilon. pdf^{MHV} and pdf^{GS} are values of the PDF of wind power output ramping for MHVs and generated scenarios, respectively. N is the sampling size. Once the indicator epsilon is calculated, the epsilon is chosen based on the smallest value of the indicator epsilon, I_ε. Epsilon for the value of I_ε less than 0.5 is adequate [27].

$$I_\varepsilon = \frac{1}{N} \sum_{is \in SMP} \left| pdf^{MHV}(is) - pdf^{GS}(is) \right| \tag{10.95}$$

10.4.3.2 Calculation of covariance matrix

As explained earlier, the parameter ε, representing the correlation between the two random time variables, is used in the calculation of the covariance matrix. Thus, the temporal correlation (i.e., correlation between wind power outputs at two periods) is represented by a covariance matrix. The covariance matrix is characterized as a positive definite matrix whose size is equal to the maximum forecasting period. Eqs. (10.96) and (10.97) present the calculation of the covariance matrix:

$$\Sigma = \begin{bmatrix} \sigma_{1,1} & \sigma_{1,2} & \cdots & \sigma_{1,H} \\ \sigma_{2,1} & \sigma_{2,2} & \cdots & \sigma_{2,H} \\ \vdots & \vdots & \ddots & \vdots \\ \sigma_{H,1} & \sigma_{H,1} & \cdots & \sigma_{H,H} \end{bmatrix} \tag{10.96}$$

$$\sigma_{Ri,Rj} = cov(Ri, Rj) = e^{-\frac{|i-j|}{\varepsilon}}, \ 0 \le i, j \le H \tag{10.97}$$

where H denotes the maximum forecasting period and R denotes random variables.

Finally, the Gaussian random vector is generated based on the normal distribution with zero mean and covariance matrix, $\sim N(0, \Sigma)$. The number of scenarios is set to 500, whereas the forecasting period equals 96. For each point forecast, the number of scenarios generated is set to 500. Hence, the size of the Gaussian random vector equals to [500,96].

10.4.4 Inverse transformation

Finally, the last step of the scenario generation method is the inverse transformation. The inverse transformation samples the scenarios around the point forecasts and includes four steps. First, a day is chosen from a historical dataset, and the FHVs for the chosen day are normalized. Second, for the chosen day, the power bins to which each point forecast belongs to are obtained. The power bin number indicates which ECDF of the MHV is used for the inverse transformation. Third, a cumulative distribution function for standard normal distribution evaluated at the multivariate random scenarios values is considered. Finally, for every point forecast and for the number of the scenarios of the random vectors, the inverse transformation is applied. At the end, the normalized wind power scenarios are converted to the power unit.

As elaborated previously, a large number of scenarios creates a high computational burden. Therefore, the scenario reduction techniques can be used. To reduce the initial number of generated scenarios, the backward scenario reduction technique is chosen. The basic principle is to find the minimum distance of all scenarios until the reduced number of scenarios is reached. As a starting point, the probability of each scenario is equal to the

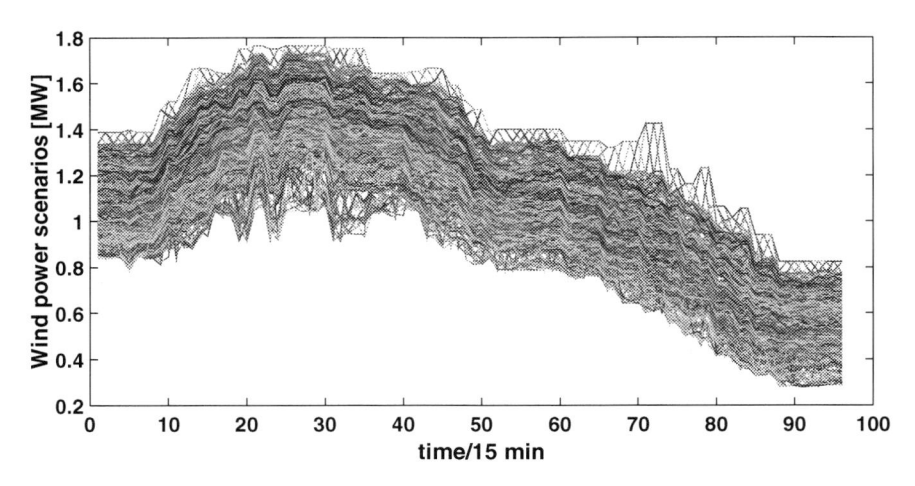

Figure 10.5 Generated scenarios by the proposed scenario generation method.

equally distributed probability for all cases. First, for all scenarios generated, the distance of every two generated scenarios is calculated for the entire forecasting period. Second, the probabilities of minimum distant scenarios are multiplied with the distance of the scenario from another scenario. Third, for the first minimum distance between the scenarios, probabilities of those scenarios that are the least distant are summed and a new probability for the first reduced scenario is achieved. From the set of probabilities, the probability of the second scenario is removed as it was previously added to the probability of the first scenario, and the second scenario in the matrix including all generated scenarios is deleted. The distance matrix is smaller by a scenario. This procedure is repeated until the number of reduced scenarios is reached. Fig. 10.5 represents the generated scenarios obtained by the proposed method. The reduced number of scenarios is shown in Fig. 10.6 with its corresponding probability of occurring.

To summarize, the overview of the proposed scenario generation method and its implementation to the two-stage stochastic IES are presented in Fig. 10.7.

10.5 Example of a case study

In this section, a case study will be illustrated and the mathematical model will be summarized.

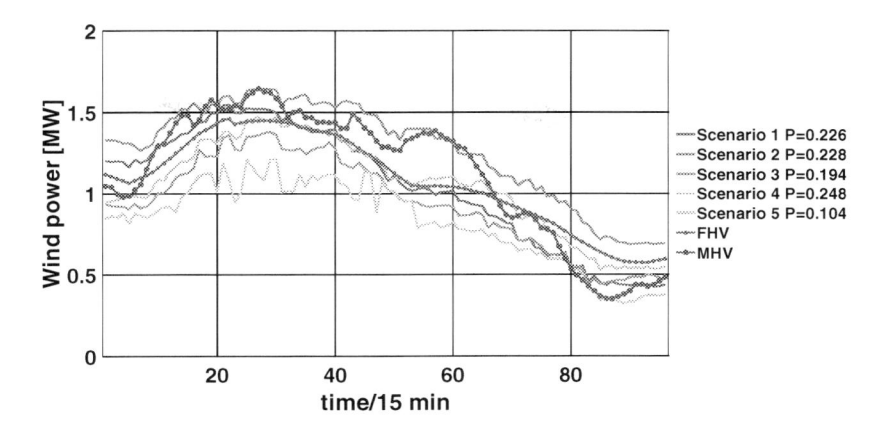

Figure 10.6 Reduced number of scenarios and its corresponding probabilities.

The test system is shown in Fig. 10.8 and parameters are given in Table 10.1. The NGS is denoted with a blue line, the EPS is denoted with a green line, and the DHS is denoted with a red line. The wind profiles can be found at the transparency platform of ENTSO-E [37]. Data from January 1, 2019, to January 1, 2020, are selected for profiles in Germany (BZN|DE-LU) and appropriately scaled. The costs of the URP and DRP are based on the work of Ordoudis et al. [38]. The remaining parameters are based on the data found elsewhere [39–43, 32, 44, 45].

Based on Section 10.3, a mixture of different units is demonstrated. To decrease the computational burden, the pu system is used for the entire mathematical model of the IES. Therefore, the NGS is converted to the pu system as follows. First, the parameters are converted to a power unit as shown in Eq. (10.98):

$$Q_{MW} = \frac{UCV}{3600} Q_{m^3/h} \tag{10.98}$$

where UCV denotes the upper calorific value. The unit of UCV is MJ/m^3 [1]. A similar description can be found in Section 2.5.3 of Chapter 2. Second, to obtain the NGS in pu, the base power and base pressure should be chosen. In Eq. (10.44), the parameter C_{nm} is in (m^3/h)/kPa. As a first step, a conversion from (m^3/h)/kPa to (MW)/kPa is performed by using Eq. (10.98). Moreover, the base power and base pressure are used to convert the parameter to the pu system. Furthermore, all parameters from Eqs. (10.47) through (10.65) are converted from m^3/h to the pu system in the same manner. The exception is Eq. (10.52). The conversion of the GC

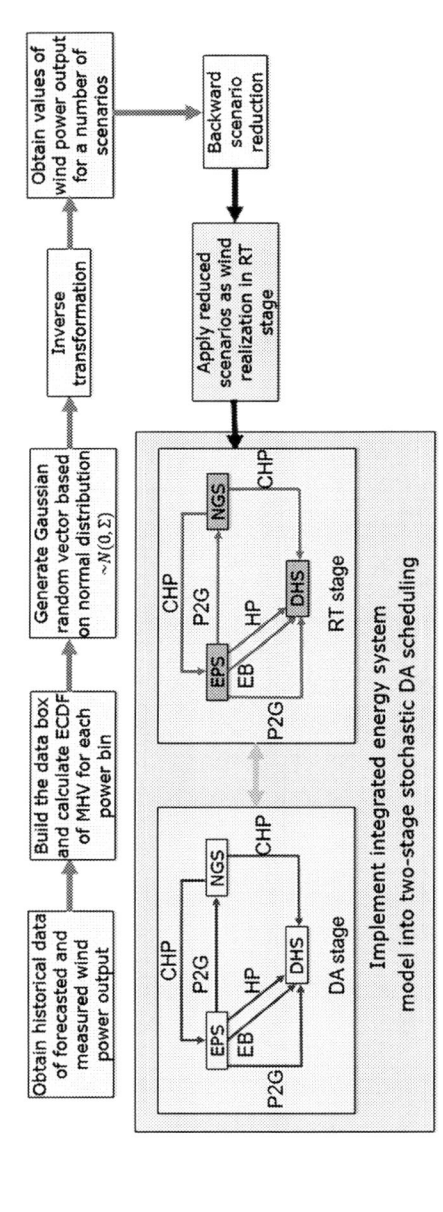

Figure 10.7 Schematic overview of implementation of proposed method into two-stage DA scheduling.

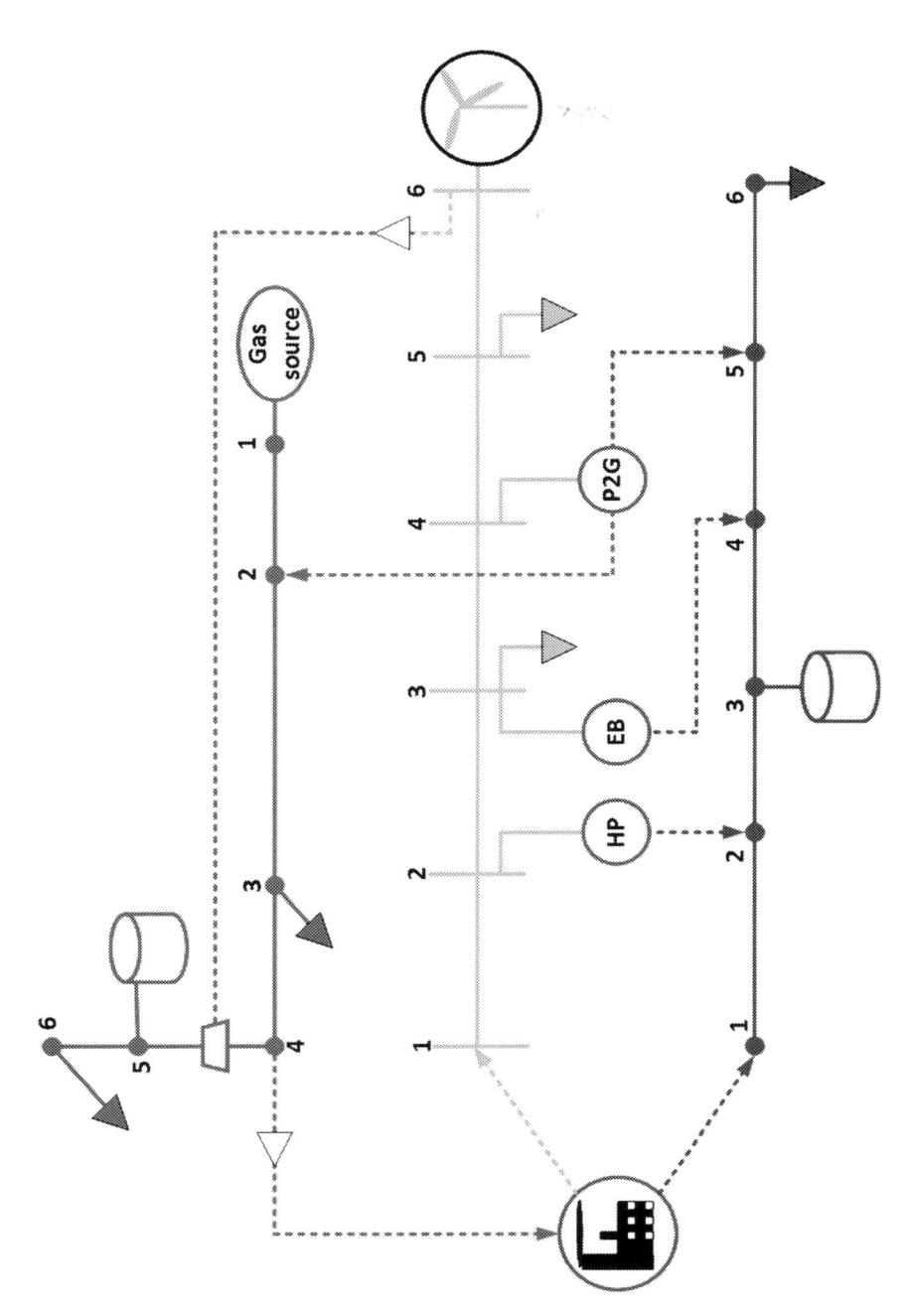

Figure 10.8 Test case: the IES.

Table 10.1 Parameters for the case study.

Parameters (EPS)	Value	Unit	Parameters (NGS)	Value	Unit
$x_{12}, x_{23}, x_{34}, x_{45}, x_{56}$	$0.2, 0.25, 0.1, 0.3, 0.2$	pu	C	47.8917e-6	
$P_j^{CHP,max/min}$	$5/0$	MW	f	0.025	
P_{nm}^{rate}	$1.6–1.8$	MW	UCV	53.72	MJ/m3
P_j^{RRL}	10	%/h	$p_n^{max/min}$	$4/1.6$	MPa
$W_w^{WF,max}$	4	MW	p_{ref}	16	(MPa)2
$D_{eb}^{EB,max/min}$	$0.6/0$	MW	T_b	273.15	K
$D_{hp}^{HP,max/min}$	$0.75/0$	MW	p_b	0.1013	MPa
$D_p^{P2G,max/min}$	$0.2/0$	MW	γ_g	0.633	
$C_j^{CHP,P}$	99	DKK/MWh	$\eta_{p,nm}$	0.9995	
C_j^{DR}	89.1	DKK/MWh	T_a	295.65	K
C_j^{UR}	108.9	DKK/MWh	Z_a	0.95	
C_e^{VOLL}	2000	DKK/MWh	c_k	1.3	
C_w^{spill}	500	DKK/MWh	$\eta^{P2G,Q}$	0.7	
$C_p^{P2G} C_{eb}^{EB}, C_{hp}^{HP}$	60	DKK/MWh	$Q_{st}^{ST,OUT,min/max}$	$0/0.75$	MW
Parameters (DHS)	**Value**	**Unit**	$Q_{st}^{ST,IN,min/max}$	$0/0.44$	MW
T^A	10	°C	$D_{12}, D_{23}, D_{34}, D_{45}, D_{56}$	355.6, 304.8, 304.8, 355.6, 304.8	mm
$\lambda_{12}, \lambda_{23}, \lambda_{34}, \lambda_{45}, \lambda_{56}$	$0.201, 0.185, 0.240,$ $0.184, 0.250$	W/mK	$Q_{12}^{max}, Q_{23}^{max}, Q_{34}^{max},$ $Q_{45}^{max}, Q_{56}^{max}$	$(0.5223,\ 0.8953,$ $0.2761,\ 0.5223,$ $1.0446) \times 10^{-3}$	MW
$L_{12}, L_{23}, L_{34}, L_{45}, L_{56}$	$250, 300, 300, 120, 850$	m	C^{GC}	4.0639×10^{-3}	

(continued on next page)

Table 10.1 (*continued*)

Parameters (EPS)	Value	Unit	Parameters (NGS)	Value	Unit
c	1.16167	Wh/kgK	$L_{12}, L_{23}, L_{34}, L_{45}, L_{56}$	18.99, 1.831, 17.792, 18.99, 34.535	km
$m_{mn}^{S,min/max}, m_{mn}^{R,min/max} mq_m^{min/max}$	0.1/36000	kg/h	T_s	295.65	K
$H_s^{HS,in/out,max}$	0.3	MW	E^{GC}	0.98	
$H_s^{SOE,min/max/0}$	1.8/18/9	MWh	η_g^{GC}	0.85	
$C_j^{CHP,H}$	150	DKK/MWh	f_{C1}^{GC}	67.0141	
$\eta^{CHP,e}$	0.4		f_{C2}^{GC}	24×10^{-6}	
$\eta^{CHP,h}$	0.38		CR	1.4	
$\eta^{P2G,H}$	0.08		$Q_g^{GS,max/min}$	3	MW
η^{EB}	0.99		$R_j^{CHP,D/U,max}$	2	MW
COP	3.6		$R_g^{GS,D/U,max}$	2	MW
$T_m^{S,fixed}$	80	°C	C^{LP}	7.8550×10^{-4}	
$T_{mn}^{S,IN,min/max}$	60/80	°C	$Q_{st}^{SOE,min/max/0}$	0/1065/532.5	MWh
$T_m^{S,min/max}$	60/80	°C	C_g^{GS}	99	DKK/MWh
$T_{mn}^{S,OUT,min/max}$	60/80	°C	C_g^{DR}	89.1	DKK/MWh
$T_{mn}^{R,IN,min/max}$	30/80	°C	C_g^{UR}	108.9	DKK/MWh
$T_m^{R,min/max}$	30/80	°C			
$T_{mn}^{R,OUT,min/max}$	30/80	°C			

consumption from Mm^3/day to the pu system should be noted as follows in Eq. (10.99):

$$D_{c,t}^{GC,DA,pu} = C^{GC} Z_a \frac{T_s}{E^{GC}\eta_{gc}^{GC}} \frac{c_k}{c_k - 1} \left(CR^{\frac{c_k-1}{c_k}} - 1 \right) Q_{c,t}^{GC,DA,pu} f_{C1}^{GC} f_{C2}^{GC},$$

$$\forall c \in \Omega^{GC}, \forall t \in T \tag{10.99}$$

where $Q_{c,t}^{GC,DA,pu}$ and $D_{c,t}^{GC,DA,pu}$ are in the pu system. f_{C1}^{GC} is a constant describing conversion from MW to m^3/h similar to Eq. (10.98) as $(3600/UCV)S_{MW}$. f_{C2}^{GC} describes conversion from m^3/h to Mm^3/day, and it is equal to 24×10^{-6}.

In addition, the mathematical model of the DHS and NGS is highly nonlinear. Hence, the linearization is performed to reach the global optimum of the entire optimization problem. The linearization process by incremental piece-wise function is used for the NGS for Eq. (10.44). The procedure can be found in Section 2.5.4 of Chapter 2. The DHS consists of a few nonlinear terms, such as the nodal mass balance equation, temperature drop, and temperature mix equations. The equations in the DHS are linearized by Taylor first-order approximation.

The simulation results are presented in later sections. For simulation purposes, the YALMIP toolbox in MATLAB is used [46]. Section 10.5.1 presents the benefits of using the stochastic approach over the deterministic approach. Further on, Section 10.5.2 demonstrates the flexibility synergy of IESs. In Section 10.5.3, the proposed scenario generation method is validated in terms of temporal correlation. Finally, a summary of the findings is given in Section 10.5.4.

10.5.1 Benefits of the stochastic programming approach

In this section, the benefits of stochastic DA scheduling are compared with deterministic DA scheduling. As mentioned earlier, the deterministic approach is based on a single deterministic forecast, and due to the variability and uncertainty of wind power output, the DA schedule can result in a nonoptimal solution in RT. Hence, the stochastic model takes uncertainties into account to decrease the difference between the DA and RT schedule, decrease the total expected system costs, and optimize reserves. The comparison is performed based on the quality metrics called *values of stochastic solution* (VSS) [47, 11]. VSS quantifies the economic advantage of the stochastic approach compared with the deterministic approach. There are five steps required to obtain VSS for a minimization problem, as shown in Fig. 10.9.

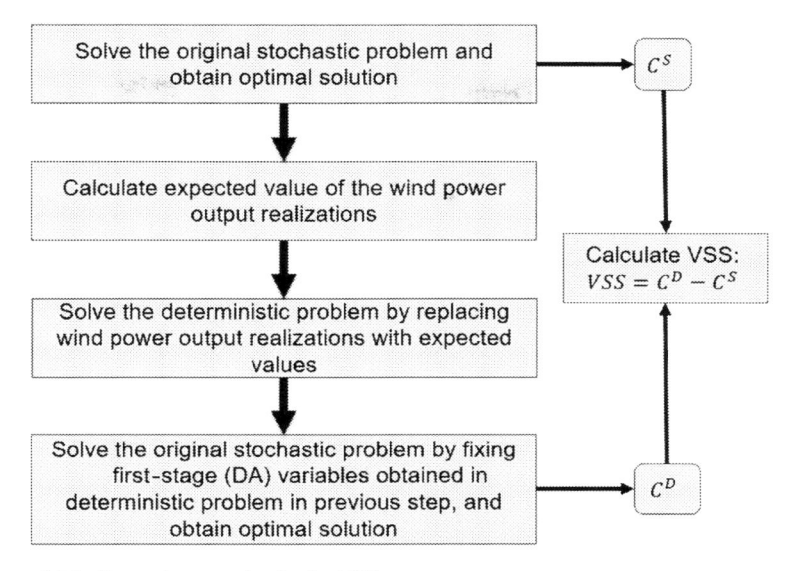

Figure 10.9 Procedure to obtain the VSS.

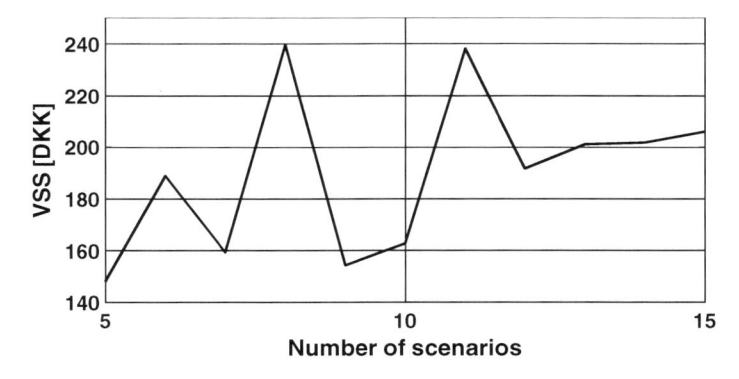

Figure 10.10 VSS for different number of reduced scenarios.

The VSS is shown in Fig. 10.10 for a different number of reduced scenarios. A number of reduced scenarios is in the range from 5 to 15. As can be observed, the economic advantage of the stochastic problem over the deterministic problem is high throughout the entire range of reduced scenarios. It can be concluded that it is not required to generate a large number of scenarios, as the benefit of the stochastic solution is shown even with a smaller number of generated scenarios. Moreover, it should be

Table 10.2 Illustration of probabilities of the scenarios and calculation of the costs in the DA and RT stages.

Different elements of the DA and RT stages	Probability	Cost (DKK)
DA scheduling operation cost	—	4114.7
RT: Scenario 1	0.1720	−3125.1
RT: Scenario 2	0.2700	−4447.2
RT: Scenario 3	0.1300	−4524.0
RT: Scenario 4	0.2260	−4467.8
RT: Scenario 5	0.2020	−4532.5
RT expected system cost	—	−4251.7
Total expected system cost	—	−137.0

noted, as mentioned earlier, that the computational burden increases with an increase in the number of scenarios.

To summarize, the positive VSS shows the economic advantage of the stochastic approach compared with the deterministic approach. Hence, by using stochastic programming, several future wind power realizations are taken into account as the scheduling is based on stochastic forecasts of the uncertain parameters. Due to reserves optimized, the imbalance in RT is lower and consequently the costs are lower. The stochastic programming approach results in minimum expected system cost compared to the deterministic approach.

10.5.2 Validation of the flexibility provision by synergy of multiple energy subsystems

In this section, an evaluation of the flexibility synergy provided by integration of several different subsystems is demonstrated. First, an illustration of DA costs, probabilities, and calculated total expected costs is presented in Table 10.2. The total expected system cost consists of DA scheduling operation cost and RT expected system cost that is calculated based on the probability of the scenario. It is obtained as presented in Eq. (10.1). Each scenario has its corresponding cost and probability. The expected system cost in each scenario is obtained by multiplication of the costs for the scenario and its corresponding probability. The expected system cost in RT summarizes each scenarios' expected system cost. To further elaborate, if a realization of Scenario 1 occurs in RT, the system cost is calculated as a sum of the DA scheduling operation cost and system cost in Scenario 1 that is equal to 989.6 Danish krone (DKK).

To evaluate the effectiveness of the flexibility provision by synergy of multiple energy sectors, three cases are listed, as shown in Table 10.3. Based

Table 10.3 Description of different cases analyzed.

Case	Case Description of Systems With Different Flexibility Synergies
1	IES based on presented case study in Fig. 10.8
2	Integrated EPS and DHS
3	Integrated EPS and NGS

on the three cases, the comparison of the operational costs and wind spillage is presented in Table 10.4. The wind curtailment rate is the percentage of wind curtailed in all scenarios in contrast to all wind power output realizations. DA scheduling operation costs, RT expected system costs dependent on the probability of the scenario, and total expected system costs were elaborated earlier. DA scheduling costs and RT expected costs are calculated and presented for each subsystem. The sum of costs for all subsystems equal to the total expected system cost. The costs are calculated based on Eq. (10.1) by only considering units belonging to a specified subsystem. As can be observed in Table 10.4, a high wind curtailment rate and correspondingly high wind spillage costs are seen in Case 2. On the contrary, in Cases 1 and 3, the wind spillage does not occur. The reasons behind high wind spillage in Case 2 are as follows. First, in Case 2, the NGS with a P2G component is not included and hence the excess electricity from wind farms is curtailed. However, Cases 1 and 3 have P2G components and excess electricity is converted to gas by the P2G unit and gas is provided to the NGS. Further on, the flexibility of CHP is limited in the DHS. Furthermore, the lowest total expected system cost is seen in Case 1 due to higher flexibility provided by integration of all three subsystems. The lowest DA costs can be observed in the EPS due to integrated flexible demand units and high wind power. On the contrary, DA costs are higher in the NGS and DHS. The summarized DA cost for all subsystems is higher in Case 1 compared with Cases 2 and 3, whereas the lowest is in Case 2. However, the RT expected system cost is the highest in Case 2, whereas it is lowest in Case 1. This is because Case 1 has more flexibility available by the EB, the HP, and P2G, and excess electricity can be converted to gas by the P2G unit. Hence, the EPS in Case 2 has the highest RT expected system cost due to wind spillage.

The comparison has shown that the main motive in obtaining low wind curtailment is the NGS with an integrated P2G unit. The DHS has limited flexibility through its CHP plant, and hence the NGS can provide higher flexibility than the DHS. Moreover, the integration of the EPS, DHS, and NGS provides high flexibility and low expected system costs. Furthermore,

Table 10.4 Comparison of operational costs and wind spillage costs for different flexibility level cases.

Case	Wind Curtailment Rate (%)	Wind Spillage Cost (DKK)	DA Scheduling Operation Cost (DKK)			RT Expected System Cost (DKK)			Total Expected System Costs (DKK)
			EPS	NGS	DHS	EPS	NGS	DHS	
1	0	0	−303.6	2989.9	1428.4	−326.4	−2536.1	−1389.2	−137.0
2	28.1	7345.4	5.4	—	1466.5	6789.3	—	−1426.1	6835.1
3	0	0	704.4	2989.9	—	−846.2	−2536.2	—	312.0

Table 10.5 Total expected system costs and wind spillage for different flexibility levels of the IES.

Case	Description of Flexibility Level of the IES	Total Expected Operational Cost (DKK)	Wind Power Curtailment Rate (%)
1	IES with initial setup as in Fig. 10.8	−137.0	0
1.1	IES without linepack	−115.07	0
1.2	IES without P2G	18.94	0
1.3	IES without P2G and gas storage	843.52	0

the NGS includes the gas storage and linepack. In the case of high excess electricity, P2G can convert the excess electricity to gas, and gas can be injected to the NGS. Moreover, the gas in the NGS can be stored in pipelines as linepack storage or in the gas storage. Both storages increase the flexibility of the entire IES.

Table 10.5 analyzes the influence of the flexibility sources on the economic efficiency and wind curtailment rate. As mentioned earlier, the NGS provides high flexibility to the IES, and therefore three cases with different flexibility levels in the NGS are compared with the initial Case 1. As can be observed, Case 1 leads to the lowest total expected operational system costs. Due to integration of three subsystems, and flexibility provided by P2G, linepack, and gas storage, the flexibility of the entire IES increases and costs are reduced. Moreover, the excess electricity is converted to gas by the P2G unit, and heat power through the EB and HP. Case 1.1 increases the total system cost due to reduced flexibility that is provided previously by the linepack in Case 1. Even higher cost can be seen in Case 1.2 due to absence of the P2G unit. The cost is highest in Case 1.3. Due to unavailability of gas storage and the P2G unit, the flexibility of the entire system decreases, and consequently the total expected system cost becomes higher. To conclude, the linepack is realistic storage in the pipelines, and linepack modeling can increase the flexibility of the system and decrease the costs. Moreover, the P2G component and storages play a major role in flexibility provision and cost reduction.

10.5.3 Validation of the temporal correlation of the proposed scenario generation method

In this section, validation of the temporal correlation of the proposed scenario generation method is illustrated. To verify the temporal correlation of the scenario set generated by the proposed scenario generation method,

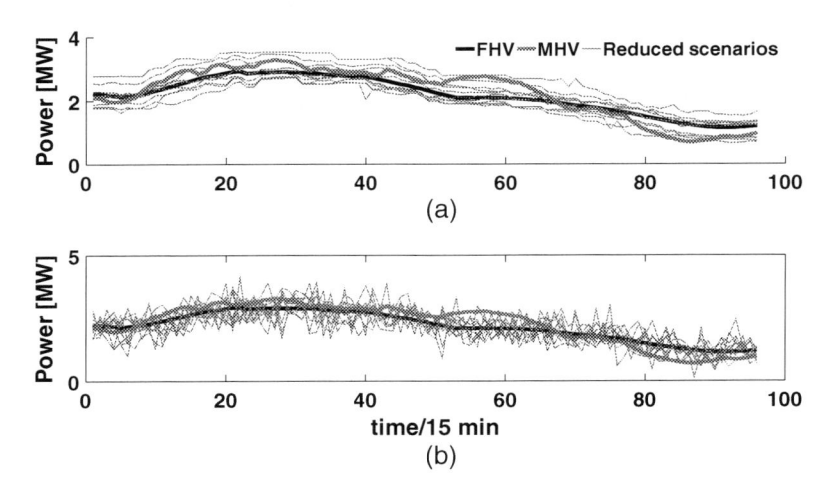

Figure 10.11 MHV, FHV, and 10 reduced scenarios by (a) the proposed method and (b) the LHS method.

Table 10.6 The optimal solution and scheduling results for scenario sets generated by the LHS and proposed methods.

Method for Generation of Scenario Set	Total Expected System Cost (DKK)	Wind Power Curtailment Rate (%)
LHS proposed	2012.4	0
Proposed method	−137.0	0

another scenario set which neglects the temporal correlation is generated by Latin hypercube sampling (LHS). A comparison between LHS and the proposed scenario generation method is performed. Both methods generate a number of scenarios, and an illustration of the reduced scenarios generated by the proposed method and the LHS method is shown in Fig. 10.11. Moreover, the optimal solution and scheduling results of both methods are compared in Table 10.6.

Based on the scheduling results, the total expected system costs are higher for the LHS method compared with the proposed method. However, both methods have zero wind curtailment. Moreover, comparison of reduced scenarios between the LHS method and the proposed method reveals the following. The proposed method considers the temporal correlation, and the reduced scenarios follow the profiles of wind power output. On the contrary, the temporal correlation is not considered in the LHS method, and the scenarios alternate above and below the forecasted and measured values. The scenario set generated by LHS method does not represent realistic scenarios

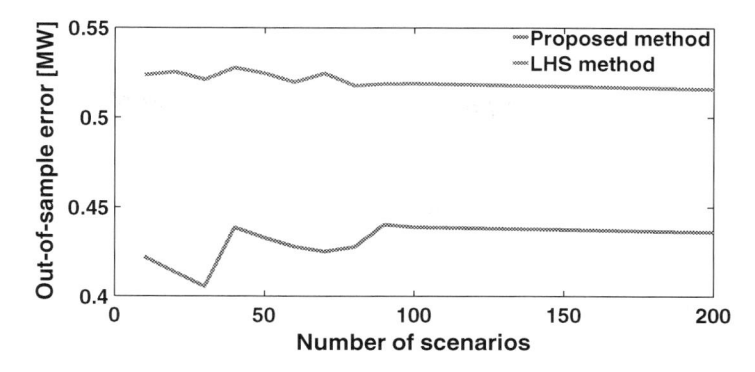

Figure 10.12 Out-of-sample-error calculated for scenario sets generated by the proposed method and the LHS method.

Table 10.7 Average, minimum, and maximum value of out-of-sample-error for the proposed and LHS methods.

Method for Generation of the Scenario Set	μ (MW)	min (MW)	max (MW)
Proposed	0.4281	0.4055	0.4402
LHS	0.5216	0.5158	0.5278

in RT. Moreover, the reserve capacity scheduled in each scenario results in nonoptimal points, as the scenarios do not follow the measured values of wind power output. Therefore, the costs are not realistic and optimal in RT.

To further evaluate the temporal correlation of the scenario set generated by the proposed method, out-of-sample error is used. Out-of-sample error is calculated for scenario sets generated by the proposed scenario generation method and the LHS method. First, the root mean square error is calculated between each of the scenarios in the scenario set and MHV. Afterward, the average of all values of root mean square error is obtained as out-of-sample error. The comparison between both methods in regard to out-of-sample error is shown in Fig. 10.12. As expected, and as can be observed, the proposed method has lower out-of-sample error. The scenario sets obtained by the proposed method are closer to the MHV. On the contrary, the out-of-sample error is quite high for the LHS method. However, out-of-sample errors from both methods saturate around the number of scenarios equaling 100. Furthermore, the overall key parameters of out-of-sample error are shown in Table 10.7. As can be observed, the LHS method has higher average value of out-of-sample error compared with the proposed method. Minimum and maximum values are large as well, due to the high fluctuating nature of the scenario set generated by the LHS method, as seen in Fig. 10.11. It can be concluded that the scenario set generated by the LHS method is not representing the realistic scenarios of wind power output.

10.5.4 Summary of the results and discussion

To summarize, the previous sections focused on three objectives.

The first objective was to present the economic advantage of the stochastic approach over the deterministic approach. It has been shown that a lower imbalance in RT can be achieved by applying the stochastic modeling approach. The stochastic programming approach has shown the minimum total expected system cost. A high economic advantage can be obtained by applying the stochastic approach compared with the deterministic approach.

The second objective focused on provision of flexibility through integration of several different energy systems. It has been shown that integration of several different subsystems provides high flexibility, economic efficiency, and wind utilization. Moreover, it has been illustrated that the highest flexibility and lower costs can be obtained by integration of linking units such as the P2G, EB, and HP. Moreover, the NGS has shown the highest flexibility provision. The gas in the NGS can be stored in a linepack or in the gas storage. Both storages are major providers of flexibility for the entire IES.

The third objective was validating the proposed method in terms of temporal correlation. The results have shown that the proposed method realistically represents possible future scenarios. Moreover, the generated scenario set by the LHS method, which does not consider temporal correlation, alternates to a large amount around the measured values and results in high out-of-sample error. It has been shown that the temporal-correlated scenario set follows the profiles of wind power output and represents the realistic scenarios and performs better than the LHS method. Finally, the proposed method results in cost and energy-efficient scheduling.

10.6 Conclusion

In this chapter, a two-stage DA scheduling of IESs was introduced. A two-stage stochastic programming approach was developed for the IES and includes the uncertainty of wind power generation. First, DA scheduling approaches were demonstrated. Then, a detailed illustration of a two-stage

DA scheduling approach was given. The first stage represents the DA market, whereas the second stage represents the RT market and accommodates the wind power output uncertainty through reserves. Wind power output uncertainties have been generated as a future realization of wind power through a scenario set. The scenario set has been generated by a scenario generation method based on historical observation of the measured values and the stochasticity of the uncertainty through realistic scenarios obtained by the scenario generation method was described. To decrease the computational burden, the scenario reduction method has been used. Last, a mathematical model of the two-stage DA scheduling approach for IESs was presented.

Some conclusions can be drawn. The stochastic approach takes uncertainties into account to decrease the difference between the DA and RT schedule, decrease the total expected system costs, and optimize reserves. The proposed method considers the temporal correlation, and the reduced scenarios follow the profiles of wind power output. The proposed scenario generation method integrated in the second stage of the proposed stochastic model provides reliable and cost–efficient solutions. Furthermore, integration and coordination of several energy subsystems lead to higher utilization of wind power and a decrease in total expected system costs. Linking units, such as P2G, can convert excess electricity to gas and decrease wind spillage. Hence, system flexibility and wind power utilization are increased by integration of the NGS, EPS, and DHS through linking components. Moreover, P2G components, storages, and linepack play an important role as providers of flexibility. The total expected system cost is decreased due to coordinated scheduling of energy and reserves.

References

[1] Q Zeng, J Fang, J Li, Z Chen, Steady-state analysis of the integrated natural gas and electric power system with bi-directional energy conversion, Appl Energy 184 (2016) 1483–1492.

[2] Q Zeng, B Zhang, J Fang, Z Chen, A bi-level programming for multistage co-expansion planning of the integrated gas and electricity system, Appl Energy 200 (2017) 192–203.

[3] S Frank, S Rebennack, An introduction to optimal power flow: theory, formulation, and examples, IIE Trans 48 (12) (2016) 1172–1197.

[4] A Seungwon, Q Li, TW Gedra, Natural gas and electricity optimal power flow, in: Proc 2003 IEEE PES Transmission and Distribution Conference and Exposition, 2003.

[5] Z Yang, C Gao, J Zhang, The interaction of gas and electricity hybrid energy Internet, in: Proc 2017 IEEE Conference on Energy Internet and Energy System Integration (EI2), 2017.

[6] J Fang, Q Zeng, X Ai, Z Chen, J Wen, Dynamic optimal energy flow in the integrated natural gas and electrical power systems, IEEE Trans Sustain Energy 99 (2017) 188–198.

[7] European Power to Gas, White Paper on Power-to-gas in a Decarbonized European Energy System based on, Renewable Energy Sources (2017) https://www.afhypac.org/documents/European%20Power%20to%20Gas_White%20Paper.pdf. (Accessed 20 April 2021).

[8] X Liu, J Wua, N Jenkins, A Bagdanavicius, Combined analysis of electricity and heat networks, Appl Energy 162 (2016) 1238–1250.

[9] EL Forsk, Final Report on Harmonized Integration of Gas, District Heating and Electric Systems, HIGHE (No. 12220), Aalborg University, Aalborg, Denmark, 2017.

[10] Q Zeng, J Fang, B Zhang, Z Chen, The coordinated operation of electricity, gas and district heating systems, in: Proc 2017 Applied Energy Symposium and Forum, Renewable Energy Integration with Mini/Microgrids, 2017.

[11] JM Morales, AJ Conejo, H Madsen, P Pinson, M Zugno, Integrating Renewables in Electricity Markets, Springer, New York, 2014.

[12] Q Zeng, AJ Conejo, Z Chen, J Fang, A two-stage stochastic programming approach for operating multi-energy systems, in: Proc 2017 IEEE Conference on Energy Internet and Energy System Integration, 2017.

[13] T Schulzea, K McKinnon, The value of stochastic programming in day-ahead and intra-day generation unit commitment, Energy 101 (2016) 592–605.

[14] J Wang, M Shahidehpour, Z Li, Security-constrained unit commitment with volatile wind power generation, IEEE Trans Power Syst 23 (3) (2008) 1319–1327.

[15] A Ghasemi, M Banejad, M Rahimiyan, Integrated energy scheduling under uncertainty in a micro energy grid, IET Gen Trans Distr 12 (12) (2018) 2887–2896.

[16] W Su, J Wang, J Roh, Stochastic energy scheduling in microgrids with intermittent renewable energy resources, IEEE Trans Smart Grid 5 (4) (2014) 1876–1883.

[17] J Zhang, JD Fuller, S Elhedhli, A stochastic programming model for a day-ahead electricity market with real-time reserve shortage pricing, IEEE Trans Power Syst 25 (2) (2010) 703–713.

[18] N Growe-Kuska, H Heitsch, W Romisch, Scenario reduction and scenario tree construction for power management problems, in: Proc 2003 IEEE Bologna Power Tech Conference, 2003.

[19] YH Wu, A Stochastic Mathematical Program with Complementary Constraints for Market-Wide Power Generation and Transmission Expansion Planning (master dissertation), Iowa State University, Ames, IA, 2014.

[20] Y Wang, Scenario Reduction Heuristics for a Rolling Stochastic Programming Simulation of Bulk Energy Flows With Uncertain Fuel Cost (Ph.D. dissertation), Iowa State University, Ames, IA, 2010.

[21] A Rabiee, A Soroudi, Stochastic multiperiod OPF model of power systems with HVDC-connected intermittent wind power generation, IEEE Trans Power Deliv 29 (1) (2014) 336–344.

[22] Y-M Atwa, EF El-Saadany, Probabilistic approach for optimal allocation of wind-based distributed generation in distribution systems, IET Renew Power Gen 5 (1) (2011) 79–88.

[23] L Shi, C Wang, L Yao, Y Ni, M Bazargan, Optimal power flow solution incorporating wind power, IEEE Syst J 6 (2) (2012) 233–241.

[24] M Bornapour, R-A Hooshmand, A Khodabakhshian, M Parastegari, Optimal coordinated scheduling of combined heat and power fuel cell, wind, and photovoltaic units in micro grids considering uncertainties, Energy 117 (2016) 176–189.

[25] M Zhang, X Ai, J Fang, W Yao, W Zuo, Z Chen, et al., A systematic approach for the joint dispatch of energy and reserve incorporating demand response, Appl Energy 230 (2018) 1279–1291.

[26] C Yıldız, M Tekin, A Gani, OF Keçecioglu, H Acıkgoz, M Sekkeli, A day-ahead wind power scenario generation, reduction, and quality test tool, Sustainability 9 (5) (2017) 1–15.

[27] X-Y Ma, Y-Z Sun, H-L Fang, Scenario generation of wind power based on statistical uncertanty and variablity, IEEE Trans Sustain Energy 4 (4) (2013) 894–904.

[28] ILR Gomes, HMI Pousinho, R Melício, VMF Mendes, Stochastic coordination of joint wind and photovoltaic systems with energy storage in day-ahead market, Energy 124 (2017) 310–320.

[29] M Lei, J Zhang, X Dong, JJ Ye, Modeling the bids of wind power producers in the day-ahead market with stochastic market clearing, Sustain Energy Tech Assess 16 (2016) 151–161.

[30] IG Marneris, PN Biskas, AG Bakirtzis, Stochastic and deterministic unit commitment considering uncertainty and variability reserves for high renewable integration, Energies 10 (1) (2017) 1–25.

[31] Danish Energy Agency, Regulation and planning of district heating in Denmark, 2017. https://ens.dk/sites/ens.dk/files/Globalcooperation/regulation_and_planning_of_district_heating_in_denmark.pdf. (Accessed 20 April 2021).

[32] EnerginetSecurity of Gas Supply 2018, Energinet, Fredericia, Denmark, 2018.

[33] S Mokhatab, WA Poe, JG Speight, Handbook of Natural Gas Transmission and Processing, Gulf Professional Publishing, Oxford, 2006.

[34] E Shashi Menon, Gas Pipeline Hydraulics, CRC Press, Boca Raton, FL, 2005.

[35] W Wei, J Wang, Modeling and Optimization of Interdependent Energy Infrastructures, Springer, Cham, Switzerland, 2020.

[36] A Turk, Q Wu, M Zhang, J Østergaard, Day-ahead stochastic scheduling of integrated multi-energy system for flexibility synergy and uncertainty balancing, Energy 196 (2020) 117130.

[37] ENTSO-E, Transparency platform, 2020. https://transparency.entsoe.eu/dashboard/show. (Accessed 20 April 2021).

[38] C Ordoudis, S Delikaraoglou, P Pinson, J Kazempour, Exploiting flexibility in coupled electricity and natural gas markets: a price-based approach, in: Proc 2017 IEEE PES PowerTech Conference, 2017.

[39] Energinet, Danish Energy AgencyTechnology Data: Energy Storages—Technology Descriptions and Projections for Long-Term Energy System Planning, Energinet, Danish Energy Agency, Erritso, Denmark, 2018.

[40] Energinet, Analysis assumptions 2017, 2017. https://en.energinet.dk/Analysis-and-Research/Analysis-assumptions/Analysis-assumptions-2017. (Accessed 20 April 2021).

[41] EnerginetFuture Natural Gas Qualities—Fact Sheet, Energinet, Fredericia, Denmark, 2017.

[42] EnerginetTechnology Data for Renewable Fuels, Energinet, Fredericia, Denmark, 2018.

[43] Energinet, Danish Energy Agency, Technology Data for Generation of Electricity and District Heating, Danish Energy Agency, 2020, https://ens.dk/sites/ens.dk/files/Statistik/technology_data_catalogue_for_el_and_dh_-_0009.pdf. (Accessed 20 April 2021).

[44] European Commission, EurostatCombined Heat and Power (CHP) Generation, Eurostat, Luxembourg City, 2017.

[45] Danish Gas Technology Centre (DGC)Future Gas, WP1: perspectives on utilization of biogas without upgrading, DGC, Horsholm, Denmark, 2019.

[46] Yalmip, YALMIP, 2020. https://yalmip.github.io/?n=Main.License. (Accessed 20 April 2021).

[47] Antonio J Conejo, M Carrión, JM Morales, Decision Making Under Uncertainty in Electricity Markets, Springer, New York, 2010.

CHAPTER 11

MPC-based real-time operation of integrated energy systems

11.1 Introduction

Due to the increasing level of renewable energy sources (RESs), higher uncertainties and fluctuations in the power system are expected, leading to unstable operation in real time (RT). To integrate a higher amount of renewable energy and operate the power system in a secure manner, the flexibility that can be provided by different energy sectors should be explored. This chapter takes a new perspective on energy scheduling and proposes a model predictive control (MPC)-based scheduling for the RT operation for integrated energy systems (IESs) taking into account a number of uncertainties, such as load and wind power, to achieve a cost–efficient energy system. MPC is jointly optimizing the electric power system (EPS), natural gas system (NGS), and district heating system (DHS) to obtain higher flexibility and penalizes the deviation from the prescheduled values. The proposed online scheduling is taking into account current and future states of the IES. Future states of the energy system are provided by the online learning method (OLM) resulting in time–efficient and accurate forecasts. Simulation results are tested on a small-scale IES showing an optimal economic operation and an increase in flexibility in RT. Computational efficiency is adequate for RT scheduling. The simulation results have shown an increase in energy efficiency and cost savings.

Section 11.2 provides the background and overview of RT scheduling. Section 11.3 introduces RT scheduling based on the MPC strategy. Moreover, elaboration of the forecasting method is given. Section 11.4 gives the mathematical model of the MPC-based RT scheduling approach for IESs. Moreover, the solution process and the case study are demonstrated in Section 11.5, and simulation results are presented in Section 11.6. Finally, a summary of the chapter is given in Section 11.7.

Optimal operation of integrated multi-energy systems under uncertainty. Copyright © 2022 Elsevier Inc.
DOI: https://doi.org/10.1016/B978-0-12-824114-1.00002-0 All rights reserved.

11.2 Background and RT scheduling

This section introduces the background and RT scheduling for IESs and gives an overview of RT scheduling based on MPC approaches.

11.2.1 Background

In recent years, the installed capacity of RESs is increasing at a high rate. One of the fast-growing RESs is wind power [1]. With respect to the Paris Agreement, the long-term goal of limiting the increase of global average temperature below $2°$ is pushing toward the decrease of CO_2 emissions and an increase of RESs [2]. Toward the future, more fluctuating RES-based units will be integrated into the system and the system will face major operation and stability challenges. To contribute to the reduction of CO_2 and maintain stability of the system with a high share of RESs, integration of different energy sectors and the flexibility that can be provided by integration of different energy sectors is proposed. It is expected that integration of different energy sectors will provide the needed flexibility to the future energy system, and power-to-gas (P2G) will have a major role in balancing of the energy systems [3].

The integration of multiple energy systems has been a major topic and has been investigated to a large extent. The modeling and integration of the EPS and NGS can be found in the work of Zeng et al. [4], Correa-Posada and Sánchez-Martín [5], and Ordoudis et al. [6]. Zeng et al. [4] present steady-state modeling of the NGS and EPS with an integrated P2G unit to increase flexibility and reduce total energy losses in the system. Correa-Posada and Sánchez-Martín [5] provide unified and linearized modeling formed as mixed-integer linear programming (MILP) to decrease the computational burden and assure the global optimum and EPS reliability in the short-term, respectively. Moreover, in the work of Correa-Posada and Sánchez-Martín [5], storage in terms of linepack is taken into account to provide lower operating costs and higher flexibility. Zeng et al. [7] and Liu et al. [8] investigate an integrated EPS and DHS, and Zeng et al. [9] integrate the EPS, NGS, and DHS. The authors show that coordination and joint operation provide high flexibility and efficiency and low wind curtailment.

In the IES, a number of uncertainties can be listed, such as demand, prices, and RESs. The fluctuating nature of RESs is the larger uncertainty and can lead to high system imbalance. In the day-ahead (DA) market, scheduling is performed based on the forecasted values of RESs and demand. The demand has slight changes in RT compared to DA values, whereas RES generation has higher forecast error. Therefore, in RT, there is a mismatch between the forecasted and actual value of the RES, and in such case, the RT market is balancing the demand and generation in the EPS due to imbalances caused by RESs [10]. As a consequence, costs are increasing and the balancing price is usually higher than the DA price [11]. Therefore, a great importance is placed on RT scheduling, as it is the last opportunity to balance the generation and demand [12]. A promising solution that can be used in RT is MPC. RT scheduling performed in the traditional manner is a single-period optimization, which only considers one timestep at a time. On the contrary, MPC can deal with the uncertainties of RESs in a way that it is optimizing the hourly schedule based on the measured current status of the system and future predictions [13]. MPC can be applied in both a centralized manner and distributed manner [14]. The centralized economic dispatch in a distributed manner is shown in the work of Velasquez et al. [14]. On the contrary, centralized MPC is studied in the work of Du et al. [13]. A centralized MPC approach with the objective to minimize the operational costs taking into account uncertain load, prices, and wind power has been developed for the microgrid. The MILP model was developed to reach the global optimum, and the forecasts for MPC were based on the forecast error following the probability distribution with zero mean and known standard deviation. MPC has shown better economic efficiency over traditional scheduling. Furthermore, a nonlinear model of the EPS and NGS with an objective to minimize the balancing costs in RT with the purpose of exploiting energy storage has shown that MPC can reduce the operation cost, and the longer prediction horizon is advantageous in case of large energy storages in the system [15]. In addition, adaptive and corrective scheduling based on MPC for the microgrid have an objective to minimize operational costs and penalize the deviation from the prescheduled values to evaluate the flexibility available in the system, as shown in other works [16, 17]. Intra-day adjustments based on MPC have shown the possibility of balanced RESs and demand uncertainty [16]. Besides deterministic approaches for MPC, stochastic approaches have been explored by Guo et al. [18] and Li et al. [19]. In those works [18, 19], the two-stage stochastic approach for the EPS and DHS in buildings and the microgrid have been proposed. Both methods used scenario generation and scenario reduction techniques.

As mentioned earlier, MPC optimizes based on current measurements and future predictions. In addition, MPC incorporates the prediction horizon, and a single-period optimization becomes a multiperiod optimization. Hence, the prediction horizon is a key parameter that influences the computational burden and optimization decisions [15]. The prediction horizon length and the resolution are quite important when MPC is applied online. Selection of the prediction horizon should take into account the computational efficiency and storage capacities. A large prediction horizon leads to a high computational burden. However, a large prediction horizon consists of more information regarding the future. Hence, the balance between the computational burden and future information should be found.

To summarize, MPC has been extensively applied to the operation of smart home building systems and microgrids. However, the literature regarding RT scheduling based on MPC for the IES is scarce. Hence, the main objective of this chapter is to present RT scheduling for the integrated EPS, DHS, and NGS that is based on MPC strategy. MPC-based RT scheduling reduces the regulation costs in RT and increases the flexibility of the entire energy system. Moreover, the MPC strategy penalizes the deviation from the DA schedule and therefore explores the possibility of providing the regulating power by storage devices and the P2G unit. The OLM provides accurate forecasts one step ahead and performs inside the designated limits of computational burden requested for RT operation. The simulations were conducted using the YALMIP toolbox in MATLAB [20].

11.2.2 Overview of traditional RT scheduling

The scheduling can be performed in three markets. The first market is the DA market. DA scheduling is performed the day before the day of operation based on the predicted values of the uncertainties. The market following the DA market is the intra-day market. The intra-day market closes 1 hour before the operational hour, and it is used to change the planned production from the DA. The additional explanations on the scheduling in the DA and intra-day market are not given but can be found in other works [10-12, 21, 22]. For simplicity, the intra-day market is neglected in this chapter. The last market is the RT market. The RT market in different regions of the world differs in regard to settlements and time of dispatch. However, the RT market maintains the balance between the generation and demand during the operational hour and acts as the last market opportunity to balance the

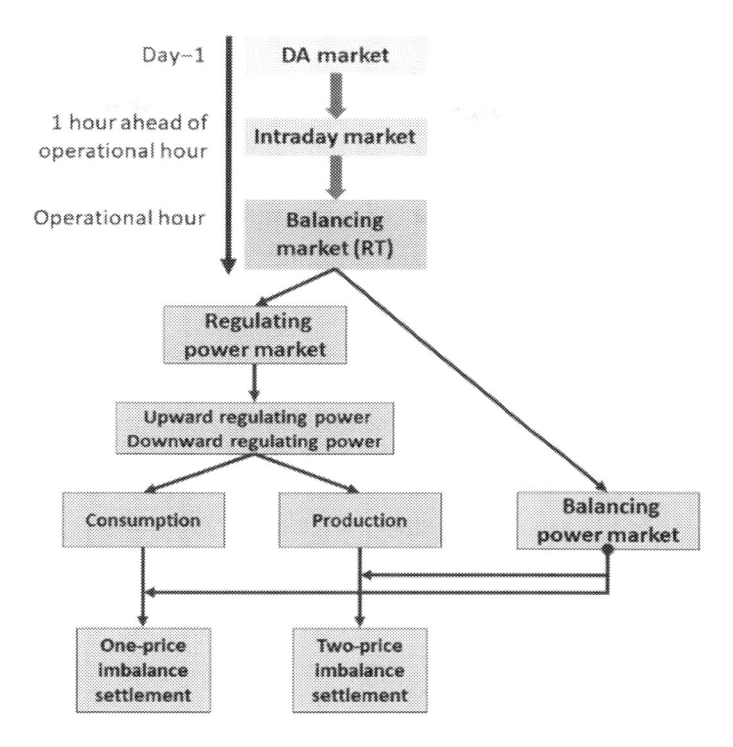

Figure 11.1 Overview of the RT scheduling used in Nordic electricity markets.

prediction errors [22]. For the purposes of understanding the RT market, the Nordic RT electricity market is elaborated [10-12, 21, 22].

The Nordic RT electricity market is called the *balancing market*. The balancing market is a single-period optimization. The division of the balancing market is shown in Fig. 11.1. The balancing market is divided into the regulating power market and the balancing power market [21]. The regulating power market is the last opportunity to deal with the deviations in RT due to the forecasting error of demand and RESs. The regulating power is purchased by the transmission system operator and consists of upward and downward regulating power usually provided by conventional generators. However, the consumption can also submit bids for the regulating power [11, 21]. By activating the upward and downward regulating power, the imbalance in RT decreases. The minor imbalances in range of a few megawatts are handled by reserves (e.g., frequency containment reserves) [11, 23].

The balancing power market is used for the settlement of the imbalances. The transmission system operator must compensate for the imbalances of the market players and purchase the balancing power. As mentioned, the consumption and production can provide bids for the regulating power and the imbalance settlements are separately handled as described in the following. The main difference is in the price settlement scheme. Two kinds of settlement schemes are used in the balancing power market. The imbalances of the production side are settled according to the two-price balancing scheme, whereas the imbalances of the consumption side are settled according to the one-price balancing scheme [21].

First, the two-price balancing scheme for the producer's imbalance is presented in Fig. 11.2. In the case when the system imbalance is negative and the upward regulation has been requested, two scenarios must be looked at. The production exceeding the scheduled production receives the DA market price for the excess production, whereas the production in deficit of power compared to the scheduled will pay the upward regulation price. Such upward regulation price is usually higher compared to the DA market price. In the case when the system imbalance is positive and downward regulation is provided, similarly, the two scenarios are elaborated. In the case when the production is in deficit compared to the schedule, the producer will pay the DA market price for the lack of production. On the contrary, the production exceeding the scheduled production will receive the downward regulation price that is less than or equal to the DA market price. Hence, in a two-price balancing settlement, the deviations contributing to the stability and to the decrease of the imbalances will be traded at the DA market price. However, the deviations that caused the higher system imbalance will be penalized and traded at the regulation prices [11].

Last, the one-price balancing scheme is elaborated as follows and shown in Fig. 11.3. As mentioned, the one-price balancing scheme is used to settle imbalances caused by consumption. In the case when the consumption contributes to the system balance, the consumers receive the balancing price. On the contrary, if the consumption contributes to the system imbalance, the consumers pay the DA market price [11].

In a case where the one-price balancing scheme is used for the production imbalance settlements, the following applies. In the case when upward regulation is provided, the balancing price is higher than the DA price. On the contrary, in the case of downward regulation, the balancing price is lower compared to the DA price. The generators providing balancing power to provide stability to the system are being awarded with the bonus, and

Request for regulating power	Producer's imbalance		Imbalance settlement
	Scheduled	Measured	
(Negative imbalance)			Receives DA market price for excess production
			Pays upward regulation price for the lack of production
(Positive imbalance)			Receives downward regulation price for excess production
			Pays DA market price for the lack of production

Figure 11.2 Imbalance settlement for the production side based on a two-price settlement scheme.

Request for regulating power	Consumer's imbalance		Imbalance settlement
	Scheduled	Measured	
(Negative imbalance)			Pays the DA market price
			Receives the balancing price
(Positive imbalance)			Receives the balancing price
			Pays the DA market price

Figure 11.3 Imbalance settlement for the consumption side based on a one-price settlement scheme.

otherwise the generator is penalized. Hence, in the two–price settlement, the bonus for the generator does not exist [10]. Moreover, the two–price balancing scheme is used as an incentive to the producers to proactively participate in contributing to the system balance [11].

The main principle of MPC strategy and the proposed MPC–based RT scheduling for the IES are elaborated in the next section.

11.3 MPC-based RT scheduling

11.3.1 MPC strategy

In this section, the main principle of the MPC strategy is introduced.

Two aspects are discussed for MPC-based RT optimization of the IES. The first aspect is related to the implementation process. The MPC controller determines the optimal control actions in a prediction horizon, and only the first control action of the prediction horizon is executed [14]. The measured and predicted values of the system are constantly updated as the timestep moves forward. Once new predicted values are provided, MPC moves to the next timestep and repeats the procedure. Fig. 11.4 demonstrates the main principle of the MPC strategy. As shown in Fig. 11.4(a), the MPC strategy is based on a closed loop with four modules, including system state measurement, updating of the forecast information, optimization over the prediction horizon, and execution of first control measures. The four modules interact with each other and are implemented sequentially. The process of the closed loop is repeated with the timestep forward, as illustrated in Fig. 11.4(b). With such a principle, the deviation of RT scheduling caused by the prediction errors can be minimized by the MPC controller [13].

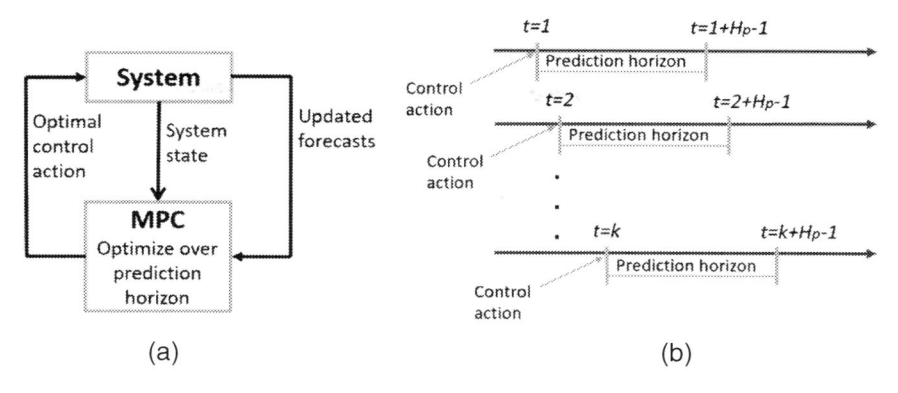

Figure 11.4 MPC (a) strategy and (b) principle.

The second aspect of the MPC strategy is its application in the RT market. The traditional RT market is a single-period optimization, and the production and consumption are usually balanced by the upward or downward regulations from conventional generators. The traditional RT scheduling does not take into account the future information of the uncertainties. However, when the MPC strategy is applied to the RT market, a multiple-period optimization problem will replace the original single-period one. The time horizon of the optimization (i.e., the prediction horizon) needs to be selected. The MPC approach receives and updates the system values based on measured and future forecasts and finds the optimal solutions for the selected horizon. Only the first control step is implemented. MPC moves a step forward and repeats the procedure.

11.3.2 MPC-based RT scheduling for IESs

The overview of the MPC strategy for RT scheduling of the IES is illustrated in Fig. 11.5. The main principle of the MPC strategy is based on a closed loop as mentioned earlier [14]. MPC is applied in RT and replaces the traditional RT scheduling based on single-period optimization with multiperiod optimization. Additionally, to create a multiperiod optimization for each timestep, the prediction horizon is used. The prediction horizon length needs to be selected based on the requirements of the system, such as storage capacities and computational burden. The prediction horizon is created based on the current measurements and historical values, and the OLM is used to predict one step ahead and update forecasts in each timestep. The characteristics of OLM are low computational burden and

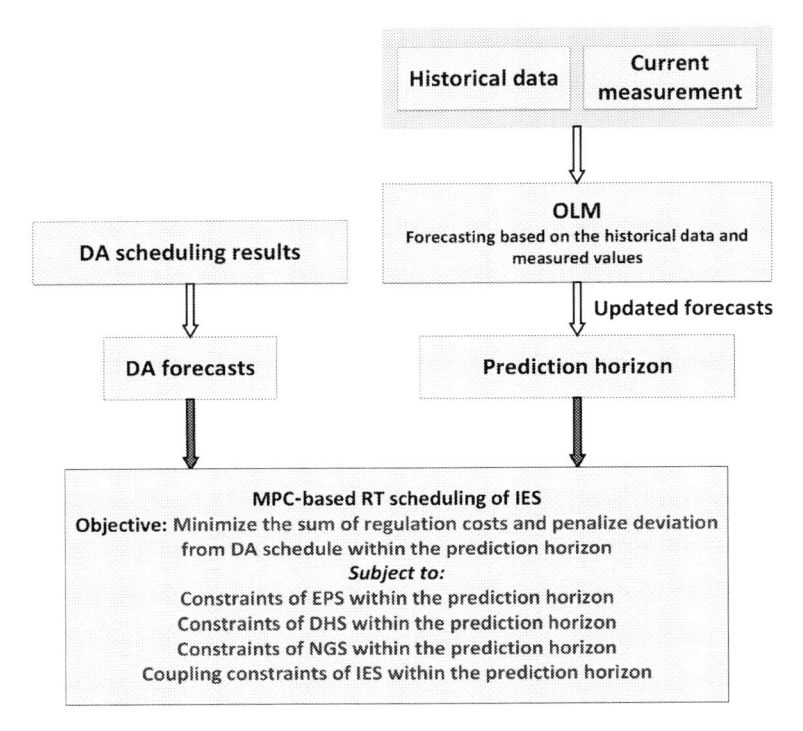

Figure 11.5 Overview of the main steps in MPC-based RT scheduling for the IES.

high accuracy and the inputs to OLM are historical values of uncertainties. The uncertainties considered are wind power output and electric demand. Moreover, the MPC approach takes into account the values obtained by the DA schedule. The DA schedule takes into account the forecasted values of the uncertainties and schedules the energy for the next day. The MPC strategy uses the prescheduled values from the DA to decrease the deviation of the conventional generators in RT compared to DA. The penalty represents a simplified two-price balancing scheme. Moreover, by taking the penalty into account, the deviations in RT are mainly compensated by the storage devices and P2G units. Providing the regulating power from the consumption side plays a major role in the future, as it is expected that all conventional units based on the consumption of gas, oil, and coal will be phased out completely.

Finally, the main objective of the MPC strategy is to minimize the sum of the regulation costs and penalize any deviation from the DA schedule. While optimization is performed, the constraints of each subsystem must be taken into account, as well as the coupling constraints. The scheduling performed by the MPC strategy provides economic efficiency and flexibility increase.

11.3.3 Principle and application of the OLM

In this section, the forecasting method used in MPC is presented. The forecasting methods and further explanations regarding the method used in this work can be found elsewhere [12, 24]. However, a brief introduction and overview of the method is given in this section. First, the introduction to regression is given. Second, the reason behind using the regression and OLM is given. Last, the procedure of the OLM is presented. The OLM will be used to predict the uncertain parameters of the IES model, which include wind power output and electric demand.

The regression method is used to forecast in order to create the relationship between the response and explanatory variables for any given time. As an example, an explanatory variable such as wind speed helps to predict the response variable such as wind power. By considering continues variables, such as wind speed and wind power, and a set of observations in the past of response and explanatory variables, the relationship between wind speed and wind power can be written through linear regression. Linear regression introduces regression parameters in which values can be found by using least square estimation. By creating a relationship between those two sets of variables and obtaining the regression parameters, the previous timesteps are ignored, and linear relationship can be found for each time instant between the wind speed and wind power. Therefore, the obtained relationship can be used to forecast in the future. However, the linear regression method used to forecast the future leads to three concerns that should be taken into account. First, to forecast in the future, the wind speed for the future is acquired. Thus, wind speed must be forecasted to further obtain future forecasted wind power values. By using forecasting values to forecast the future, the forecasting error is increased. Second, uncertain parameters can vary with time. For example, seasonal effects can be seen in wind power. Seasonal effects can be taken into account through the estimation of model parameters on the sliding window. In other words, the least square estimation is performed for a season or a specified window size. For each timestep, the regression parameters are recalculated. That leads to the third concern, which is increased computational burden.

There are two linear regression models. The first model considers simple mapping of an explanatory variable to a response variable, or in other words, mapping of wind speed to wind power as explained earlier. The second model is the autoregressive model. The autoregressive model is called the *OLM*. The OLM is based on recursivity, which means that the new update is based on the relationship between the last and the new data point [24]. Due to highlighted concerns, like computational burden and forecast error, the OLM is implemented in this study based on the research given by DTU CEE Summer School [24]. By using the OLM, the computational burden is not increasing, and the explanatory variable, such as wind speed forecast, is not needed, as will be shown in the following.

The model for the OLM is as follows. The autoregressive model is similar to the simple mapping model and is shown in Eq. (11.1). The more compact formulation is shown in Eq. (11.2). The difference is that the explanatory variable is a response variable in the previous or present time. β denotes model parameters, and ε stands for noise, which is the deviation between the measured and modeled relationship. Moreover, y is the response variable, whereas x_t is the explanatory variable. For example, y_t can be the wind power output at time t. x_t and β_t can be written as $x_t = [1 \; y_{t-1} \ldots y_{t-p}]^T$ and $\beta_t = [\beta_{t,0} \; \beta_{t,1} \ldots \beta_{t,p}]^T$, respectively. p denotes the order of the model that controls the recursivity. Hence, the recursivity of the autoregressive model means that the model parameters β_t, for $p = 1$ are estimated at time $t-1$ and the data before timestep $t-1$ can be deleted.

$$y_t = \beta_{t,0} + \beta_{t,1}y_{t-1} + \beta_{t,2}y_{t-2} + \; \ldots \; + \beta_{t,p}y_{t-p} + \varepsilon_t, \quad \forall t \qquad (11.1)$$

$$y_t = \beta_t^T x_t + \varepsilon_t, \quad \forall t \qquad (11.2)$$

To estimate the model parameters, βt, the regressive least square method is applied. The procedure to obtain the update scheme for regressive least square and the further explanation of the OLM can be found elsewhere [25, 24]. Finally, once the parameters for the update scheme for the OLM are available, the historical values can be used as an input to the learning process of the OLM.

Each uncertainty has different parameters such as standard deviation and different model parameters. Based on the historical data, the OLM predicts wind power output and electric demand. The time resolution is 15 minutes. However, the OLM can predict one timestep ahead. Hence, to create a prediction horizon longer than one timestep, the rest of the prediction horizon (H_p - 1) is generated based on the forecast error distribution. The forecast errors follow the normal distribution with zero mean and known

Table 11.1 Parameters for uncertainties used in the OLM and forecasting.

Uncertainty	Learning Process Duration	Standard Deviation
Wind power	8 days	0.0555
Electric demand	6 days	0.0119

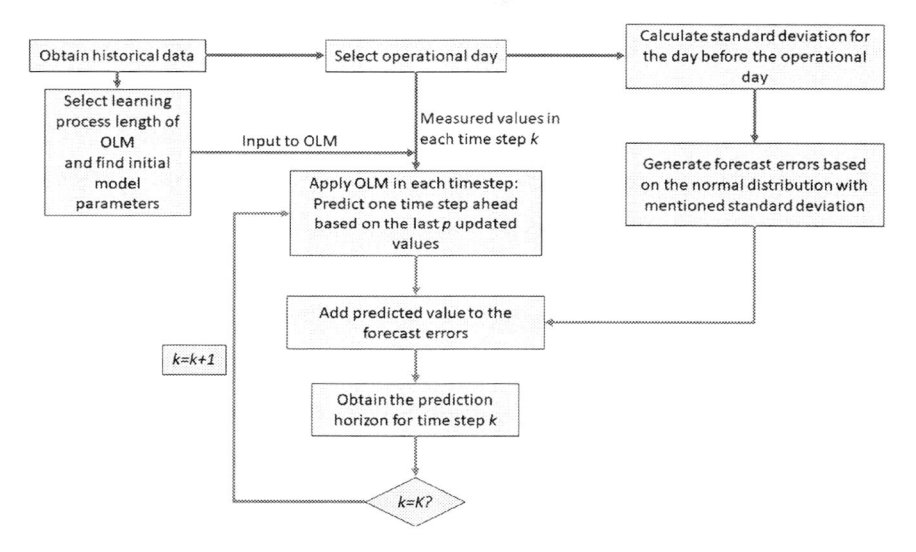

Figure 11.6 Application of the OLM for RT scheduling based on MPC.

standard deviation. Hence, in Table 11.1, a summary of the parameters is given. The learning process duration between wind power output and electric demand differs. It was obtained by increasing the learning process duration until the smallest deviation between the measured and predicted values is obtained and it no longer decreases. Furthermore, the standard deviation from Table 11.1 is calculated for the day before the day of operation. Afterward, the forecast errors are generated based on the normal distribution with the mentioned standard deviation and summed with the predicted value obtained by the OLM. Hence, the prediction horizon is created, and finally MPC can be applied to minimize the entire regulation cost over the prediction horizon.

The scheme to obtain the prediction horizon is shown in Fig. 11.6 based on the explanations given earlier.

11.4 Mathematical models of the IES for MPC-based RT scheduling

In this section, a mathematical model of the integrated and coordinated multi-energy system is designed. The IES combines the EPS, DHS, and NGS along with the linking components and considers the network constraints for each of the energy systems. The mathematical models are presented for RT scheduling based on MPC. Two uncertainties are considered as mentioned earlier: wind power output and electric demand. It is assumed that DA scheduling has been performed beforehand based on the forecasts of these uncertainties and the operating points of units for DA scheduling are available. The optimization problem is solved for the entire day, and the time resolution is 15 minutes.

To summarize, the objective function and four sets of constraints are given. First, the objective function is elaborated, whereas the four sets of constraints include the EPS, NGS, DHS, and linking constraints. The objective of the entire model is to minimize total regulation costs for IES while being subjected to security constraints. The notation for sets used in the following sections is described next.

Symbol	Description
$c \in \Omega^{GC}$	Gas compressors
$d \in \Omega^{GD}$	Gas demand
$e \in \Omega^{ED}$	Electric demand
$eb \in \Omega^{EB}$	Electric boilers (EBs)
$g \in \Omega^{GS}$	Gas sources
$hp \in \Omega^{HP}$	Heat pumps (HPs)
$j \in \Omega^{CHP}$	Combined heat and power (CHP)
$n, m \in \Omega^{EPS/DHS/NGS}$	Set of nodes in EPS/DHS/NGS
$p \in \Omega^{P2G}$	P2G
$s \in \Omega^{HS}$	Heat storages
$st \in \Omega^{ST}$	Gas storages
$w \in \Omega^{WF}$	Set of wind farms at node n
$m \in \Omega^{DHS,mix, S/R}$	Set of nodes with the mixing temperatures in the supply and return system
$l \in \Omega^{HL}$	Heat demand

11.4.1 Objective function

The objective of RT scheduling based on MPC is to minimize total regulation costs for the entire IES and penalize the conventional generators for any deviation from the DA schedule. MPC–based RT scheduling finds the optimal solution based on the measured system values and future predictions in the specified prediction horizon. The objective is presented in Eq. (11.3):

$$\min \left\{ \sum_{t=k}^{k+H_{p1}-1} \left\{ \begin{array}{l} \sum_{j=1}^{n^{CHP}} \left(C_j^{UR} R_{j,t}^{CHP,U} - C_j^{DR} R_{j,t}^{CHP,D} \right) \\ + \sum_{j=1}^{n^{CHP}} C_j^{CHP,H} \left(H_{j,t}^{CHP,RT} - H_{j,t}^{CHP,DA} \right) \\ - \sum_{p=1}^{n^{P2G}} C_p^{P2G} \left(D_{p,t}^{P2G,RT} - D_{p,t}^{P2G,DA} \right) \\ - \sum_{eb=1}^{n^{EB}} C_{eb}^{EB} \left(D_{eb,t}^{EB,RT} - D_{eb,t}^{EB,DA} \right) \\ - \sum_{hp=1}^{n^{HP}} C_{hp}^{HP} \left(D_{hp,t}^{HP,RT} - D_{hp,t}^{HP,DA} \right) \\ + \sum_{g=1}^{n^{GS}} \left(C_g^{UR} R_{g,t}^{GS,U} - C_g^{DR} R_{g,t}^{GS,D} \right) \\ + \sum_{e=1}^{n^{ED}} \left(C_e^{VOLL} D_{e,t}^{ED,shed} \right) + \sum_{w=1}^{n^{WF}} \left(C_w^{spill} W_{w,t}^{spill} \right) \\ + \sum_{j=1}^{n^{CHP}} \left(C^{penalty} \left| P_{j,t}^{CHP,RT} - P_{j,t}^{CHP,DA} \right| \right) \\ + \sum_{j=1}^{n^{CHP}} \left(C^{penalty} \left| H_{j,t}^{CHP,RT} - H_{j,t}^{CHP,DA} \right| \right) \end{array} \right\} \right\} \quad (11.3)$$

where parameters and variables are denoted as follows:

Parameters

$C_j^{CHP,H} =$ Marginal cost of heat production by CHP (DKK/MWh),

$C_g^{DR} =$ Marginal cost of downward regulating power of the gas source unit g (DKK/MWh),

$C_j^{DR} =$ Marginal cost of downward regulating power of the CHP unit j (DKK/MWh),

$C_{eb}^{EB} =$ Marginal cost of electricity for producing heat by the EB (DKK/MWh)

$C_{hp}^{HP} =$ Marginal cost of electricity for producing heat by the HP (DKK/MWh),

$C^{penalty} =$ Penalty cost for deviation from the DA schedule (DKK/MWh),

$C_p^{P2G} =$ Cost of electricity for producing gas and heat by P2G (DKK/MWh),

$C_g^{UR} =$ Marginal cost of upward regulating power of the gas source unit g (DKK/MWh),

$C_j^{UR} =$ Marginal cost of upward regulating power of the CHP unit j (DKK/MWh),

$C_e^{VOLL} =$ Value of the lost load (DKK/MWh), and

$C_w^{spill} =$ Cost of wind spillage (DKK/MWh).

Variables

$D_{eb}^{EB,RT/DA} =$ Electric consumption of the EB in RT/DA (MW),

$D_e^{ED,shed} =$ Electric load shedding (MW),

$D_{hp}^{HP,RT/DA} =$ Electric consumption of the HP in RT/DA (MW),

$D_p^{P2G,RT/DA} =$ Electric consumption of P2G in RT/DA (MW),

$H_j^{CHP,RT/DA} =$ Heat power output from CHP in RT/DA (MW),

$R_j^{CHP,U/D} =$ Upward/downward regulation power from CHP (MW),

$R_g^{GS,U/D} =$ Upward/ downward regulation power from the gas source (MW)

$W_w^{spill} =$ Wind spillage of wind farms (MW).

The first term describes the cost of upward regulation power and downward regulation power provided by CHP. The second term shows the cost of the regulation heat provided by CHP. The third, fourth, and fifth terms present the costs for providing the regulating power by P2G, the electric boiler (EB), and the heat pump (HP), respectively. The regulation power is calculated as the difference in power between the RT stage and DA stage. Moreover, the sixth term represents the regulation cost of gas supply. In the seventh and eighth terms, the value of lost load and wind power spillage cost are described, respectively. Finally, the last two terms represent the penalty for deviating from the DA schedule values for the thermal and electric power provided by CHP. Such measure increases regulation power provided by storage and P2G units, and the flexibility of the system increases.

As mentioned earlier, MPC schedules at each timestep and takes into account the measured and forecasted values that create the prediction horizon. MPC optimizes throughout the selected prediction horizon. Hence, in Eq. (11.3), the total number of timesteps in the scheduling horizon is denoted as K and the current timestep of the scheduling horizon is denoted as k.

The index for the timestep in the prediction horizon is denoted as t, and the length of the prediction horizon is equal to $k + H_p - 1$. To summarize, the optimization problem is solved for each timestep k for all constraints in the prediction horizon. This can be denoted as $\forall t \in [k, k + H_p - 1]$, $\forall k \in K$. The notation in the following constraints is omitted for ease of reading. However, it must be accounted for.

11.4.2 Electric power system

Constraints for the EPS are summarized in Eqs. (11.4) through (11.20). Eq. (11.4) introduces the nodal power balance equation for RT. It takes into account the regulation power from the CHP, the difference between the power consumed in RT and DA and the difference between the generated power in RT and DA. The regulating power can be provided by multiple units, and both production and consumption sides participate in provision of the regulating power. Additional notation compared to Eq. (11.3) is as follows. $W_w^{WF,RT}$ denotes the measured wind power output in RT and $W_w^{WF,DA}$ denotes the forecasted value of wind power output in DA. $D_c^{GC,RT}$ and $D_c^{GC,DA}$ stand for gas compressor consumption in the RT and DA, respectively. The electric demand in RT and DA is denoted as $D_e^{ED,RT}$ and $D_e^{ED,DA}$. B_{nm} denotes susceptance of the transmission line, whereas δ_n denotes the phase angle of bus n. Due to regulation power of CHP, the total output power of CHP is demonstrated in Eq. (11.5). Correspondingly, the total output power of CHP is limited by its maximum and minimum operating limits as shown in Eq. (11.6). The capacity limits of the upward and downward regulating power are given in Eqs. (11.7) through (11.11). The wind spillage and load shedding cannot exceed the measured values as demonstrated in Eqs. (11.12) and (11.13). Moreover, Eqs. (11.14) through (11.16) limit the consumption of P2G, the EB, and the HP in RT, respectively, by its maximum consumption levels. Furthermore, Eqs. (11.17) represents the limitations for the transmission line capacity, whereas Eq. (11.18) gives the reference angle. Eq. (11.19) describes the non-negative variable of the electric consumption of the gas compressor. Finally, Eq. (11.20) illustrates the upward and downward ramping rate limits expressed by P_j^{RRL}.

$$\sum_{j \in \Omega_n^{CHP}} \left(R_{j,t}^{CHP,U} - R_{j,t}^{CHP,D} \right) + \sum_{w \in \Omega_n^{WF}} \left(W_{w,t}^{WF,RT} - W_{w,t}^{WF,DA} - W_{w,t}^{spill} \right)$$
$$- \sum_{p \in \Omega_n^{P2G}} \left(D_{p,t}^{P2G,RT} - D_{p,t}^{P2G,DA} \right) - \sum_{eb \in \Omega_n^{EB}} \left(D_{eb,t}^{EB,RT} - D_{eb,t}^{EB,DA} \right)$$

$$- \sum_{hp \in \Omega_n^{HP}} \left(D_{hp,t}^{HP,RT} - D_{hp,t}^{HP,DA} \right) - \sum_{c \in \Omega_n^{GC}} \left(D_{c,t}^{GC,RT} - D_{c,t}^{GC,DA} \right)$$

$$+ \sum_{e \in \Omega_n^{ED}} D_{e,t}^{ED,shed} - \sum_{e \in \Omega_n^{ED}} \left(D_{e,t}^{ED,RT} - D_{e,t}^{ED,DA} \right) \tag{11.4}$$

$$= \sum_{m \in \Omega_n^{EPS}} B_{nm} \left(\delta_n^{RT} - \delta_m^{RT} - \delta_n^{DA} + \delta_m^{DA} \right), \quad \forall n \in \Omega^{EPS}$$

$$P_{j,t}^{CHP,RT} = P_{j,t}^{CHP,DA} + \left(R_{j,t}^{CHP,U} - R_{j,t}^{CHP,D} \right), \quad \forall j \in \Omega^{CHP} \tag{11.5}$$

$$P_j^{min} \le P_{j,t}^{CHP,RT} \le P_j^{max}, \quad \forall j \in \Omega^{CHP} \tag{11.6}$$

$$0 \le R_{j,t}^{CHP,U} \le R_j^{CHP,U,max} x_{j,t}^{CHP,U}, \quad \forall j \in \Omega^{CHP} \tag{11.7}$$

$$0 \le R_{j,t}^{CHP,D} \le R_j^{CHP,D,max} x_{j,t}^{CHP,D}, \quad \forall j \in \Omega^{CHP} \tag{11.8}$$

$$x_{j,t}^{CHP,U} + x_{j,t}^{CHP,D} \le 1, \quad \forall j \in \Omega^{CHP} \tag{11.9}$$

$$x_{j,t}^{CHP,U} \in \{0, 1\}, \quad \forall j \in \Omega^{CHP} \tag{11.10}$$

$$x_{j,t}^{CHP,D} \in \{0, 1\}, \quad \forall j \in \Omega^{CHP} \tag{11.11}$$

$$0 \le W_{w,t}^{spill} \le W_{w,t}^{WF,RT}, \quad \forall w \in \Omega^{WF} \tag{11.12}$$

$$0 \le D_{e,t}^{ED,shed} \le D_{e,t}^{ED,RT}, \quad \forall e \in \Omega^{ED} \tag{11.13}$$

$$0 \le D_{p,t}^{P2G,RT} \le D_p^{P2G,max}, \quad \forall p \in \Omega^{P2G} \tag{11.14}$$

$$0 \le D_{eb,t}^{EB,RT} \le D_{eb}^{EB,max}, \quad \forall eb \in \Omega^{EB} \tag{11.15}$$

$$0 \le D_{hp,t}^{HP,RT} \le D_{hp}^{HP,max}, \quad \forall hp \in \Omega^{HP} \tag{11.16}$$

$$-P_{nm}^{rate} \le B_{nm} \left(\delta_{n,t}^{DA} - \delta_{m,t}^{DA} \right) \le P_{nm}^{rate}, \quad \forall (nm) \in \Omega^{EPS} \tag{11.17}$$

$$\delta_{ref,t}^{RT} = 0 \tag{11.18}$$

$$D_{c,t}^{GC,RT} \ge 0, \quad \forall c \in \Omega^{GC} \tag{11.19}$$

$$\left| P_{j,t-1}^{CHP,RT} - P_{j,t}^{CHP,RT} \right| \leq P_j^{RRL}, \quad \forall j \in \Omega^{CHP} \tag{11.20}$$

11.4.3 District heating system

The model of the DHS is similar to the one in Chapter 10. First, the thermal and hydraulic model are presented. The nodal thermal balance is presented in Eq. (11.21). It is based on the thermal energy difference between the RT values and DA values for multiple units and the temperature difference between RT and DA. The temperature difference results in either generation or consumption of thermal energy. A thorough explanation can be found in Chapter 10. The nodal balance equation considers CHP, EB, P2G, and HP regulating thermal power in RT. H_{eb}^{EB}, H_{hp}^{HP} and H_p^{P2G} denote the thermal power output from the EB, the HP, and P2G, respectively. In addition, the thermal energy storage is considered. $H_s^{HS,in}$ denotes the heat power input to the thermal energy storage, whereas $H_s^{HS,out}$ stands for the heat power output from the thermal energy storage. All units are in MW. The continuity of the mass flow is demonstrated in Eqs. (11.22) through (11.24). m_{mn}^{RT} denotes the mass flow entering or leaving the node in kg/s, whereas mq_{in}^{RT} denotes the mass flow at the consumption or production node in kg/s. A is the network incidence matrix consisting of values $\{-1, 0, +1\}$, and A_q is a vector of similar values. The superscript S and R stand for supply and return pipelines, respectively. Furthermore, due to several pipelines being connected, the temperature mixing occurs at the nodes. Hence, Eqs. (11.25) and (11.26) illustrate the temperature mixing at the nodes in the supply and return pipelines respectively. The temperature at a node in the supply or return pipelines is denoted as $T_m^{S/R}$, whereas $T_{mn}^{S,OUT}$ and $T_{mn}^{S,IN}$ are the outlet and inlet temperatures of the supply pipelines, respectively. Similarly, $T_{mn}^{R,OUT}$ and $T_{mn}^{R,IN}$ are the outlet and inlet temperatures of the return pipelines, respectively. In addition, while the heat is transported in the pipelines, the temperature drop happens due to the temperature difference between the ambient temperature and water temperature. The temperature drop equations for supply and return pipelines are demonstrated in Eqs. (11.27) and (11.28), respectively. As can be observed, the temperature drop depends also on the parameters of the pipelines. The diameter of the pipeline and heat transfer coefficient are taken into account through parameter λ, which is the thermal conductivity of the pipeline in W/mK. Moreover, the length of the pipeline is denoted as L_{mn}, whereas T^A and c denote the ambient temperature (°C) and specific heat capacity of water (Wh/kgK). Last,

Eqs. (11.29) through (11.34) present the additional temperature constraints.

$$
\begin{pmatrix}
\displaystyle\sum_{j\in\Omega_n^{CHP}} \left(H_{j,t}^{CHP,RT} - H_{j,t}^{CHP,DA}\right) \\[2ex]
+ \displaystyle\sum_{hp\in\Omega_n^{HP}} \left(H_{hp,t}^{HP,RT} - H_{hp,t}^{HP,DA}\right) \\[2ex]
+ \displaystyle\sum_{eb\in\Omega_n^{EB}} \left(H_{eb,t}^{EB,RT} - H_{eb,t}^{EB,DA}\right) \\[2ex]
+ \displaystyle\sum_{p\in\Omega_n^{P2G}} \left(H_{p,t}^{P2G,RT} - H_{p,t}^{P2G,DA}\right) \\[2ex]
+ \displaystyle\sum_{s\in\Omega_n^{HS}} \left(\begin{array}{c} H_{s,t}^{HS,out,RT} - H_{s,t}^{HS,in,RT} \\ -H_{s,t}^{HS,out,DA} + H_{s,t}^{HS,in,DA} \end{array}\right)
\end{pmatrix}
$$

$$
= \left\{ \begin{array}{c}
T_{m,t}^{final} > T_{m,t}^{initial}: \\[1ex]
\left[mq_{m,t}^{RT} c\left(T_{m,t}^{S,RT} - T_{m,t}^{R,RT}\right)\right] - \left[mq_{m,t}^{DA} c\left(T_{m,t}^{S,DA} - T_{m,t}^{R,DA}\right)\right] \\[3ex]
T_{m,t}^{initial} > T_{m,t}^{final}: \\[1ex]
\left[mq_{m,t}^{RT} c\left(T_{m,t}^{R,RT} - T_{m,t}^{S,RT}\right)\right] - \left[mq_{m,t}^{DA} c\left(T_{m,t}^{R,DA} - T_{m,t}^{S,DA}\right)\right]
\end{array} \right\},
$$

$$
\forall m \in \Omega^{DHS} \tag{11.21}
$$

$$
A^S\, m_{mn,t}^{S,\,RT} = A_q^S \circ mq_{m,t}^{RT}, \quad \forall m \in \Omega^{DHS} \tag{11.22}
$$

$$
A^R\, m_{mn,t}^{R,\,RT} = A_q^R \circ mq_{m,t}^{RT}, \quad \forall m \in \Omega^{DHS} \tag{11.23}
$$

$$
m_{mn,t}^{S,\,RT} = m_{mn,t}^{R,\,RT}, \quad \forall (mn) \in \Omega^{DHS} \tag{11.24}
$$

$$
\sum \left(m_{nm,t}^{S,\,RT}\, T_{nm,t}^{S,OUT,RT}\right) + \sum \left(mq_{m,t}^{RT}\, T_{m,t}^{S,RT}\right) = T_{nm,t}^{S,IN,RT} \sum \left(m_{nm,t}^{S,\,RT}\right),
$$
$$
\forall m \in \Omega^{DHS,mix,S} \tag{11.25}
$$

$$
\sum \left(m_{mn,t}^{R,\,RT}\, T_{mn,t}^{R,OUT,RT}\right) + \sum \left(mq_{m,t}^{RT}\, T_{m,t}^{R,RT}\right) = T_{nm,t}^{R,IN,RT} \sum \left(m_{nm,t}^{R,\,RT}\right),
$$
$$
\forall m \in \Omega^{DHS,mix,R} \tag{11.26}
$$

$$
T_{mn,t}^{S,OUT,RT} - T^A = \left(T_{mn,t}^{S,IN,RT} - T^A\right) e^{-\frac{\lambda_{mn} L_{mn}}{cm_{mn,t}^{S,\,RT}}}, \quad \forall (mn) \in \Omega^{DHS} \tag{11.27}
$$

$$
T_{mn,t}^{R,OUT,RT} - T^A = \left(T_{mn,t}^{R,IN,RT} - T^A\right) e^{-\frac{\lambda_{mn} L_{mn}}{cm_{mn,t}^{R,\,RT}}}, \quad \forall (mn) \in \Omega^{DHS} \tag{11.28}
$$

$$T_{m,t}^{S,RT} = T_{mn,t}^{S,IN,RT} = T_{m,t}^{S,fixed}, \quad m = 1 \tag{11.29}$$

$$T_{m,t}^{R,RT} = T_{mn,t}^{R,IN,RT}, \quad m = max(m) \tag{11.30}$$

$$T_{mn,t}^{S,OUT,RT} = T_{mn,t}^{S,IN,RT}, \quad \forall m \notin \Omega^{DHS,mix,S} \tag{11.31}$$

$$T_{mn,t}^{S,OUT,RT} = T_{m,t}^{S,RT}, \quad \forall m \notin \Omega^{DHS,mix,S} \tag{11.32}$$

$$T_{mn,t}^{R,OUT,RT} = T_{m,t}^{R,RT}, \quad \forall m \notin \Omega^{DHS,mix,R} \tag{11.33}$$

$$T_{mn,t}^{R,OUT,RT} = T_{mn,t}^{R,IN,RT}, \quad \forall m \notin \Omega^{DHS,mix,R} \tag{11.34}$$

Furthermore, the scheduling constraints are presented in the following to ensure that the DHS operates inside the operational limits. The mass flow limits are presented in Eqs. (11.35) through (11.37). Inflow and outflow temperatures of the pipeline in the supply and return networks, as well as the node temperatures in the supply and return networks, are given in Eqs. (11.38) through (11.43). Eqs. (11.44) and (11.45) represent the state of energy of the thermal energy storage, H_s^{SOE} (MWh), whereas Eq. (11.46) limits the capacity of the thermal energy storage. The injection and extraction rates of thermal energy storage are limited as shown in Eqs. (11.47) and (11.48). Moreover, additional binary variables denoted as $x_s^{HS,out}$ and $x_s^{HS,in}$ ensure that only one process is available at a time. The operational constraints with additional variables are presented in Eqs. (11.47) through (11.51).

$$m_{mn}^{S,min} \leq m_{mn,t}^{S,RT} \leq m_{mn}^{S,max}, \quad \forall (mn) \in \Omega^{DHS} \tag{11.35}$$

$$m_{mn}^{R,min} \leq m_{mn,t}^{R,RT} \leq m_{mn}^{R,max}, \quad \forall (mn) \in \Omega^{DHS} \tag{11.36}$$

$$mq_m^{min} \leq mq_{m,t}^{RT} \leq mq_m^{max}, \quad \forall m \in \Omega^{DHS} \tag{11.37}$$

$$T_{mn}^{S,IN,min} \leq T_{mn,t}^{S,IN,RT} \leq T_{mn}^{S,IN,max}, \quad \forall (mn) \in \Omega^{DHS} \tag{11.38}$$

$$T_{mn}^{S,OUT,min} \leq T_{mn,t}^{S,OUT,RT} \leq T_{mn}^{S,OUT,max}, \quad \forall (mn) \in \Omega^{DHS} \tag{11.39}$$

$$T_{mn}^{R,IN,min} \leq T_{mn,t}^{R,IN,RT} \leq T_{mn}^{R,IN,max}, \quad \forall (mn) \in \Omega^{DHS} \tag{11.40}$$

$$T_{mn}^{R,OUT,min} \leq T_{mn,t}^{R,OUT,RT} \leq T_{mn}^{R,OUT,max}, \quad \forall (mn) \in \Omega^{DHS} \tag{11.41}$$

$$T_m^{S,min} \leq T_{m,t}^{S,RT} \leq T_m^{S,max}, \quad \forall m \in \Omega^{DHS} \tag{11.42}$$

$$T_m^{R,min} \leq T_{m,t}^{R,RT} \leq T_m^{R,max}, \quad \forall m \in \Omega^{DHS} \tag{11.43}$$

$$H_{s,t}^{SOE,RT} = H_{s,t-1}^{SOE,RT} + H_{s,t}^{HS,in,RT} - H_{s,t}^{HS,out,RT}, \quad \forall s \in \Omega^{HS}, t > 1 \tag{11.44}$$

$$H_{s,t}^{SOE,RT} = H_s^{SOE0,RT} + H_{s,t}^{HS,in,RT} - H_{s,t}^{HS,out,RT}, \quad \forall s \in \Omega^{HS}, \ t = 1 \tag{11.45}$$

$$H_s^{SOE,min} \leq H_{s,t}^{SOE,RT} \leq H_s^{SOE,max}, \quad \forall s \in \Omega^{HS}, t > 1 \tag{11.46}$$

$$0 \leq H_{s,t}^{HS,in,RT} \leq H_s^{HS,in,max} x_{s,t}^{HS,in,RT}, \quad \forall s \in \Omega^{HS} \tag{11.47}$$

$$0 \leq H_{s,t}^{HS,out,RT} \leq H_s^{HS,out,max} x_{s,t}^{HS,out,RT}, \quad \forall s \in \Omega^{HS} \tag{11.48}$$

$$x_{s,t}^{HS,out,RT} + x_{s,t}^{HS,in,RT} \leq 1, \quad \forall s \in \Omega^{HS} \tag{11.49}$$

$$x_{s,t}^{HS,out,RT} \in \{0, 1\}, \quad \forall s \in \Omega^{HS} \tag{11.50}$$

$$x_{s,t}^{HS,in,RT} \in \{0, 1\}, \quad \forall s \in \Omega^{HS} \tag{11.51}$$

11.4.4 Natural gas system

The nodal gas balance equation is demonstrated as a difference between the values in RT and DA as presented in Eq. (11.52). $Q_{st}^{ST,IN/OUT}$ denotes the gas injection or extraction to or from gas storage (MW), whereas Q_j^{DCHP} denotes the gas consumption of CHP (MW). The upward and downward regulating gas power of P2G are denoted as $R_p^{P2G,U}$ and $R_p^{P2G,D}$, respectively. The gas flow is denoted as Q_{nm} (MW). In addition, the steady–state gas flow is shown in Eq. (11.53). p_n and p_m denote inlet and outlet pressures in the pipeline (MPa), whereas C_{nm} denotes the pipeline parameter in $(m^3/h)/MPa$. The calculation of C_{nm} is presented in Eq. (11.54) [26]. A thorough explanation of the parameters can be found in Chapter 2 and Chapter 10. Due to regulation gas at the entry point of the gas system in RT, the total gas injected is calculated as demonstrated in Eq. (11.55) and limited by the maximum injection rate in Eq. (11.56). The capacity of the gas injected in RT is limited in Eqs. (11.57) through (11.61). In a similar manner, the upward and downward regulation from P2G is activated in RT and shown in

Eqs. (11.62) through (11.67). The operational limits for the pressure are expressed in Eq. (11.68), whereas the pressure at the reference node is accounted for in Eq. (11.69). Eq. (11.70) limits the gas flow in the pipelines. As mentioned earlier, the gas compressors are integrated in the gas system and consume the electric power from the EPS. The control of the flow to compensate for losses is demonstrated by the compression ratio, CR, in Eq. (11.71), whereas the electric consumption of the gas compressor is expressed in Eq. (11.72). Explanations on the parameters in the equations can be further reviewed in Chapter 2 and Chapter 10. The state of energy of the gas storage, Q_{st}^{SOE}, is shown in Eqs. (11.73) and (11.74). The gas storage capacity is limited in Eq. (11.75), whereas the injection and withdrawal rates of the gas storages are limited in Eqs. (11.76) through (11.80). $x_{st}^{ST,IN}$ and $x_{st}^{ST,OUT}$ are introduced to ensure that only one process is available at a time: either injection to the gas storage or extraction from the gas storage.

$$\sum_{g\in\Omega_n^{GS}}\left(R_{g,t}^{GS,U}-R_{g,t}^{GS,D}\right)$$

$$+\sum_{st\in\Omega_n^{ST}}\left(Q_{st,t}^{ST,OUT,RT}-Q_{st,t}^{ST,IN,RT}-Q_{st,t}^{ST,OUT,DA}+Q_{st,t}^{ST,IN,DA}\right)$$

$$+\sum_{p\in\Omega_n^{P2G}}\left(R_{p,t}^{P2G,U}-R_{p,t}^{P2G,D}\right)-\sum_{j\in\Omega_n^{CHP}}\left(Q_{j,t}^{DCHP,RT}-Q_{j,t}^{DCHP,DA}\right)$$

$$=\sum_{m\in\Omega_n^{NGS}}\left(Q_{mn,t}^{RT}-Q_{mn,t}^{DA}\right),\quad\forall n,\,m\in\Omega^{NGS}$$

$$\tag{11.52}$$

$$\left(\left(p_{n,t}^{RT}\right)^2-\left(p_{m,t}^{RT}\right)^2\right)C_{nm}^2=\left(Q_{nm,t}^{RT}\right)^2,\quad\forall(nm)\in\Omega^{NGS}\tag{11.53}$$

$$C_{nm}=C\frac{T_b}{p_b}D_{nm}^{2.5}\left(\frac{1}{L_{nm}\gamma_g T_a Z_a f}\right)^{0.5}\eta_{p,nm},\quad\forall(nm)\in\Omega^{NGS}\tag{11.54}$$

$$Q_{g,t}^{GS,RT}=Q_{g,t}^{GS,DA}+R_{g,t}^{GS,U}-R_{g,t}^{GS,D},\quad\forall g\in\Omega^{GS}\tag{11.55}$$

$$Q_g^{GS,min}\le Q_{g,t}^{GS,RT}\le Q_g^{GS,max},\quad\forall g\in\Omega^{GS}\tag{11.56}$$

$$0\le R_{g,t}^{GS,U}\le R_g^{GS,U,max}x_{g,t}^{GS,U},\quad\forall g\in\Omega^{GS}\tag{11.57}$$

$$0\le R_{g,t}^{GS,D}\le R_g^{GS,D,max}x_{g,t}^{GS,D},\quad\forall g\in\Omega^{GS}\tag{11.58}$$

$$x_{g,t}^{GS,U}+x_{g,t}^{GS,D}\le 1,\quad\forall g\in\Omega^{GS}\tag{11.59}$$

$$x_{g,t}^{GS,U} \in \{0, 1\}, \quad \forall g \in \Omega^{GS} \tag{11.60}$$

$$x_{g,t}^{GS,D} \in \{0, 1\}, \quad \forall g \in \Omega^{GS} \tag{11.61}$$

$$Q_{p,t}^{P2G,RT} = Q_{p,t}^{P2G,DA} + R_{p,t}^{P2G,U} - R_{p,t}^{P2G,D} , \quad \forall p \in \Omega^{P2G} \tag{11.62}$$

$$0 \leq R_{p,t}^{P2G,U} \leq R_{p}^{P2G,U,max} x_{p,t}^{P2G,U}, \quad \forall p \in \Omega^{P2G} \tag{11.63}$$

$$0 \leq R_{p,t}^{P2G,D} \leq R_{p}^{P2G,D,max} x_{p,t}^{P2G,D}, \quad \forall p \in \Omega^{P2G} \tag{11.64}$$

$$x_{p,t}^{P2G,U} + x_{p,t}^{P2G,D} \leq 1 , \quad \forall p \in \Omega^{P2G} \tag{11.65}$$

$$x_{p,t}^{P2G,U} \in \{0, 1\}, \quad \forall p \in \Omega^{P2G} \tag{11.66}$$

$$x_{p,t}^{P2G,D} \in \{0, 1\} , \quad \forall p \in \Omega^{P2G} \tag{11.67}$$

$$\left(p_{m}^{min}\right)^2 \leq \left(p_{m,t}^{RT}\right)^2 \leq \left(p_{m}^{max}\right)^2 , \quad \forall m \in \Omega^{NGS} \tag{11.68}$$

$$\left(p_{ref,t}^{RT}\right)^2 = p_{ref} \tag{11.69}$$

$$-Q_{nm}^{max} \leq Q_{nm,t}^{RT} \leq Q_{nm}^{max} , \quad \forall (nm) \in \Omega^{NGS} \tag{11.70}$$

$$\left(p_{m,t}^{RT}\right)^2 \leq CR^2 \left(p_{n,t}^{RT}\right)^2 , \quad \forall n, m \in \Omega^{NGS} \tag{11.71}$$

$$D_{c,t}^{GC,RT} = C^{GC} Z_a \frac{T_s}{E^{GC}\eta_{gc}^{GC}} \frac{c_k}{c_k-1} \left(CR^{\frac{c_k-1}{c_k}} - 1\right) Q_{c,t}^{GC,RT}, \quad \forall c \in \Omega^{GC} \tag{11.72}$$

$$Q_{st,t}^{SOE,RT} = Q_{st,t-1}^{SOE,RT} + Q_{st,t}^{ST,IN,RT} - Q_{st,t}^{ST,OUT,RT} , \quad \forall st \in \Omega^{ST}, t > 1 \tag{11.73}$$

$$Q_{st,t}^{SOE,RT} = Q_{st}^{SOE0,RT} + Q_{st,t}^{ST,IN,RT} - Q_{st,t}^{ST,OUT,RT} , \quad \forall st \in \Omega^{ST}, t = 1 \tag{11.74}$$

$$Q_{st}^{SOE,min} \leq Q_{st,t}^{SOE,RT} \leq Q_{st}^{SOE,max} , \quad \forall st \in \Omega^{ST} \tag{11.75}$$

$$Q_{st}^{ST,IN,min} \leq Q_{st,t}^{ST,IN,RT} \leq Q_{st}^{ST,IN,max} x_{st,t}^{ST,IN,RT} , \quad \forall st \in \Omega^{ST} \tag{11.76}$$

$$Q_{st}^{ST,OUT,min} \leq Q_{st,t}^{ST,OUT,RT} \leq Q_{st}^{ST,OUT,max} x_{st,t}^{ST,OUT,RT} \quad, \quad \forall st \in \Omega^{ST} \tag{11.77}$$

$$x_{st,t}^{ST,IN,RT} + x_{st,t}^{ST,OUT,RT} \leq 1 \quad, \quad \forall st \in \Omega^{ST} \tag{11.78}$$

$$x_{st,t}^{ST,IN,RT} \in \{0, 1\}, \quad \forall st \in \Omega^{ST} \tag{11.79}$$

$$x_{st,t}^{ST,OUT,RT} \in \{0, 1\}, \quad \forall st \in \Omega^{ST} \tag{11.80}$$

11.4.5 Linking components

P2G, the EB, the HP, and CHP link the three subsystems. The relationships between the gas consumption and the electricity generated by CHP is described by generated electricity efficiency, $\eta^{CHP,e}$, in Eq. (11.81). Moreover, the relationship between the gas consumption and the heat generated by the CHP is expressed through generated heat efficiency, $\eta^{CHP,h}$, in Eq. (11.82). Eqs. (11.83) and (11.84) demonstrate the conversion of the excess electricity by P2G to the gas and heat, respectively. The relationship between the electricity consumed and heat generated by the HP is presented in Eq. (11.85), whereas the relationship between the electricity consumption and heat generated from the EB is given in Eq. (11.86).

$$P_{j,t}^{CHP,RT} = \eta^{CHP,e} Q_{j,t}^{DCHP,RT}, \quad \forall j \in \Omega^{CHP} \tag{11.81}$$

$$H_{j,t}^{CHP,RT} = \eta^{CHP,h} Q_{j,t}^{DCHP,RT}, \quad \forall j \in \Omega^{CHP} \tag{11.82}$$

$$Q_{p,t}^{P2G,RT} = \eta^{P2G,Q} D_{p,t}^{P2G,DA}, \quad \forall p \in \Omega^{P2G} \tag{11.83}$$

$$H_{p,t}^{P2G,RT} = \eta^{P2G,H} D_{p,t}^{P2G,RT}, \quad \forall p \in \Omega^{P2G} \tag{11.84}$$

$$H_{hp,t}^{HP,RT} = COP\, D_{hp,t}^{HP,RT}, \quad \forall hp \in \Omega^{HP} \tag{11.85}$$

$$H_{eb,t}^{EB,RT} = \eta^{EB} D_{eb,t}^{EB,RT}, \quad \forall eb \in \Omega^{EB} \tag{11.86}$$

11.5 Solution process and case study

In this section, an overview of the solution process and case study are given.

11.5.1 Solution process

In this section, an overview of the solution process is provided. Three main concerns regarding the mathematical model of IES are introduced and explained. Furthermore, the solution process is elaborated.

Initially, the mathematical model of the IES is nonlinear. Hence, the linearization procedure must be performed for the nonlinear terms to reach the global optimum. In the NGS, the nonlinearity appears in Eq. (11.53). The gas flow equation is linearized by the piece-wise function. The linearization procedure can be found in Section 2.5.4 of Chapter 2. Moreover, Eqs. (11.21) and (11.25) through (11.28) are linearized by Taylor first-order approximation. Finally, the absolute term in Eq. (11.3) is linearized as well. The absolute term is replaced by an auxiliary variable in the objective function. The values of absolute term create two new constraints that are limited with the auxiliary variable. The attained mathematical model of the IES is from mixed-integer linear programming (MILP). In addition, the second concern is the units used in the IES. According to Section 11.4, the units in the NGS largely differ compared to the EPS and DHS. Initially, all units regarding the flow in the NGS are in m^3/h. Hence, due to such differences, the entire model of the IES is converted to the pu system. Moreover, the simulations performed for a system in pu results in lower computational burden [4]. The thorough procedure to convert the systems to pu can be found in Section 2.5.3 of Chapter 2 and in Chapter 10. Finally, the last concern discussed is how to obtain the feasibility of the entire scheduling horizon. As mentioned, MPC is applied for every timestep of the scheduling horizon and optimizes the RT schedule for the entire prediction horizon, H_p. Hence, additional constraints are added to the model of IES to provide feasibility inside the scheduling horizon. The constraints highly dependent on the previous and current state are shown in Eqs. (11.20), (11.44), and (11.73). By introducing the additional variables, the optimization problem solved for $\forall k \in K$ becomes feasible for both scheduling and prediction horizons.

Once the modified linearized model of the IES is obtained for the entire scheduling and prediction horizons, the MPC strategy can be applied. The simulations are carried out in MATLAB by using the YALMIP toolbox [20]. The solver used for the optimization problem is MOSEK [27]. The solution process for MPC–based RT scheduling for the IES model is shown in Fig. 11.7. The DA schedule is assumed to already be determined, and the focus is mainly on RT scheduling. RT scheduling is performed every 15 minutes

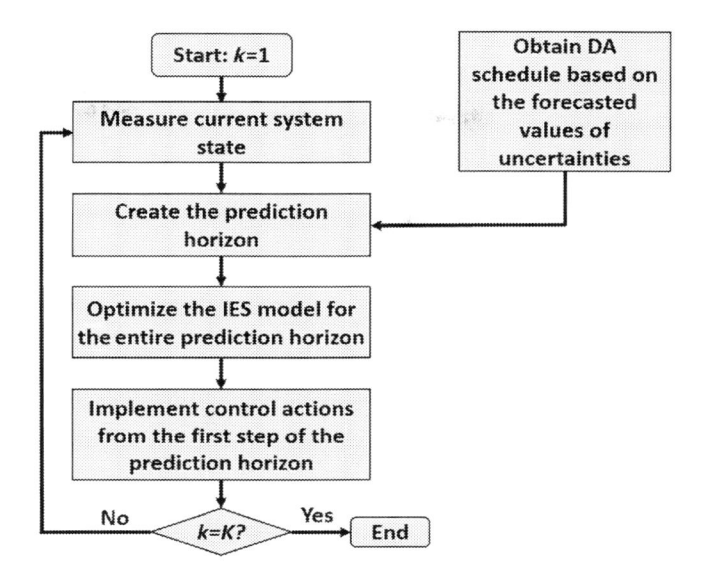

Figure 11.7 The solution process for MPC-based RT scheduling of the IES.

for a period of 24 hours. In every timestep of the scheduling horizon, new measurements are provided to MPC. The OLM that is implemented in MPC forecasts for the next timestep, and the forecasting error distribution is used for the rest of the forecasted value. This creates the prediction horizon. MPC is applied and the optimal control actions are calculated for each timestep. This procedure is repeated until the final timesteps of the scheduling horizon are reached.

To verify the efficiency of MPC, traditional RT scheduling is used for comparison. Traditional RT scheduling is based on the explanations in Section 11.2.2 and implemented as a single-period optimization. The time resolution is set to 15 minutes to compare the results with the MPC approach.

Finally, as mentioned earlier, the RT balancing market consists of balancing power and regulating power markets. The balancing power market is used for the settlements of the imbalances for the production and consumption side. For simplicity, the cost of traditional RT scheduling will be calculated after the scheduling has been performed by including the penalty for the deviation, similar to the penalty included in the MPC strategy. However, the settlement results of the balancing power market will be briefly presented to validate the proposed MPC strategy.

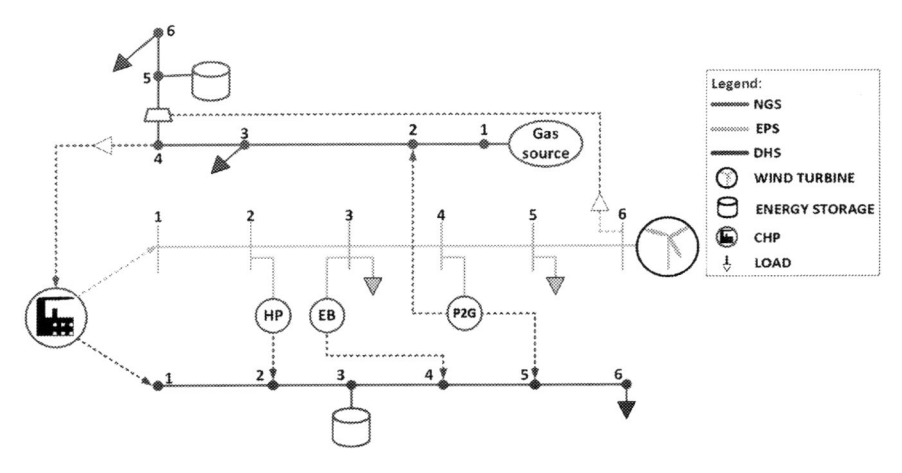

Figure 11.8 Test case: the IES.

11.5.2 Case study

The case system is presented in Fig. 11.8, whereas the parameters of the IES are given in Table 11.2 [6, 28–35]. The historical data of the uncertainties are used from January 1, 2019, to January 1, 2020 [36]. In the next sections, the simulation results are presented.

11.6 Simulation results

In this section, the simulations results will be presented and the MPC approach will be evaluated. Section 11.6.1 evaluates the benefits of the OLM. Section 11.6.2 presents the computational burden of the MPC approach. In addition, the validation of the MPC strategy for RT scheduling of the IES is investigated in Section 11.6.3. Finally, the summary of the simulations results is given in Section 11.6.4.

11.6.1 Benefits of the OLM combined with MPC

In this section, the benefits of the OLM are presented. Three cases are compared in Table 11.3. The first case, Case A, describes the measured uncertainties of the system. The last two cases represent different methods that can be used to create the prediction horizon that is later input to MPC. Case B is the OLM that is based on the explanations in Section 11.3.3. On the contrary, to present the benefits of the OLM, Case C is not based on the OLM. Case C is the conventional method that is based on the generation of

Table 11.2 Parameters for the case study.

Parameters (EPS)	Value	Unit	Parameters (NGS)	Value	Unit
$x_{12}, x_{23}, x_{34}, x_{45}, x_{56}$	0.2, 0.25, 0.1, 0.3, 0.2	pu	C	47.8917e-6	
$P_j^{CHP,max/min}$	5/0	MW	f	0.025	
P_{nm}^{rate}	1.6 − 1.8	MW	UCV	53.72	MJ/m3
P_j^{RRL}	10	%/h	$P_n^{max/min}$	4/1.6	MPa
$W_w^{WF,max}$	4	MW	p_{ref}	16	(MPa)2
$D_{eb}^{EB,max/min}$	0.6/0	MW	T_b	273.15	K
$D_{hp}^{HP,max/min}$	0.75/0	MW	p_b	0.1013	MPa
$D_p^{P2G,max/min}$	0.4/0	MW	γ_g	0.633	
$C_j^{CHP,P}$	99	DKK/MWh	$\eta_{p,nm}$	0.9995	
C_j^{DR}	89.1	DKK/MWh	T_a	295.65	K
C_j^{UR}	108.9	DKK/MWh	Z_a	0.95	
C_e^{VOLL}	2000	DKK/MWh	c_k	1.3	
C_w^{spill}	500	DKK/MWh	$\eta^{P2G,Q}$	0.7	
$C_p^{P2G} C_{eb}^{EB}, C_{hp}^{HP}$	60	DKK/MWh	$Q_{st}^{ST,OUT,min/max}$	0/0.1875	MW
Parameters (DHS):	Value	Unit	$Q_{st}^{ST,IN,min/max}$	0/0.11	MW
T^A	10	°C	$D_{12}, D_{23}, D_{34}, D_{45}, D_{56}$	355.6, 304.8, 304.8, 355.6, 304.8	mm
$\lambda_{12}, \lambda_{23}, \lambda_{34} \lambda_{45}, \lambda_{56}$	0.201, 0.185, 0.240, 0.184, 0.250	W/mK	$Q_{12}^{max}, Q_{23}^{max}, Q_{34}^{max}, Q_{45}^{max}, Q_{56}^{max}$	(0.5223, 0.8953, 0.2761, 0.5223, 1.0446) $\times$ 10^{-3}	MW
$L_{12}, L_{23}, L_{34}, L_{45}, L_{56}$	250, 300, 300, 120, 850	m	C^{GC}	4.0639 $\times$ 10^{-3}	

(continued on next page)

Table 11.2 (*continued*)

Parameters (EPS)	Value	Unit	Parameters (NGS)	Value	Unit
c	1.16167	Wh/kgK	$L_{12}, L_{23}, L_{34}, L_{45}, L_{56}$	18.99, 1.831, 17.792, 18.99, 34.535	km
$m_{mn}^{S,min/max}, m_{mn}^{R,min/max}$ $mq_m^{min/max}$	0.1/36000	kg/h	T_s	295.65	K
$H_s^{HS,in/out,max}$	0.3	MW	E^{GC}	0.98	
$H_s^{SOE,min/max/0}$	1.8/18/9	MWh	η_{gx}^{GC}	0.85	
$C_j^{CHP,H}$	150	DKK/MWh	f_{C1}^{GC}	67.0141	
$\eta^{CHP,e}$	0.4		f_{C2}^{GC}	24×10^{-6}	
$\eta^{CHP,h}$	0.38		CR	1.4	
$\eta^{P2G,H}$	0.08		$Q_g^{GS,max/min}$	3	MW
η^{EB}	0.99		$R_j^{CHP,D/U,max}$	2	MW
COP	3.6		$R_g^{GS,D/U,max}$	2	MW
$T_m^{S,fixed}$	80	°C	C^{LP}	7.8550×10^{-4}	
$T_{mn}^{S,IN,min/max}$	60/80	°C	$Q_{st}^{SOE,min/max/0}$	0/1065/532.5	MWh
$T_m^{S,min/max}$	60/80	°C	C_g^{GS}	99	DKK/MWh
$T_{mn}^{S,OUT,min/max}$	60/80	°C	C_g^{DR}	89.1	DKK/MWh
$T_{mn}^{R,IN,min/max}$	30/80	°C	C_g^{UR}	108.9	DKK/MWh
$T_m^{R,min/max}$	30/80	°C			
$T_{mn}^{R,OUT,min/max}$	30/80	°C	$C^{peanlty}$	500	DKK/MWh

Table 11.3 Description of cases to create prediction horizon in MPC.

Case	Description of Cases
A	Measured wind power and demand
B	Prediction horizon based on the OLM
C	Prediction horizon based on the normal distribution with a known standard deviation of forecast errors

Table 11.4 RMSE values for Case B and Case C in comparison to Case A.

Case	RMSE of Wind Power (pu)	RMSE of Electric Demand (pu)
B	8.4957e−4	4.1202e−4
C	0.0687	0.0183

the forecast error based on the normal distribution with a known standard deviation taken from Table 11.1 and zero mean. Moreover, it takes into account the last measured value of the uncertainty and adds the forecast error values to the last measured value.

The comparison results between three cases are shown in Fig. 11.9. Only the first timestep of the prediction horizon is shown for every step in the scheduling horizon. As mentioned, Case A is a reference point, and it represents the actual measured values of wind power and demand. The root mean square error (RMSE) is calculated for two different schemes in regard to Case A and is shown in Table 11.4. First, the predicted values in Case B follow the measured values in Case A. It can be seen that Case B is close to Case A and the accuracy is very high. Moreover, the RMSE value is close to zero. However, Case C has large deviations compared to the measured values in Case A. The wind power output in Case C has larger deviations compared to the electric demand in Case C, as can be seen in Fig. 11.9(a) and Fig. 11.9(b), respectively. A similar observation for Case C can be observed in RMSE values. Due to high forecast errors of wind power, the RMSE value is larger for wind power compared to electric demand. To summarize, the OLM method provides forecasts at high accuracy compared to the conventional method.

11.6.2 Impact of the prediction horizon length on the computational efficiency

The regulating power must be activated inside the 15-minute interval from the time when the order for the upward or downward regulation has been sent. Hence, the computational burden of the MPC approach must be low and meet the requirement of 15 minutes. The simulations are performed on a

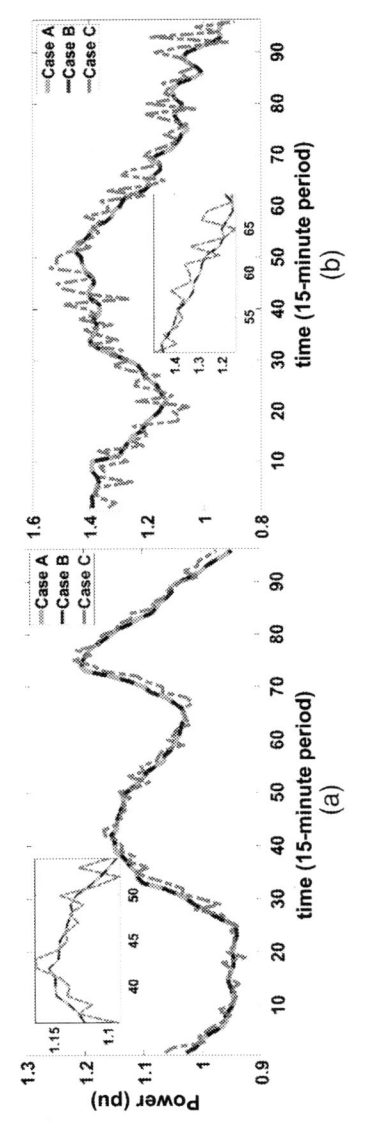

Figure 11.9 Comparison of two different schemes to construct the prediction horizon for (a) electric demand and (b) wind power.

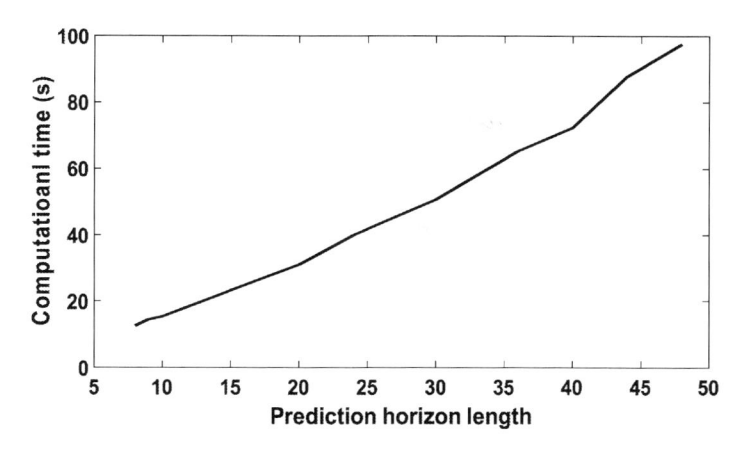

Figure 11.10 Computational time for different prediction horizon lengths.

computer with a 2.7–GHz CPU and 16 GB of RAM. A very long prediction horizon can lead to a high computational burden. On the contrary, a short prediction horizon does not fully capture the future information. As mentioned earlier, the prediction horizon must be chosen on the basis of the balance among computational burden, forecast errors, and storage capacities in the system. Hence, the MPC approach has been performed for one timestep in the scheduling horizon while considering different lengths of the prediction horizon. The prediction horizon is chosen in the range from 8 to 48 timesteps. The computational time is shown in Fig. 11.10. As expected, the computational time is increasing as the length of the prediction horizon increases. Moreover, it is not exceeding the limits of the required time. Due to higher forecast values in a long prediction horizon, a prediction horizon longer than 48 timesteps has not been investigated. The MPC approach is validated and can be used for RT scheduling.

To summarize, a higher prediction horizon length contains a higher amount of future information. Due to a longer prediction horizon, a control action can be taken in the first step based on future information. On the contrary, a shorter prediction horizon lacks information and therefore cannot reach the best optimal solution. However, a long prediction horizon leads to higher forecasts errors, and hence the length of the prediction horizon should be chosen wisely based on the forecasting method available. Both computational time and forecast error should be taken into account when selecting the length of the prediction horizon.

Table 11.5 Description of cases.

Case	Description of Cases
1	MPC-based RT scheduling with the length of prediction horizon $H_p = 8$
2	Traditional RT scheduling

11.6.3 Validation of the MPC strategy

In this section, the effectiveness of the MPC approach for RT scheduling is verified. As a benchmark model, traditional RT scheduling is used. Traditional RT scheduling is a single-period optimization occurring near RT. The time resolution will be equal to the time resolution of the MPC-based approach, which is 15 minutes. However, the results for the conventional method to obtain the prediction horizon in Section 11.6.1 contain large deviations. Due to such large deviations in the conventional method, the forecasts of the uncertainties in traditional RT scheduling will not be based on generation of the forecast error based on the normal distribution with a known standard. The deviation obtained for the conventional method is very high and unrealistic. Hence, the measurements of the uncertainties from the previous timestep will be used in traditional RT scheduling. This will provide higher accuracy and values that are more realistic inputs to traditional RT scheduling. Traditional RT scheduling will receive the last measured values of the uncertainties and perform single-period optimization similar to Eq. (11.3). The difference in the objective function of traditional RT scheduling compared to Eq. (11.3) is the fact that there is no incentive given to decrease the deviation of the producers in RT compared to the DA schedule. Therefore, the absolute terms are ignored.

Two cases are introduced in Table 11.5 to show the benefit of MPC-based RT scheduling. Case 1 is the MPC-based approach, whereas Case 2 is based on the traditional RT scheduling scheme as earlier elaborated.

In Fig. 11.11(a), the imbalance in RT for the described cases is presented. The imbalance summarizes the wind power imbalance and electric demand imbalance between RT and DA. The values for the MPC approach based on the OLM and the RT traditional approach are shown for the first timestep of the prediction horizon. This first timestep is applied to the system. As mentioned earlier, there might be some minor imbalances left after RT scheduling is performed. The minor imbalances after RT scheduling is performed are shown in Fig. 11.11(b). Minor imbalances are calculated as a difference between the predicted values in the first timestep of the prediction horizon and measured values. Such imbalances are to

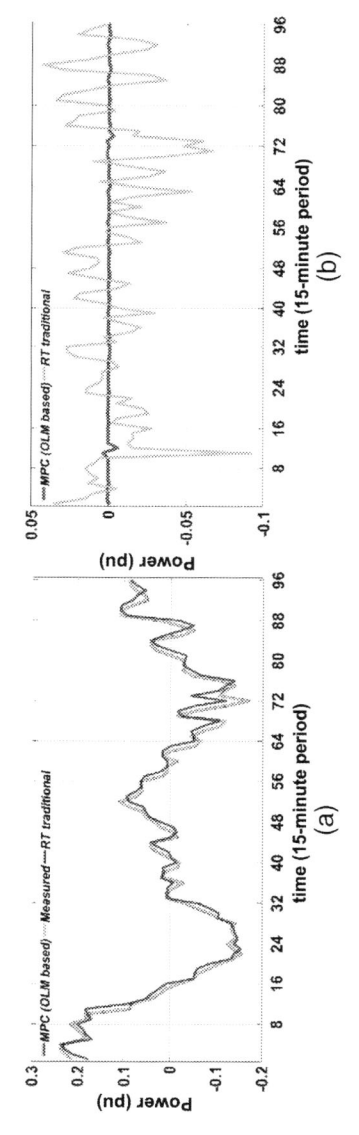

Figure 11.11 (a) Imbalance for different cases and (b) imbalances for MPC and RT scheduling compared to measured values.

Table 11.6 Comparison of system cost obtained by optimization problem and wind spillage cost.

Case	System Cost ($)	Penalty ($)	Wind Spillage Cost ($)	Total System Cost ($)
1	3612.65	—	0	3612.65
2	−2128.34	12,971.40	0	15,099.74

be met by the frequency reserves. It can be observed that the predicted imbalance by MPC-based RT scheduling based on the OLM is close to the actual measured values of the imbalances of systems. On the contrary, traditional RT scheduling deviates from the measured values. Hence, the minor imbalances left in RT are very high after traditional RT scheduling is performed compared to the MPC approach. To summarize, the MPC approach based on the OLM ensures the minimum imbalances compared to the traditional RT approach.

Furthermore, the results of regulating power scheduling are shown in Fig. 11.12(a) and Fig. 11.12(b) for Case 1 and Case 2, respectively. The values above zero represent the downward regulation power from the production side and upward regulation power from the consumption side. On the contrary, the values below zero represent the upward regulation provided from the production side and downward regulation provided from the consumption side. It can be observed that Case 1 has low downward regulation power provided by the conventional generators compared to Case 2. Moreover, P2G units provided a larger amount of upward regulation in Case 1 compared to Case 2. Hence, the MPC approach explores the possibility to provide a higher amount of regulating power from the flexible units, such as P2G, compared to traditional RT scheduling. It should be mentioned that the balancing price for the consumption side is usually lower compared to the production side.

Table 11.6 summarizes the results of the optimization problem for two different cases. As mentioned earlier, the cost of traditional RT scheduling is obtained as a sum of the solution of the optimization problem and the penalty for the deviation, similar to the penalty included in the MPC strategy. However, this does not represent the balancing settlement in the balancing power market. The calculations represent the simplified two–price scheme where a penalty is included in MPC-based optimization, whereas in the traditional RT schedule it is accounted for after the optimization problem is solved. As observed, the wind spillage in both cases is zero. The system costs are lower in Case 2 compared to Case 1. However, the additional penalty in Case 2 contributes to the total system cost increase as every deviation from

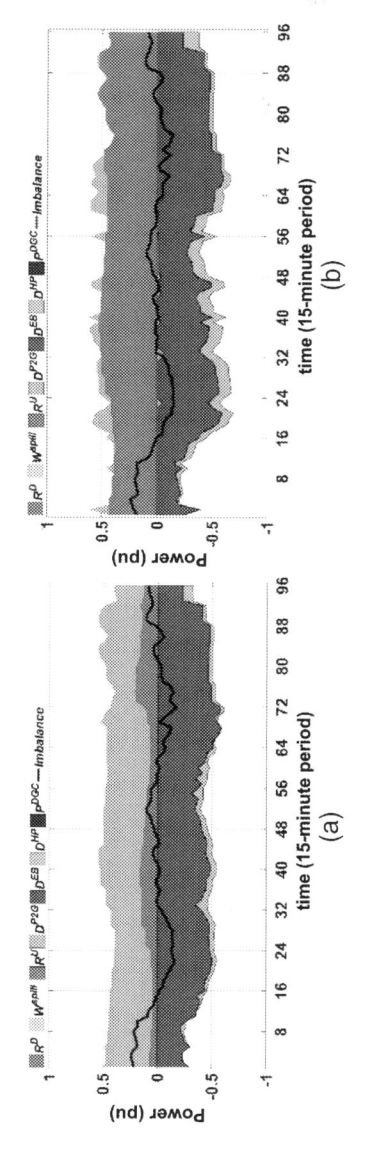

Figure 11.12 Results of the regulating power market from (a) Case 1 and (b) Case 2.

the DA schedule is penalized. The total system costs are therefore higher in Case 2. To summarize, the MPC-based approach schedule generation and demand more optimally and give opportunity to the linking units and storages to provide the regulating power required in RT.

Finally, the balancing power market provides the settlement of the imbalances for the production and consumption side. Based on Fig. 11.12, it is evident that units are both contributing to system balance and causing system imbalance. Hence, the units helping to restore the system balance are called *helpers*, whereas the units causing further imbalance of the system are called *causers*. It can be noticed that the MPC approach decreases the regulating power provided by the conventional generator and explores the possibility of the storages and P2G unit to provide regulating power. The participation of linking units such as P2G, the EB, and the HP in the regulating power market is highly important for future scenarios. As mentioned, it is expected that all conventional units based on the consumption of gas, oil, and coal will be phased out completely in the future. For simplicity, only the daily amount of the regulating power of the helper and causer that participate in the regulating power market is presented in Table 11.7. It can be observed that Case 1 limits the regulating power from the conventional generator compared to Case 2. Moreover, high participation is seen from the consumption side. The consumption side provides more regulating power throughout the day to help balance the system rather than increase system imbalance. On the contrary, energy provided by the conventional generator in Case 2 is very high and similar to the energy of the consumption side. Moreover, the daily regulating power provided by consumption is quite similar for the units helping to keep the balance and units increasing the imbalance of the system. Finally, the total daily regulating power is high in Case 2 compared to Case 1. To summarize, the MPC approach includes future information of the uncertainties and tries to minimize the provision of the regulating power from the conventional generators.

As described earlier, MPC-based RT scheduling provides a cost and energy efficient solution for RT scheduling. Moreover, it gives an opportunity to linking units to participate in the regulating power market.

11.6.4 Summary of the results and discussion

Three main conclusions were presented in previous sections.

The first part was validation of the OLM. The OLM is the forecasting method based on recursivity, which is combined with the MPC strategy.

Table 11.7 Accumulated regulating power for entire day (15-minute resolution) for two different cases.

Player	Case 1	Case 2	Settlement
PRODUCTION—accumulated daily			
Causer	5.56	20.41	Pays upward regulation price for the lack of production
Helper	5.29	24.13	Pays DA market price for the lack of production
CONSUMPTION—accumulated daily			
Total helper	42.76	26.97	Receives the balancing price
Total causer	35.30	23.74	Pays the DA market price

The results have shown that the OLM provides high accuracy and low computational burden compared to the traditional forecasting method.

In addition, the second part investigated the computational burden in regard to different lengths of the prediction horizon. A longer prediction horizon has a higher amount of future information available compared to the shorter prediction horizon. It has been shown that the computational burden increases with an increase of the prediction horizon. However, the time did not exceed the limit for time to provide the regulation power. It has been concluded that the prediction horizon length should be chosen based on the forecast error and computational burden.

Finally, the last part focused on the validation of the MPC strategy compared to traditional RT scheduling. It has been observed that there are minor imbalances after the MPC strategy is applied that will be met by the other balancing services. The MPC approach combined with the OLM ensures an accurate forecast for the next timestep. On the contrary, the imbalances were very high in the traditional RT scheduling approach. Hence, it can be concluded that MPC-based RT scheduling provides minimum imbalances in RT. Furthermore, the MPC approach allows linking units to participate in the regulating market in a larger amount compared to traditional RT scheduling.

11.7 Conclusion

In this chapter, MPC-based RT scheduling of IESs was introduced. The MPC strategy has been applied to the IES and includes the uncertainty of wind power and electric demand. The system utilized the P2G unit, EB, and HP, as well as gas storage and thermal energy storage, to increase flexibility and utilize excess wind power. First, traditional RT scheduling was elaborated. Later on, the MPC strategy was introduced. The important part of the MPC strategy was the OLM, which can predict into the future with high accuracy and low computational burden. In addition, mathematical models for the IES were presented. Last, the solution process and case study were presented, and the results were shown.

MPC-based RT scheduling improves both economic and energy efficiency compared to traditional RT scheduling by predicting future information of the system. The OLM that predicted the future information was integrated in the MPC and resulted in high accuracy and energy efficiency. The limitations of the work are as follows. The linepack in the gas system has not been considered. Linepack can improve energy efficiency, decrease wind curtailment, and decrease the costs of the entire system. Furthermore, an online forecasting method can be developed to forecast few steps ahead more accurately to decrease the forecast error. In addition, the detailed balancing power market can be used to realistically represent the settlements for the entire system.

References

[1] IRENA, Wind energy, 2019. https://www.irena.org/wind. (Accessed 21 February 2019).
[2] European Commision, Paris Agreement, 2020. https://www.ec.europa.eu/clima/policies/international/negotiations/paris_en. (Accessed 21 February 2020).
[3] Energinet, PtX in Denmark before 2030: short term potential of PtX in Denmark from a system perspective, 2019. https://www.energinet.dk/-/media/8BF0CD597E1A457C8E9711B50EC2782A.PDF. (Accessed 22 February 2020).
[4] Q Zeng, J Fang, J Li, Z Chen, Steady-state analysis of the integrated natural gas and electric power system with bi-directional energy conversion, Appl Energy 30 (6) (2015) 1483–1492.
[5] CM Correa-Posada, P Sánchez-Martín, Integrated power and natural gas model for energy adequacy in short-term operation, IEEE Trans Power Syst 30 (6) (2015) 3347–3355.
[6] C Ordoudis, S Delikaraoglou, P Pinson, J Kazempour, Exploiting flexibility in coupled electricity and natural gas markets: a price-based approach, IEEE Manchester PowerTech (2017).
[7] Q Zeng, B Zhang, J Fang, Z Chen, A bi-level programming for multistage co-expansion planning of the integrated gas and electricity system, Appl Energy 200 (2017) 192–203.
[8] X Liu, J Wua, N Jenkins, A Bagdanavicius, Combined analysis of electricity and heat networks, Appl Energy 162 (2016) 1238–1250.
[9] Q Zeng, AJ Conejo, Z Chen, J Fang, A two-stage stochastic programming approach for operating multi-energy systems, in: IEEE Conference on Energy Internet and Energy System Integration, 2017.

[10] N Mazzi, P Pinson, Wind power in electricity markets and the value of forecasting, in: G Kariniotakis (Ed.), Renewable Energy Forecasting, Woodhead Publishing Series in Energy, Elsevier, Duxford, 2017, pp. 259–278. https://doi.org/10.1016/B978-0-08-100504-0.00010-X (Accessed 20 April 2021).

[11] Energinet, Danish Energy Agency, Nordic Power Market Design and Thermal Power Plant Flexibility, Energinet, Danish Energy Agency, Energinet, Danish Energy Agency, Fredericia, Syddanmark, Denmark, 2018.

[12] JM Morales, AJ Conejo, H Madsen, P Pinson, M Zugno, Integrating Renewables in Electricity Markets, Springer, New York, 2014.

[13] Y Du, W Pei, N Chen, X Ge, H Xiao, Real-time microgrid economic dispatch based on model predictive control strategy, J Mod Power Syst Clean Energy 5 (5) (2017) 787–796.

[14] MA Velasquez, J Barreiro-Gomez, N Quijano, AI Cadena, M Shahidehpour, Distributed model predictive control for economic dispatch of power systems with high penetration of renewable energy resources, Int. J Electr Power Energy Syst 113 (2019) 607–617.

[15] M Arnold, G Andersson, Model Predictive Control of Energy Storage Including Uncertain Forecasts, Power System Computation Conference 23 (2011) 24–29.

[16] N Holjevac, T Capuder, I Kuzle, Adaptive control for evaluation of flexibility benefits in microgrid systems, Energy 92 (3) (2015) 487–504.

[17] N Holjevac, T Capuder, N Zhang, I Kuzle, C Kang, Corrective receding horizon scheduling of flexible distributed multi-energy microgrids, Appl Energy 207 (2017) 176–194.

[18] X Guo, Z Bao, W Yan, Stochastic model predictive control based scheduling optimization of multi-energy system considering hybrid CHPs and EVs, Appl Sci 9 (2) (2019) 356.

[19] Z Li, C Zang, P Zeng, H Yu, Combined two-stage stochastic programming and receding horizon control strategy for microgrid energy management considering uncertainty, Energies 9 (7) (2016) 499.

[20] Yalmip, YALMIP, 2020. https://www.yalmip.github.io/?n=Main.License. (Accessed 20 April 2021).

[21] Energinet, Rules and regulations in the Danish electricity Market—Market regulations archive, 2017. https://en.energinet.dk/Electricity/Rules-and-Regulations/Archive-Market-Regulations. (Accessed 20 April 2021).

[22] Q Wang, C Zhang, Y Ding, G Xydis, J Wang, J Østergaard, Review of real-time electricity markets for integrating distributed energy resources and demand response, Appl Energy 138 (2014) 695–706.

[23] Danish Energy Agency, Energy policy toolkit on system integration of wind power: experiences from Denmark, 2015. https://www.ens.dk/sites/ens.dk/files/Globalcooperation/system_integration_of_wp.pdf. (Accessed 20 April 2021).

[24] DTU CEE Summer School, Data-driven analytics and optimization for energy systems, 2019. https://www.energy-markets-school.dk/summer-school-2019/. (Accessed 20 April 2020).

[25] A. Turk, Electronic companion, 2020. https://www.github.com/Ana-Turk/Electronic-companion-Datasets-parameters-and-methods. (Accessed 20 April 2021).

[26] E Shashi Menon, Gas Pipeline Hydraulics, CRC Press, Boca Raton, FL, 2005.

[27] MOSEK, MOSEK optimization toolbox for MATLAB 9.0, 2019. https://www.docs.mosek.com/9.0/toolbox/index.html. (Accessed 20 April 2021).

[28] European Commission, Eurostat, Combined Heat and Power (CHP) Generation, Eurostat, Luxembourg City, 2017.

[29] Energinet, Future Natural Gas Qualities—Fact Sheet, Energinet, Fredericia, Denmark, 2017.

[30] Energinet, Security of Gas Supply 2018, Energinet, Fredericia, 2018.

[31] Energinet, Danish Energy Agency, Technology Data: Energy storages—Technology Descriptions and Projections for Long-Term Energy System Planning, Energinet, Danish Energy Agency, Erritso, Denmark, 2018.

[32] Energinet, Analysis assumptions 2017, 2017. https://www.en.energinet.dk/Analysis-and-Research/Analysis-assumptions/Analysis-assumptions-2017. (Accessed 20 April 2021).

[33] Energinet, Danish Energy Agency, Technology Data for Generation of Electricity and District Heating, Danish Energy Agency, 2020, https://ens.dk/sites/ens.dk/files/Statistik/technology_data_catalogue_for_el_and_dh_-_0009.pdf.

[34] Energinet, Technology Data for Renewable Fuels, Energinet, Fredericia, Denmark, 2018.

[35] Danish Gas Technology Centre (DGC), Future Gas, WP1: perspectives on utilization of biogas without upgrading, DGC, Horsholm, Denmark, 2019.

[36] ENTSO-E, Transparency platform, 2020. https://www.transparency.entsoe.eu/dashboard/show. (Accessed 20 April 2021).

APPENDIX A

Basics of stochastic optimization

As an important branch of optimization theory, stochastic optimization becomes an important tool for decision-making problems with unperfect or uncertain information in many fields such as engineering, economics, and finance.

Decision-making problems in optimal operation of multi-energy systems are no exception. In fact, uncertainty is present in most decision-making problems faced by multi-energy systems. For example, the day-ahead forecast of wind and solar power, electricity demand, gas load, and heat load is uncertain when system operators have to make an operational schedule. However, decisions have to be made the day before the operating day even without perfect information. This is what motivates the use of stochastic optimization in the modeling and optimal operation of multi-energy systems.

For optimization problems, if all input parameters are exactly known, such optimization problems are called deterministic optimization. However, more often than not, some input parameters are uncertain but describable through probability distribution functions. Optimization problems based on probability distribution functions of uncertain parameters are not tractable in computation. In this situation, a discrete representation of uncertain parameters is necessary.

Scenario representation of uncertain parameters is the most commonly used technique in stochastic optimization. The basic idea is to generate a large number of scenarios that can approximately reflect the true probability distribution of the uncertain parameters. In the scenario set, each scenario represents a possible realization of uncertain parameters with an associated probability of occurrence.

The key to scenario generation is the sampling technique, which decides whether the generated scenarios can properly reflect the true probability distribution. Since energy optimization usually involves multiple time periods and multiple locations of renewable energy sites, not only the marginal probability distribution of a single uncertain parameter but also the temporal and spatial correlation of different uncertain parameters need to be

Optimal operation of integrated multi-energy systems under uncertainty. Copyright © 2022 Elsevier Inc.
DOI: https://doi.org/10.1016/B978-0-12-824114-1.00012-3 All rights reserved.

considered in the sampling. The temporal and spatial correlation modeling of uncertain parameters is an important aspect of the probability distribution function modeling.

Based on the generated scenarios, the stochastic optimization problem can be formulated. Different from deterministic optimization, stochastic optimization is to minimize or maximize the expected value of the objective function by weighting all considered scenarios with the occurrence probability. The constraints of stochastic optimization need to ensure the feasibility of each scenario. In addition, stochastic optimization usually involves two or multiple stages. Two-stage stochastic optimization is the most commonly used, where the first stage represents the current stage and the second stage represents the future stage related to the realization of uncertain parameters. The basic idea of two-stage stochastic optimization is to make decisions of the first stage with uncertainties of the second stage incorporated. However, only the first-stage results will be used.

The dilemma for using a stochastic optimization approach is the tradeoff between solution quality and computational burden. Intuitively, the solution quality increases with the number of scenarios. However, the computational burden will surge with a dramatic increase in the size of the optimization problem, which may lead to intractability if unhandled. To this end, the scenario reduction techniques have been proposed in the literature to reduce the number of scenarios without sacrificing the accuracy to a large extent.

A.1 Stochastic optimization fundamentals

A.1.1 Random variables and probability distribution

In probability theory and statistics, a random variable, denoted by ξ, is a variable whose possible values are numerical outcomes of a random experiment.

A probability distribution is a mathematical function that gives the probabilities of occurrence of different possible outcomes for an experiment. It is a mathematical description of a random phenomenon in terms of its sample space and the probabilities of events.

The probability distributions are distinguished for specific types of variables. In stochastic optimization, random variables are classified into two types:

(1) discrete random variable;
(2) continuous random variable.

For the discrete case, a discrete random variable takes a finite or countable number of different values with associated probabilities. It is necessary to specify a probability mass function (PMF), which assigns a probability to each possible outcome. The mathematical description of discrete random variables and their associated probabilities are expressed as follows:

$$f(\xi^k) = P(\xi = \xi^k) \tag{A.1}$$

$$\sum_{k=1}^{K} f(\xi^k) = 1 \tag{A.2}$$

where $\xi^k (k \in K)$ is the list of possible values, the probability of each possible value is between 0 and 1, and the sum of all the probabilities is equal to 1.

In contrast, for a continuous case, a random variable takes values from a continuum and any individual outcome has probability zero, and only events that include infinitely many outcomes, such as intervals, can have a probability. Continuous probability distributions can be described in several ways.

The probability density function (PDF) describes the infinitesimal probability of any given value, and the probability that the outcome lies in a given interval can be computed by the integral over that interval.

An alternative description of the distribution is by means of the cumulative distribution function (CDF), which describes the probability that the random variable is no larger than a given value x. The CDF is the area under the PDF from $-\infty$ to x.

For a continuous random variable, its probability of being in an interval [a, b] can be described by (A.3), or equivalently described by (A.4) [1]:

$$P(a \leq \xi \leq b) = \int_a^b f(\xi) d\xi \tag{A.3}$$

$$P(a \leq \xi \leq b) = \int_a^b dF(\xi) \tag{A.4}$$

where $f(\xi)$ is the PDF of ξ and $F(\xi)$ is the CDF of ξ. For the CDF $F(\xi)$, it must be such that $\int_{-\infty}^{+\infty} dF(\xi) = 1$.

Apart from the preceding PMF, PDF, and CDF, the moment-generating function also serves to identify a probability distribution, as they uniquely determine an underlying CDF. The nth moment of a random variable ξ is defined to be $E[\xi^n]$. The nth central moment of ξ is defined to be $E[(\xi - E[\xi])^n]$. The first moment and the second central moment are the most commonly used. The former is called the *expected value*, denoted by μ, whereas the latter is called the *variance*, denoted by σ^2. These two key parameters are expressed as follows:

$$\mu = \int_{-\infty}^{+\infty} \xi \, dF(\xi) \tag{A.5}$$

$$\sigma^2 = \int_{-\infty}^{+\infty} (\xi - \mu)^2 dF(\xi) \tag{A.6}$$

A probability distribution whose sample space is one-dimensional is called *univariate*, whereas a distribution whose sample space is a vector space of dimension 2 or more is called *multivariate*. A univariate distribution gives the probabilities of a single random variable taking on various alternative values; a multivariate distribution, such as a joint probability distribution, gives the probabilities of a random vector (i.e., a list of two or more random variables) taking on various combinations of values.

A.1.2 Multivariate probability distributions

In energy optimization problems, a number of probability distributions are utilized to characterize the uncertainties of renewables, electricity demand, heat load, and gas load.

The commonly encountered univariate probability distributions include Gaussian distribution, beta distribution, Weibull distribution, empirical distribution, and versatile distribution, among others.

In contrast, the multivariate probability distributions include the Gaussian mixture distribution, Copula distribution, and multivariate normal distribution, among others.

To characterize the temporal and spatial correlations of uncertainties, different multivariate probability distributions are utilized to model the uncertainties in previous chapters based on stochastic optimization. Therefore, only multivariate probability distributions are detailed in this section.

- *Multivariate normal distribution.*

  The multivariate normal distribution [2], multivariate Gaussian distribution, or joint normal distribution is a generalization of the one-dimensional (univariate) normal distribution to higher dimensions. The multivariate normal distribution of a K-dimensional random vector ξ $=[\xi_1, \xi_2, ..., \xi_K]^T$ can be expressed by the K-dimensional mean vector μ and $K \times K$ covariance matrix Σ as follows:

$$\xi \sim N(\mu, \Sigma) \tag{A.7}$$

$$\mu = E[\xi] = (E[\xi_1], E[\xi_2], ..., E[\xi_K])^T \tag{A.8}$$

$$\Sigma_{i,j} = E\big[(\xi_i - \mu_i)(\xi_j - \mu_j)\big] = \text{Cov}\big(\xi_i, \xi_j\big) \tag{A.9}$$

The joint PDF for K-dimensional random vector ξ is expressed as follows:

$$f(\xi_1, \xi_2, ..., \xi_K) = \frac{e^{-\frac{1}{2}(\xi - \mu)^T (\Sigma)^{-1} (\xi - \mu)}}{\sqrt{(2\pi)^K |\Sigma|}} \tag{A.10}$$

where $|\Sigma|$ is the determinant of Σ.

According to expressions in (A.7) through (A.10), when the mean vector μ and covariance matrix Σ are obtained, the multivariate normal distribution will be uniquely determined.

- *Multivariate Gaussian mixture distribution.*

Multivariate Gaussian mixture distribution is a convex combination of multiple multivariate Gaussian distributions. The PDF of Gaussian mixture distribution is a finite weighted sum of Gaussian PDFs. It is characterized by an adjustable parameter set $\Gamma = \{\mu_m, \Sigma_m, \omega_m; m = 1, 2, ..., M\}$, where M is the number of mixture components; ω_m is the weight of the mth component; and μ_m and Σ_m are the mean vector and covariance matrix of the mth component, respectively. A mathematical expression of the Gaussian mixture distribution is given as follows [3]:

$$f(\xi_1, \xi_2, ..., \xi_K) = \sum_{m=1}^{M} \omega_m N_m(\mu_m, \Sigma_m) \tag{A.11}$$

$$\sum_{m=1}^{M} \omega_m = 1, \quad \omega_m > 0 \tag{A.12}$$

$$N_m(\mu_m, \Sigma_m) = \frac{e^{-\frac{1}{2}(\xi - \mu_m)^T (\Sigma_m)^{-1} (\xi - \mu_m)}}{\sqrt{(2\pi)^K |\Sigma_m|}} \tag{A.13}$$

The main advantage of multivariate Gaussian mixture distribution is that it can approximately represent non-Gaussian distribution by properly fitting the parameter set Γ.

- *Copula distribution.*

Copula theory provides an effective way of modeling stochastic dependence. A Copula function is a multivariate distribution function expressed in terms of marginally uniform random variables on the unit interval. A Copula separates the joint distribution into two factors: the marginal distributions of the individual variables and their mutual dependency.

Suppose F is the joint distribution of random vector $\xi = [\xi_1, \xi_2, ..., \xi_K]^T$ and $F_1, F_2, ..., F_K$ are the CDF of each random variable $\xi_1 \xi_2, ..., \xi_K$. If F_k is regarded as random variable, then it follows the uniform distribution $F_k(\xi_k) \sim U(0, 1)$.

In Copula theory, the joint distribution F can be written as

$$F(\xi_1, \xi_2, ..., \xi_K) = C(F_1(\xi_1), F_2(\xi_2), ..., F_K(\xi_K)) \qquad (A.14)$$

The function C in (A.14) is named a Copula function. It is a special kind of multivariate CDF that has uniform margins. Copula theory transforms the modeling of F into the modeling of F_k and C separately. It takes advantage of the fact that stochastic dependence is more easily recognized for uniform variables than for other arbitrarily distributed variables [4].

A.2 Scenario generation and reduction

A.2.1 Scenario generation

To make the optimization tractable, a convenient way to represent stochastic processes is through scenarios [5]. In stochastic optimization, a scenario is a single realization of a stochastic process. The scenario quality is an important determinant of the quality of solutions.

The sampling technique is the basis for scenario generation. The inverse transform method has been widely used for sampling. Assume $F(\xi)$ is the known CDF of random variable ξ. To sample from $F(\xi)$—that is, to sample random variable ξ with the property that $\Pr(\xi \leq p) = F(\xi)$ for all p. The inverse transform sampling is expressed as follows:

$$P = F^{-1}(U), \ U \sim \text{Unif}[0, 1] \qquad (A.15)$$

where F^{-1} is the inverse function of $F(\xi)$, U is the probability of sampling point P, and $\text{Unif}[0, 1]$ is the uniform distribution on $[0, 1]$.

However, in energy optimization problems, the stochastic process comprising multiple random variables usually has temporal or spatial correlations. Only depending on the inverse transform sampling in (A.15) cannot ensure

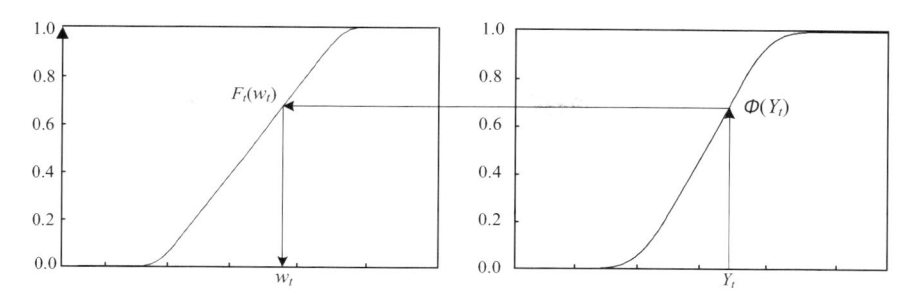

Figure A.1 Nataf transformation.

the correlations between different random variables in the stochastic process. In such a situation, Nataf transformation is utilized to achieve the sampling of a multi-dimensional stochastic process with correlations.

Consider a temporal related stochastic process $[y_1, y_2, ..., y_t, ..., y_T]$ obeying non-Gaussian distribution. Since it is difficult to model the temporal correlation for the multivariate non-Gaussian distribution function. The temporal correlation is therefore modeled in the Gaussian distribution function space and then mapped into the original distribution function space by Nataf transformation. The temporal correlation modeling of multivariable Gaussian distribution has been detailed in Chapter 3 and is thus not detailed here.

Take the tth random variable in the stochastic process as an example. The Nataf transformation is illustrated in Fig. A.1 [6]. The arrows indicate the transformation process. Starting from a sampling point Y_t of Gaussian distribution, the associated Gaussian distribution probability is $\Phi(Y_t)$. This probability value is then assigned as the value of F_t, which is the CDF of the tth random variable. Given the value of F_t, the inverse transformation based on (A.15) is implemented to get the sampling point w_t for the tth random variable y_t.

In summary, it is much easier to represent the temporal correlated stochastic process with the multivariate Gaussian probability distribution instead of working with the empirical CDF. Nataf transformation can achieve the sampling transformation from the Gaussian function space to the original CDF space.

Mathematically, the procedure of Nataf transformation can be expressed as follows:

$$\Phi(Y_t) = F_t(w_t) \tag{A.16}$$

$$w_t = F_t^{-1}(\Phi(Y_t)) \tag{A.17}$$

A.2.2 Scenario reduction

As mentioned earlier, it is critical to generate plausible scenarios to adequately describe the stochastic process and guarantee the solution quality of stochastic optimization. However, since a multi-energy system usually has a large scale of varibles and constraints in a scenario, the plausible scenarios may render the stochastic optimization of multi-energy systems computationally intractable.

Hence, it is necessary to develop algorithms to reduce the number of originally generated scenarios. The reduction procedure is to select a few representative scenarios that can retain most of the relevant information of the original set while reducing the size of the original set significantly.

The scenario reduction techniques are usually based on certain probabilistic metrics that are used to bundle similar scenarios. A number of scenario reduction techniques have been developed to make the stochastic optimization computationally tractable, such as unsupervised k-means clustering method, deep learning, and backward reduction based on probability distance. This section only details the procedure of the backward reduction technique, which is used in previous chapters.

Consider a scenario set that is represented by Ω, wherein the initial number of scenarios is S. Each scenario ξ_s, $\forall s \in S$ in the set Ω has a probability of π_s. The number of scenarios to be retained is N. The procedure for the backward reduction technique is detailed as follows [7]:

Step 1: For each scenario ξ_i, identify the related scenario $\xi_j, (j \neq i)$ that has the minimum probability distance with ξ_i. The identification is based on the following calculation:

$$D_{i,\min} = \min_{\substack{j \\ j \neq i, j = 1,2,\dots,S}} \pi_j \cdot d(\xi_i, \xi_j) \tag{A.18}$$

where $d(\xi_i, \xi_j)$ is the Euclidean distance between scenario ξ_i and scenario ξ_j.

Step 2: Determine the scenario ξ_i to be deleted according to (A.18). Delete scenario ξ_i from the current scenario set:

$$D_{\min} = \min_{i=1,2,\dots,S} \pi_i \cdot D_{i,\min} \tag{A.19}$$

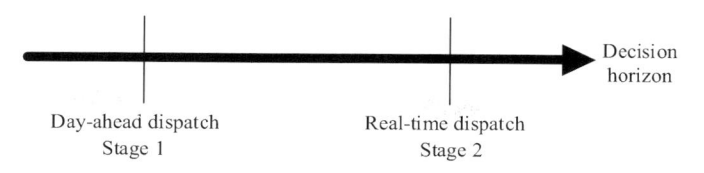

Figure A.2 Two stages of stochastic optimization.

Step 3: Update the number of remaining scenarios as $S = S - 1$. To ensure that the sum of the probabilities of all scenarios is 1, the probability of the deleted scenario is added to the scenario closest to it.

Step 4: Repeat Step 2 and Step 3 until the number of remaining scenarios reaches N.

A.3 General formulation of two-stage optimization

In energy optimization problems based on stochastic optimization, the decision making is usually divided into different stages according to the decision horizon in practice. For instance, the day-ahead dispatch and real-time dispatch are usually distinguished as two different stages, as shown in Fig. A.2.

When the system operator makes day-ahead dispatch, the uncertainty information in the real-time stage will be accounted for. Stochastic optimization problems can be distinguished between two-stage and multistage optimizations according to the number of stages. Two-stage stochastic optimization is the most widely adopted in optimization. This section only details the general formulation of two-stage stochastic optimization.

According to the different stages, the decision variables are also divided into two groups, distinguished as follows [5]:

- *First-stage or here-and-now decision variables*: These decisions are made before the realization of random variables such as wind power and electricity demand. The decisions do not rely on any scenarios in the second stage.
- *Second-stage or wait-and-see decision variables*: These decisions are made after the observation of random variables. Each scenario of stochastic process corresponds to a set of second-stage decisions. Hence, the second-stage decisions are dependent on the actual realization of stochastic processes.

The general expression of two-stage stochastic optimization are expressed as follows:

$$\min\ c^T x + E\{Q(\xi)\}$$
$$s.t.\ Ax \leq b \tag{A.20}$$

In the day-ahead stage, the unit commitment of the power sector and energy scheduling of all sectors, $\mathbf{x}$, are made ahead of time. $Ax \leq b$ represents the set of feasible decisions defined by all day-ahead constraints of all energy sectors, and A and b are known the matrix and vector, respectively. Various costs are included in the vector c. The second term in (A.20) is the expected cost of real-time operations, where ξ is the uncertain vector with a known joint probability distribution.

It should be noted that for the sake of clarity, the first part of the objective function in (A.20) is linear. However, a nonlinear version could also be utilized.

For each realization, s, of the uncertain vector ξ, the second-stage problem can be expressed as follows [8]:

$$Q(\xi) = \min_{y_s}\ q^T y_s$$
$$s.t.\ A_s x + B_s y_s \leq z_s \tag{A.21}$$

Problem (A.21), wherein the decisions are made after uncertainty is cleared, is referred to as recourse problem. In (A.21), various energy regulation costs are captured in q; y_s is the decision vector of real-time energy regulation corresponding to scenario s; A_s, B_s, and z_s are known matrices and vectors with appropriate dimensions.

According to (A.20) and (A.21), the two-stage stochastic optimization problem can be equivalently expressed as follows:

$$\min_{x,\ y_s}\ c^T x + \sum_{s \in S} \pi_s q^T y_s$$
$$s.t.\quad Ax \leq b \tag{A.22}$$
$$A_s x + B_s y_s \leq z_s$$

where π_s is the probability of scenario s. Note that for all decisions of the first and second stages to be optimal, they need to be derived simultaneously by solving a single optimization problem in (A.22) so that the relationships among the decision variables are properly accounted for.

The decision-making process based on two-stage stochastic optimization includes four steps, described as follows (Fig. A.3):

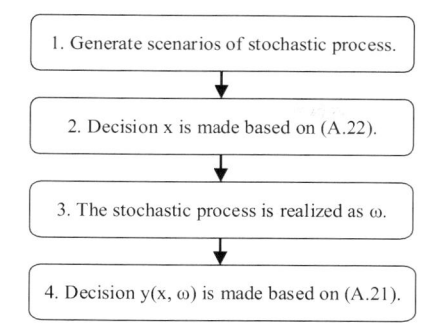

1. Generate scenarios of stochastic process.

↓

2. Decision x is made based on (A.22).

↓

3. The stochastic process is realized as ω.

↓

4. Decision y(x, ω) is made based on (A.21).

Figure A.3 Decision-making process of stochastic optimization.

References

[1] JR Birge, F Louveaux, Introduction to Stochastic Programming,, Springer Science & Business Media, 2011.

[2] YL Tong, The Multivariate Normal Distribution, Springer Science & Business Media, 2012.

[3] Z Wang, C Shen, F Liu, X Wu, C-C Liu, F Guo, Chance-constrained economic dispatch with non-Gaussian correlated wind power uncertainty,, IEEE Trans. Power Syst. 32 (6) (2017) 4880–4893.

[4] N Zhang, CQ Kang, Q Xia, N Zhang, CQ Kang, Q Xia, J Liang, Modeling conditional forecast error for wind power in generation scheduling,, IEEE Trans. Power Syst. 29 (3) (2014) 1316–1324.

[5] AJ Conejo, M Carrión, JM Morales, Decision Making Under Uncertainty in Electricity Markets,, Springer, New York, 2010.

[6] XY Ma, YZ Sun, HL Fang, Scenario generation of wind power based on statistical uncertainty and variability,, IEEE Trans. Sustain. Energy 4 (4) (2013) 894–904.

[7] J Dupačová, N Gröwe-Kuska, W Römisch, Scenario reduction in stochastic programming, Math. Program. 95 (3) (2003) 493–511.

[8] QP Zheng, J Wang, AL Liu, Stochastic optimization for unit commitment—A review, IEEE Trans. Power Syst. 30 (4) (2014) 1913–1924.

Introduction to adaptive robust optimization

Optimal operation of power systems are often characterized as decision making under uncertainty models, since the input data is always uncertain due to inexact measurements or forecast errors. For instance, the uncertainty in power system operation mainly includes the generation of renewable energy such as wind power and photovoltaic power, power demands, and market prices. Since the operation solutions are highly sensitive to these uncertain parameters, it is of great significance to develop mathematical methods taking into account these uncertainties. There are two prevalent optimization approaches to deal with uncertainty, namely stochastic optimization (SO) and robust optimization (RO). In SO, the uncertainty is represented by a true probability distribution and massive scenarios are adopted to describe the most plausible realizations. The objective is to minimize the expected cost. However, RO describes the uncertainty with an uncertainty set and the objective is to minimize the cost in the worst case. It does not require the assumption of distribution and is computationally efficient.

Based on the constructed uncertainty set, the RO problem is feasible for all possible realizations of the uncertainty within the predefined set and optimal for the worst-case realization. Thus, the formulation of the uncertainty set determines the conservativeness of the robust solution. A general overview of RO and different construction of uncertainty set can be found in the work of Bertsimas et al. [1]. According to different application mechanisms, RO can be categorized into static and adaptive RO—that is, with or without recourse [2]. In reality, there are usually multiple stages in power system operation, such as the dependence of day-ahead operation and real-time operation [3]. Some decisions are decided in the first stage (i.e., here-and-now) before the uncertainty revelation, and the other decisions are regulated in the second stage (i.e., wait-and-see) responding to the revealed uncertainty and here-and-now decisions. Here, we focus on two-stage adaptive robust optimization (ARO). This appendix provides a conceptual and concise introduction to ARO and its solution algorithm.

Optimal operation of integrated multi-energy systems under uncertainty. Copyright © 2022 Elsevier Inc.
DOI: https://doi.org/10.1016/B978-0-12-824114-1.00013-5   All rights reserved.

B.1 Formulation of ARO with resource

Here, we introduce an ARO problem with linear formulation where the first and second stage problems are linear models and the uncertainty set is a polyhedron. Let $\mathbf{x}$ be the decision vector in the first stage, $\mathbf{y}$ be the decision vector in the second stage, and $\mathbf{u}$ be the uncertain variables. All decision variables can be either continuous or discrete. The ARO is to determine the optimal values of $\mathbf{x}$ under the worst case of $\mathbf{u}$ considering the recourse decisions $\mathbf{y}$. The compact form of a two-stage ARO formulation is described as follows:

$$\min_{\mathbf{x}} \left\{ \mathbf{c}^T \mathbf{x} + \max_{\mathbf{u} \in \mathbf{U}} \min_{\mathbf{y} \in F(\mathbf{x}, \mathbf{u})} \mathbf{d}^T \mathbf{y} \right\}, \tag{B.1}$$

$$\text{s.t. } \mathbf{A}\mathbf{x} \le \mathbf{b}, \quad \mathbf{B}\mathbf{x} = \mathbf{a}, \quad \mathbf{x} \in \mathbb{R}^n_+, \tag{B.2}$$

where $\mathbf{A}, \mathbf{B}, \mathbf{a}$, and $\mathbf{b}$ are constant matrices; $\mathbf{U}$ is the predefined uncertainty set; and F is the feasible region of $\mathbf{y}$. The easiest and most commonly formulation of $\mathbf{U}$ is a polyhedral uncertainty set

$$\mathbf{U} = \left\{ \mathbf{u} | \mathbf{P}\mathbf{u} \le \mathbf{q} \right\}, \tag{B.3}$$

where $\mathbf{P}$ and $\mathbf{q}$ are constant matrices.

Given that y is optimized in the second stage where the value of first-stage variable x cannot be changed and the value of uncertain variable u is revealed, the general form of F is

$$F(\mathbf{x}, \mathbf{u}) = \{\mathbf{y} | \mathbf{C}\mathbf{y} + \mathbf{D}\mathbf{x} \le \mathbf{h}, \tag{B.4}$$

$$\mathbf{G}\mathbf{y} + \mathbf{M}\mathbf{u} = \mathbf{g}, \tag{B.5}$$

$$\mathbf{y} \in \mathbb{R}^n_+\}, \tag{B.6}$$

where $\mathbf{C}, \mathbf{D}, \mathbf{G}, \mathbf{M}, \mathbf{h}$, and $\mathbf{g}$ are constant matrices.

B.2 Solution methodology

Since the dependence function F of $\mathbf{y}$ is implicit, it is difficult to solve the aforementioned tri-level ARO. There is a lot of work been done such as affine approximations [4] and decomposition algorithms [5]. The affine approximation method assume that the wait-and-see variable y is an explicit affine function of uncertain variables, and then the ARO problem can be formulated as a one-stage problem. However, it is completed at the cost of adaptability. To solve the fully adaptive robust model, some decomposition methods are used. One is the Benders-dual cutting plane algorithm [5], which decomposes the original tri-level problem into a master problem (i.e., the first-level problem) and a subproblem (i.e., the second and third-level problems). Given that the third-level problem is linear, it is equivalently replaced by its dual problem and then the second-level problem and the third-level problem are emerged as one problem. In addition, the master problem and the reformulated subproblem are solved iteratively until an acceptable error between the upper bound and lower bound is reached.

To improve the computation efficiency of the Benders-dual cutting plane algorithm, a new algorithm named the *Constraint-and-Column Generation* (C&CG) algorithm is introduced [6]. Different from the Benders-dual cutting plane algorithm, the C&CG algorithm adds a set of variables and constraints from the subproblem into the master problem and converts the max-min subproblem into a mixed-integer linear programming using Karush-Kuhn-Tucker (KKT) conditions. The master problem is described as follows, which minimizes the total system operation cost under the worst-case realization $\mathbf{u}_l^*$ obtained from the subproblem in the previous iteration:

$$\text{MP} : \min_{\mathbf{x}, \eta} \mathbf{c}^T \mathbf{x} + \eta, \tag{B.7}$$

$$\text{s.t. } \mathbf{Ax} \leq \mathbf{b}, \tag{B.8}$$

$$\eta \geq \mathbf{d}^T \mathbf{y}^l, \ \forall l \in O, \tag{B.9}$$

$$\mathbf{Cy}^l + \mathbf{Dx} \leq \mathbf{h}, \ \forall l \in O, \tag{B.10}$$

$$\mathbf{Gy}^l + \mathbf{Mu}_l^* = \mathbf{g}, \ \forall l \in O, \tag{B.11}$$

$$\mathbf{x} \in \mathbb{R}_+^n, \mathbf{y}^l \in \mathbb{R}_+^n, \ \forall l \in O, \tag{B.12}$$

where η is the auxiliary variable representing the lower bound, $\mathbf{y}^l$ are the new variables generated from the subproblem and added to the master problem, O is the index set for wind uncertainty scenarios l, and $\mathbf{u}_l^*$ is the optimal value obtained from the subproblem in the last iteration, which is considered as the current worst-case realization.

The subproblem is formulated with fixed $\mathbf{x}^*$ given by the master problem, which is to identify the worst-case scenario $\mathbf{u}_l^*$:

$$\text{SP} : \max_{\mathbf{u} \in \mathbf{U}} \min_{\mathbf{y} \in F(\mathbf{x}^*, \mathbf{u})} \mathbf{d}^T \mathbf{y}, \tag{B.13}$$

$$\text{s.t. } \mathbf{Cy} + \mathbf{Dx}^* \leq \mathbf{h} : \quad \boldsymbol{\lambda}, \tag{B.14}$$

$$\mathbf{Gy} + \mathbf{Mu} = \mathbf{g} : \quad \boldsymbol{\mu}, \tag{B.15}$$

$$\mathbf{y} \in \mathbb{R}^n_+ : \quad \boldsymbol{\nu}, \tag{B.16}$$

where $\boldsymbol{\lambda}, \boldsymbol{\mu}, \boldsymbol{\nu}$ are dual variables corresponding to constraints (B.14) through (B.16). Reformulated with KKT conditions, the max–min subproblem is transferred into an equivalent single–level problem as follows. It includes the primary feasibility constraints (B.14) through (B.16), stationarity constraints (B.18), and complementary slackness conditions (B.19) through (B.21):

$$\text{KKT} - \text{SP} : \max_{\mathbf{u} \in U, \mathbf{y} \in F(\mathbf{x}^*, \mathbf{u}), \boldsymbol{\lambda}, \boldsymbol{\mu}, \boldsymbol{\nu}} \mathbf{d}^T \mathbf{y}, \tag{B.17}$$

s.t. (B.14) through (B.16),

$$\mathbf{d} + \mathbf{C}\boldsymbol{\lambda} + \mathbf{G}\boldsymbol{\mu} - \boldsymbol{\nu} = 0, \tag{B.18}$$

$$(\mathbf{Cy} + \mathbf{Dx}^* - \mathbf{h})_i \lambda_i = 0, \quad \forall i, \tag{B.19}$$

$$y_j \upsilon_j = 0, \quad \forall j, \tag{B.20}$$

$$\lambda_i \geq 0, \ \upsilon_j \geq 0, \ \boldsymbol{\mu} \text{ is free}, \tag{B.21}$$

where i and j are the indices of the corresponding constraints. In the KKT-SP problem, most of constraints are linear except the bilinear complementary constraints (B.19) and (B.20), which are linearized by the big-M method. Thus, the KKT-SP is converted into a mixed–integer linear programming problem.

The reformulated MP and SP are solved by the C&CG algorithm iteratively, which is summarized in Algorithm B.1.

In the aforementioned ARO problem, the second-stage problem is linear since the wait-and-see variables $\mathbf{y}$ in the second stage are continuous, and thus it can be solved successfully using the reformulation based on the dual problem or strong duality. However, solving the ARO problem will be challenging with discrete variables appearing in the recourse problem, which is thus nonconvex and the duality theory is inapplicable. Therefore, a nested C&CG algorithm is needed. The detailed introduction can be found in the work of Zhao and Zheng [7].

Algorithm B.1 C&CG algorithm.

1:	Set the upper bound as ∞ and lower bound as $-\infty$, convergence error $[= 0.01$, iteration index $l = 0$, and $O = \varnothing$;
2:	Solve the master problem (B.7) through (B.12), and derive the optimal solution $(x_{l+1}^*, \eta_{l+1}^*, y_1^*, y_2^*, \ldots, y_l^*)$ and update the lower bound LB $= c^T x_{l+1}^* + \eta_{l+1}^*$;
3:	Solve the KKT-SP (B.17), (B.19) through (B.21), with the optimal solution x_{l+1}^* obtained in Step **2**, and get the optimal solution (y_{l+1}^*, u_{l+1}^*), then update the upper bound UB $= \{UB, c^T x_{l+1}^* + d^T y_{l+1}^*\}$;
4:	Convergence check. If UB-LB $\leq [$, return y_{l+1}^* and terminate. Otherwise, generate new variables y_{l+1} and add corresponding new constraints (B.9) through (B.12) to the MP, and update $l = l + 1$, $O = O \cup \{l+1\}$, then go to Step **2**.

References

[1] D Bersimas, DB Brown, C Caramanis, Theory and applications of robust optimization, SIAM Rev. 53 (3) (2011) 464–501.

[2] V Gabrel, C Murat, A Thiele, Recent advances in robust optimization: An overview, Eur. J. Oper. Res. 235 (2014) 471–483.

[3] J Tan, Q Wu, Q Hu, W Wei, F Liu, Adaptive robust energy and reserve co-optimization of integrated electricity and heating system considering wind uncertainty, Appl. Energy 260 (2020) 114230.

[4] D Bertsimas, H Bidkhori, On the performance of affine policies for two-stage adaptive optimization: A geometric perspective, Math. Program. 153 (2015) 577–594.

[5] D Bertsimas, E Litvinov, XA Sun, J Zhao, T Zheng, Adaptive robust optimization for the security constrained unit commitment problem, IEEE Trans. Power Syst. 28 (2013) 52–63.

[6] B Zeng, L Zhao, Solving two-stage robust optimization problems using a column-and-constraint generation method, Oper. Res. Lett. 41 (2013) 57–61.

[7] L Zhao, B Zeng, An exact algorithm for two-stage robust optimization with mixed integer recourse problems, Working Paper, 2012. http://www.optimization-online.org/DB_FILE/2012/01/3310.pdf. (Accessed August 17, 2021).

Index